PERIODIC TABLE
OF THE
ELEMENTS

1 H																	2 He
3 Li	4 Be											5 B	6 C	7 N	8 O	9 F	10 Ne
11 Na	12 Mg											13 Al	14 Si	15 P	16 S	17 Cl	18 Ar
19 K	20 Ca	21 Sc	22 Ti	23 V	24 Cr	25 Mn	26 Fe	27 Co	28 Ni	29 Cu	30 Zn	31 Ga	32 Ge	33 As	34 Se	35 Br	36 Kr
37 Rb	38 Sr	39 Y	40 Zr	41 Nb	42 Mo	43 Tc	44 Ru	45 Rh	46 Pd	47 Ag	48 Cd	49 In	50 Sn	51 Sb	52 Te	53 I	54 Xe
55 Cs	56 Ba	57 La	72 Hf	73 Ta	74 W	75 Re	76 Os	77 Ir	78 Pt	79 Au	80 Hg	81 Tl	82 Pb	83 Bi	84 Po	85 At	86 Rn
87 Fr	88 Ra	89 Ac	104 Unq	105 Unp	106 Unh												

58 Ce	59 Pr	60 Nd	61 Pm	62 Sm	63 Eu	64 Gd	65 Tb	66 Dy	67 Ho	68 Er	69 Tm	70 Yb	71 Lu
90 Th	91 Pa	92 U	93 Np	94 Pu	95 Am	96 Cm	97 Bk	98 Cf	99 Es	100 Fm	101 Md	102 No	103 Lw

Materials Science,
Testing,
and Properties
for Technicians

Materials Science, Testing, and Properties for Technicians

WILLIAM O. FELLERS

American River College
Sacramento, California

WITHDRAWN

PRENTICE HALL CAREER & TECHNOLOGY
Englewood Cliffs, New Jersey 07632

Library of Congress Cataloging-in-Publication Data

Fellers, William O.
  Materials science, testing, and properties for
technicians.

  Includes index.
  1. Materials.   I. Title.
TA403.F45   1990   620.1′1   88-32536
ISBN 0-13-560764-7

Editorial/production supervision and
  interior design: Marcia Krefetz
Cover design: Wanda Lubelska Design
Manufacturing buyer: Dave Dickey

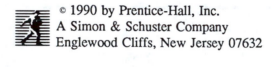 © 1990 by Prentice-Hall, Inc.
A Simon & Schuster Company
Englewood Cliffs, New Jersey 07632

Printed in the United States of America

10 9 8 7 6 5 4

ISBN 0-13-560764-7

Prentice-Hall International (UK) Limited, *London*
Prentice-Hall of Australia Pty. Limited, *Sydney*
Prentice-Hall Canada Inc., *Toronto*
Prentice-Hall Hispanoamericana, S.A., *Mexico*
Prentice-Hall of India Private Limited, *New Delhi*
Prentice-Hall of Japan, Inc., *Tokyo*
Simon & Schuster Asia Ptd. Ltd., *Singapore*
Editora Prentice-Hall do Brasil, Ltda., *Rio de Janeiro*

Contents

4 STRUCTURE OF MATTER   88

5 STEEL   134

6 HEAT TREATMENT OF STEELS   154

Preface

You are embarked on the study of engineering technology. Congratulations! Everything an engineer, engineering technician, or drafter designs must be made from materials. To make intelligent designs and decisions about materials, it is imperative that you gain some knowledge of materials and the way in which they are tested. This book is designed to help you achieve that end. The book, by itself, is not going to make you learn about materials. You must do that yourself. There are a few suggestions that might help you in this course.

1. Read the book *carefully*. Have a pencil or highliner in your writing hand. Underline the important concepts as you read them. In cases of mathematical presentations, do the calculations yourself as they are presented in the book. Refer to the figures as you read about them. Remember, this is not a novel; you cannot "speed read" this book and understand it.

2. In learning a science such as materials science, you must learn an entirely new vocabulary. You will have to learn as many new words in studying a science as you would in taking a foreign language. In this book the important words are italicized. Learn the definitions of these words before you go on. Look up immediately the meaning of any words that you do not understand. There is a glossary of these terms in the back of the book. For words not in the glossary, you may have to refer to physics, chemistry, or other handbooks and manuals. Do not wait until later, or try to guess at their meanings. Remember, many words in the English language have several meanings. Different academic disciplines apply different meanings to the same word. You must learn the meanings of these words as they are applied to materials science.

3. The problems in this book are designed to help you understand what you have just read. Work them! If you truly understand the problems, they will become easy, yes, even boring. When you can demonstrate the solution to others with ease, you have mastered them.

4. Periodically, throughout this book, there are some questions called Think-

ers. These are problems or questions without specific answers. There may be many correct answers to them. They are designed to make you stretch your brain a bit. In answering these, be prepared to defend your answers and give reasons for them as well as coming up with answers. Your reasoning and method of getting the answers is just as important as the answer itself.

5. Show an interest in materials. Ask questions when necessary. Your instructor is there to help you learn but cannot learn for you. Interested and inquiring minds have more questions than there are answers.

6. It would be impossible to cover every material in any book. If you have an interest in a particular material, look it up or research the literature on that material. Do not wait for the instructor to get around to your favorite material— he or she may never get to it.

7. Remember, materials science and technology is a subject in which new discoveries are being made faster than anyone can learn about them. The study of science and technology is not a static "learn it once and you are set forever" type of subject. Once you have started learning about materials, it will be necessary to keep on learning about them. Supplement this book with articles from current journals, such as *Scientific American, Nature, Science, Science Digest, Discover,* and even *Popular Science* and *Popular Mechanics.*

Every effort has been taken to try to make this book engaging and readable. However, it is up to you to supply the interest and the drive to learn the subject. Remember also that learning is not a spectator sport. It is hoped that you will have an interesting and profitable course in your study of materials testing and properties. Good learning to you.

William O. Fellers

Tests
of
Materials

INTRODUCTION

The progress of civilization has closely followed the development of materials. Early humans used stones, wood, and other materials which they found in the sizes and shapes that they needed. It was a major technological achievement to discover that flint or chert could be chipped and sharpened to provide a cutting edge. The first metals to be used were those found in their natural state: gold, copper, and meteoritic iron. Entire civilizations developed around the use of copper and its alloys (bronzes).

Probably among the oldest manufactured materials were ceramics in the form of pottery. Archeological finds have produced ceramic objects dating from as early as 3000 B.C. Glass objects dating from 2000 B.C. are known. Closely paralleling ceramic and glass development was that of iron. Although natural iron from meteorites had been known for centuries and was even worshipped by some as being "heaven sent," it was not sufficiently abundant to be in wide use. Around 1500 B.C. (200 years before Moses) the Hittites found a method of extracting iron from its ore. The Hittites lived in the upper Euphrates valley of Mesopotamia (now Iraq and Syria). This new material, much harder and stronger than copper and bronze, was soon used in armament. Soldiers equipped with iron swords and shields enjoyed a definite advantage over opponents outfitted with bronze equipment. The first "ultimate weapon" had been invented. By the middle of the thirteenth century B.C. the manufacture of iron products had spread throughout the Mediterranean world. The Old Testament of the Bible has nearly 90 references to iron, and by the time of Christ, iron was in common use.

As time elapsed, materials were improved and new ones developed. The Romans were known to have used lead for water pipes. As chemistry and other sciences gradually produced new materials, the variety of products increased. The technical skills needed to work with these new products expanded, with a resultant elevation in the standard of living. The most modest household today displays far more luxurious furnishings than those of most of the royalty of ancient times.

The late nineteenth century brought many new materials into commercial use. Aluminum, Bakelite, magnesium, portland cement, and celluloid are examples of a few of these materials. After World War II scientific discoveries increased greatly. Whereas only a few plastics were known prior to the 1940s, they now number in the thousands. Semiconductors, adhesives (epoxies, superglues), composite materials (fiberglass, graphite–epoxies, etc.), synthetic fibers (Dacron, nylon, etc.), and new building materials have all been discovered and produced in the twentieth century.

Very often throughout history, the development of new technologies has been hampered because materials were not available to keep pace with the theory. The rotary calculator was designed by Pascal and Leibniz in the mid-seventeenth century, but the manufacturing tolerances and materials at hand were not sufficiently precise to make the calculator work properly until the mid-nineteenth century. Similarly, the development of the powered heavier-than-air airplane was delayed because the engines available were too heavy per horsepower produced to get the planes into the air. Early internal combustion engines had limited power output because the materials available to early engineers would not withstand the high temperatures produced at the high compression ratios required to build high-horsepower engines. This problem still exists to a certain extent. Development of the thermonuclear (fusion reaction) reactor is delayed because no material known can withstand the extremely high temperatures produced in the fusion reaction.

Thinker 1-1:

List at least 10 materials that have been developed or have become available commercially within the last 10 to 15 years.

Thinker 1-2:

List five items (i.e., machines, electrical items, etc.) that have depended on the development of new materials.

Thinker 1-3:

List five items that have been replaced by newer materials or that have been improved by the introduction of new materials.

THE NEED FOR MATERIALS TESTING

Since the industrial revolution in the late eighteenth century, the reliability of machinery has improved steadily. A major factor in this better reliability is growth in the knowledge of materials. Until the early twentieth century, the treatment of metals and other materials of construction and manufacture was primarily an art. Metalsmiths were trained to gauge the temperature of heated metals by the color of the metals in the fire. They forged their shapes by hand and judged their hardness by "feel." Most of the design work was at best an educated guess based on previous experience. If a 16-ft span in a building was well supported by a 6 by 12 in. wooden beam spaced every 4 ft, a 20-ft span would probably need an 8 × 12. In reality the same 16-ft span would have been adequately supported by a 4 × 12.

For engineers to design buildings or machinery efficiently, they had to understand the properties of the materials with which they worked. Ways had to be developed to test these materials. Materials are tested for at least four reasons:

1. Research
2. Quality assurance and quality control
3. Fracture or failure analysis
4. Engineering design analysis

Research on materials is done to determine the properties of materials. When a new material is developed, engineers must know all the properties of that material before they can use it in design work. Research testing will determine the tensile strength, hardness, density, thermal conductivity, electrical conductivity, and many other properties of the new materials. (Each of these tests is discussed later in the chapter.) Armed with this information, engineers can intelligently design new products using the newly developed materials. As an example, if research indicates that a steel bar of 1-in^2 cross-sectional area can support a load of 40,000 lb in tension, an engineer can calculate that a cross-sectional area of 4-in^2 is needed to support a load of 160,000 lb.

Other examples of the need for materials research can be found in the history of science. When Thomas A. Edison was developing the electric light bulb, it is said that he tested over 600 different materials to find a suitable filament. He was using the old "cut and try" technique of research. Modern techniques would have guided Edison to rule out over 90% of the materials he tried as being theoretically unworkable for light-bulb filaments and would have allowed him to concentrate on the other 10% to determine the most suitable material.

Quality assurance comprises tests performed by a company receiving new supplies to make certain that the shipment meets the standards called for in the order. *Quality control* involves tests done by a company on their own products prior to shipment to ensure that the product meets the manufacturer's standards. Large companies will test almost every product they purchase before accepting shipment or paying for it. It is, for example, better to determine that ball bearings that a company buys to put in their electric motors are of the proper size, within tolerance, and have the specified hardness before they are installed rather than to find later that these parts did not fit, or wear out prematurely, and thus jeopardize the quality of their product. Similarly, a reliable company will test its own products before marketing them.

Fracture or failure analysis involves tests performed on parts that have failed, or on new parts which are similar to parts that have failed, to determine why the failure occurred. We are all familiar with the widespread publicity that follows the crash of a commercial airplane. Politicians belatedly call for new laws to prevent future disasters. Public outcry against aircraft manufacturers is whipped to a frenzy. The metal is hardly cool at the site of the accident before inspectors are on the scene collecting every scrap of the fallen plane. Very often fracture tests are performed on critical parts of an aircraft to determine if failure of these parts could have caused the accident. Such tests can help to determine if the crash was due to a flaw in material used in vital parts of the aircraft, an improperly designed component, a poorly manufactured part, an improperly installed part, or was due to human error. The results of such testing are made available to concerned companies and governmental agencies. Rarely is the public, which soon loses interest, informed of the months of study involved in trying to prevent similar tragedies. Often, study of structural failures involves fracture analysis of materials. Materials do not "just break"—they are broken—and we now know how to determine what causes such breaks.

There are always many ways to design and manufacture an item. The engineer's job is to determine the "best" design and the "best" manufacturing techniques. Engineering design testing involves testing an experimental product to determine such things as the product's reliability, how long it will last, how well it will perform, how safe it is, and how cheaply it can be produced. (Yes, economics is an engineering function, too.) In engineering design analysis, either models or full-scale replicas of new products are tested, often to failure, to determine their performance characteristics. Many new tires are run to failure on test machines to determine their reliability and safety prior to marketing. Before the government will buy a component for a satellite, for instance, the part must be proved, by testing, to meet all the design criteria established by the space agency. Often, many designs of the same component, sometimes made by competing companies, are tested. Other areas in which products must be tested to ensure absolute infallibility occur in a number of the health fields. Heart valves, pacemakers, artificial joints, and synthetic arterial implants, for example, must all function perfectly, or the results could be fatal. In all the instances noted above, it is materials technicians who do the actual testing.

Very few students know where they will be working or the situations in which they may find themselves once they are in industry. It is therefore imperative that all engineering technicians and engineers have, prior to completing their technical training, a working knowledge of materials of manufacture and construction and the ways by which materials are tested.

Thinker 1-4:

Listen to the radio or watch television for an evening. List 10 advertised products and outline why they should be tested.

QUALITIES OF A GOOD TEST

To be useful, a test must be done exactly the same way every time by every person running the test. This is called standardizing the test. Several organizations have drawn very specific guidelines and details that must be followed in the execution of tests of materials. The American Society for Testing and Materials (ASTM), the American Iron and Steel Institute (AISI), the American Society for Metals (ASM), the American National Standards Institute (ANSI) [formerly and often still referred to as the American Standards Association (ASA)], the Portland Cement Association (PCA), and many others have developed and standardized various types of tests.

A good test must be *reliable* and *valid*. A test is said to have *reliability* when it gives the same results for the same material on repeated tests. It has *validity* if the test measures exactly what it claims to measure and not some other, often unknown property of the material. Weighing a feather on a scale might be considered a reliable test; however, trying to determine the force of gravity by measuring the distance a feather will fall in a given amount of time might not be a valid test since the fall of a feather would be greatly influenced by the air.

Tests should be *objective*, not *subjective*. *Objective tests* are those that will give the same results regardless of who performs the test. *Subjective tests* are often affected by the judgment of the person conducting the test. Multiple-choice tests which can be graded by a machine are much more objective than essay tests which must be read and evaluated by a person. Digital readouts are usually more objective than are dials.

Precision and *accuracy* are terms used with measurements in any test. *Precision* is a measure of how close together values of the same measurement lie. *Accuracy* is a measure of how close to the true value a measurement falls. Weighing a part on a miscalibrated scale may give precise results but be entirely inaccurate. Conversely, soft-boiling an egg using a 3-minute sand (hourglass type) timer may give fairly accurate results but with little precision.

Thinker 1-5:

Evaluate the following types of measurements as to their reliability, validity, objectivity, precision, and accuracy.

a. Measuring 10 gallons of gasoline from a gas pump
b. Measuring the speed of your automobile, using the odometer (speedometer)
c. Measuring the speed of your automobile by finding the time it takes, using your wristwatch, to go a measured mile
d. Measuring the speed of your automobile by finding the time it takes, using a stopwatch, to go a measured mile
e. Laying out a mile track by "pacing it off"

There is a difference between *mistakes* and *errors*. *Mistakes* are "goof-ups," "boo-boos," "wrong moves," and so on, and can usually be corrected and eliminated if given a chance. Simply "do it over and do it right." *Errors* are inaccuracies in measurements and can be minimized but never eliminated entirely. No measurement can be made exactly. If one measures the length of a table using a standard "yardstick," the table can only be measured to the nearest $\frac{1}{8}$ in. There is a possible error due to the instrument of $\pm \frac{1}{16}$ in. If greater accuracy is desired, a better-calibrated instrument must be used. If the table is measured with an instrument calibrated to the nearest thousandth of an inch, the error could still be five ten-thousandths of an inch either way. Even finer measurements will still have an error. Remember, the more accurate and finer the calibration of test instruments, the more expensive they become.

There are several terms associated with materials that must be understood prior to discussing the methods by which materials are tested. These terms are often misunderstood since they can have other meanings when applied to different disciplines.

Brittleness: A characteristic of a material that causes it to break prior to undergoing plastic deformation. *Plastic deformation* is a permanent change in material caused by a load being applied. If material will not stretch or bend, it is brittle. A brittle material is one for which the tensile strength and the breaking strength are the same.

Ductility: The ability of a material to be drawn into a wire. It is considered the opposite of brittleness. It is one type of plasticity.

Elasticity: The ability of a material to return to its original size and shape once a load has been removed. Most but not all materials have some degree of elasticity.

Malleability: The ability of a material to withstand rolling or pounding into sheets. It is another form of plasticity.

Plasticity: The ability of a material to withstand permanent deformation without tearing or rupturing.

Resilience: The ability of a material to absorb energy without being permanently deformed. The *modulus of resilience* of a material is the amount of energy in foot-pounds per cubic inch or newtons per cubic meter that can be elastically recovered by deforming the material.

Toughness: A measure of the ability of the material to absorb energy without breaking. The *modulus of toughness* of a material is the energy in foot-pounds per cubic inch or newtons per cubic meter required to break the material.

TYPES OF TESTS

Materials are usually tested by techniques that closely approximate their use. Structural members that are being pulled on are said to be in *tension*. Figure 1-1 shows cables in a suspension bridge which are in tension. Conversely, components that are being pushed together are under *compression*. In Figure 1-2, the chair legs are an example of structural members in compression.

If a rod or bar is being twisted, it is in *torsion*. Motor shafts and drive shafts in automobiles are subject to torsion. Figure 1-3 shows a rod in torsion.

Beams and other parts that are being bent are said to be in *flexure*. A load applied to the middle of a board while it is supported at both ends will put the board into flexure. Similarly, diving boards at swimming pools are supported as a cantilever beam and are in flexure. Figure 1-4 shows a bookshelf that must resist flexure.

FIGURE 1-1   Tension members.

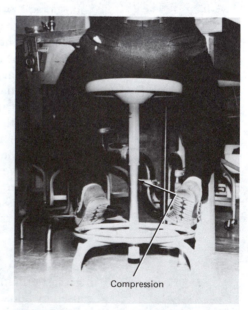

FIGURE 1-2   Compression members.

FIGURE 1-3 Torsion member.

Impact tests are used to test a material's ability to withstand shock. There are several impact tests. The *Izod* and *Charpy* tests are impact tests of notched samples in which the samples are struck with a calibrated blow by a pendulum. Tension impact tests are also conducted.

Hardness tests are measures of the ability of a material to resist surface deformation. There are many types of hardness tests, such as the *Rockwell, Brinell, Mohs, sceleroscope, Tukon, Vickers,* and others, which are discussed in Chapter 3.

Fatigue of metals is a result of their being bent back and forth many times. A wire bent double and straightened several times will eventually break as a result of metal fatigue. A material does not have to be bent severely to undergo fatigue. Often, mere vibrational strains will cause a material to break under fatigue.

FIGURE 1-4 Flexure.

FIGURE 1-5   Gear teeth in shear.

The ability of a material to resist cutting by off-axial forces is a measure of its *shear* strength. Figure 1-5 illustrates shear.

Glues and adhesives are often tested in *peel*. Peel is the breaking of one bond at a time, such as pulling a bandage from your skin. Figure 1-6 shows an example of peel.

Creep is the deformation resulting from a continuous load being applied for a long period—often many years. Bridges and other structures are sometimes instrumented to provide information as to how much a particular beam is elongating under its given loads. Records are kept for many years on these instrument readings. In the laboratory, tests for creep in a material are generally conducted at controlled temperature and other conditions. Results of these tests are important in the design of large structures.

It is rare that any structural member is in only one of the foregoing modes. Any part can be in several modes at once. The middle shaft in Figure 1-3 is in both torsion and flexure.

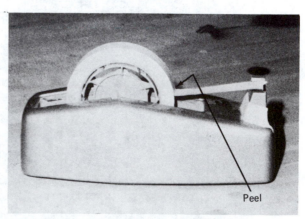

Peel

FIGURE 1-6   Peel.

Thinker 1-6:

Regarding the accompanying photograph, state whether the lettered parts are in tension, compression, torsion, shear, flexure, or combinations of these.

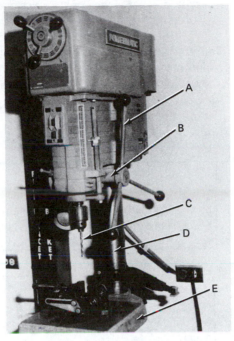

Thinker 1-7:

Look about you. List at least four examples of structural members that are in:

a. Tension

b. Compression

c. Torsion

d. Shear

e. Flexure

DESTRUCTIVE AND NONDESTRUCTIVE TESTING

The testing of materials can be classified in many ways. If a part or a sample specimen of a material is consumed, broken, or damaged to the degree that it can no longer be used for its intended purpose, the test is considered a *destructive test*. Rifle bullets that are fired to see whether they will work or to determine their accuracy provide an example of destructive testing. Similarly, samples of metal bars or concrete blocks that are broken to determine their strength, are destructive tests. Products that must be destroyed in testing can only be tested by sampling. Out of every 1000 rifle bullets, perhaps 10 will be selected at random and fired. If these 10 bullets work properly, it is assumed that the others will also work. This technique of testing by sampling obviously has its faults. It is possible that the 10 shells selected could have been the only faulty ones in the lot and the entire 1000 rounds would be rejected on the basis of those 10 bullets being bad. Conversely, only the good ones may have been chosen and the faulty lot could have been accepted. The science of statistics and probability, however, has made this method an accepted technique with very few "wrong conclusions" being reached.

Whenever possible, *nondestructive testing (NDT)* and *nondestructive evaluation (NDE)* are preferred. Automobile pistons can easily be measured to a thousandth of an inch. If they are too large or too small, they can be rejected, but the correct-size ones can still be used. Simple measuring is a nondestructive test. Paints are checked for correct color by comparing a sample with standards while they are being mixed. X-rays, sound waves, and magnetism can also be used to detect flaws in materials without damaging the parts.

Thinker 1-8:

Explain how a transmission gear for an automobile could be tested destructively and nondestructively.

Thinker 1-9:

How could you tell, using a nondestructive test, if there is a film in your camera?

Not all tests require complicated laboratory equipment. Your senses of sight, hearing, taste, and touch are very effective and often used to test instruments. An apple is rejected if a worm or soft spot is seen therein. Similarly, a crack seen in an engine block is sufficient evidence to reject it for further use. One need not send it to a laboratory for complicated tests.

Musicians test a new instrument by playing it and listening carefully to the quality of sound produced. Much to the dismay of grocery stores, we often test tomatoes, pears, cantaloupes, watermelons, or other fruit and vegetables by feeling, squeezing, or "thumping" them—a form of quality assurance of the product.

Thinker 1-10:

Explain how you could use simple household tools or instruments or your senses to test the quality of the following items.

  a. A television set
  b. Toilet tissue
  c. Light bulbs
  d. A basketball
  e. A new pair of shoes

Laboratory test equipment is just an extension of your senses. It allows you to make quantitative measurements and to increase the accuracy of the senses. Laboratory equipment also enables us to make tests not detectable by human senses. Tests of such items as electrical voltage, resistance, and amperage as well as nuclear radiation could not be tested by human sense organs, and could be dangerous to attempt without instrumentation. Laboratory test devices also make it possible to extend the range of our senses by applying greater forces, to "see" things not visible to the naked eye, and to hear vibrations beyond the audible range.

MEASUREMENTS IN TESTING

The purpose of all testing is to reduce the properties of the materials to numerical or quantitative values. Standard units must be established by which to measure the properties of the materials. In the United States the English system of measurements, using feet, pounds, gallons, and so on, has traditionally been used. However, most of the industrialized nations of the world now use the metric system. Even though most of the engineering and testing in this country is still based on English units, many companies have foreign contracts and buyers. These companies are forced to work in the metric system. For this reason, an understanding of the metric system is essential for all engineers, engineering technicians, and drafters.

In the metric system, prefixes are used to denote multiples of 10. For example, 10 meters is a dekameter, 10 dekameters is a hectometer, and 10 hectometers is a kilometer. Going down the scale, one tenth of a meter is a decimeter, one hundredth of a meter (one tenth of a decimeter) is a centimeter, and one thousandth of a meter (one tenth of a centimeter) is a millimeter. The same prefixes are used with every type of measurement. One thousand grams is a kilogram, 1000 liters is a kiloliter, and so on. Figure 1-7 lists the prefixes and their values.

In converting from one prefix to another in the metric system, all that need be done is to multiply or divide by the multipliers. For example, to determine how many centimeters there are in 3 kilometers, multiply 3 kilometers by 1000 meters per kilometer by 100 centimeters per meter:

$$(3 \text{ km})(1000 \text{ m/km})(100 \text{ cm/m}) = 300,000 \text{ cm}$$

To determine the number of grams in 7 micrograms, the problem would read

$$(7 \text{ } \mu\text{g})(0.000001 \text{ g/}\mu\text{g}) = 0.0000007 \text{ g}$$

To solve problems in engineering technology successfully, two techniques should be mastered. The first is to learn scientific notation, in which numbers are written as powers of 10. One hundred can be written as 1×10^2, 2000 as 2×10^3, and 6 million becomes 6×10^6.

The second technique is referred to as dimensional analysis or unit analysis. Unit analysis makes use of the fact that all quantities have units associated with them except such constants as π, $\sqrt{2}$, and $\frac{1}{2}$. These unitless numbers are called *numerics*. One must buy 10 *gallons* of gasoline, 50 *feet* of wire, or 25 *pounds* of potatoes. You drive down the road at 55 *miles per hour*. Note that the word "per" always indicates division. In every equation not only must the numerical values on each side of the equal sign be equivalent, but the *units* must also be the same.

FIGURE 1-7   Metric system prefixes.

Multiplier			Prefix	Standard Symbol
1,000,000,000,000	or	10^{12}	tera	T
1,000,000,000	or	10^{9}	giga	G
1,000,000	or	10^{6}	mega	M
1,000	or	10^{3}	kilo	k
100	or	10^{2}	hecto	h
10	or	10^{1}	deka	da
1	or	10^{0}	—	—
0.1	or	10^{-1}	deci	d
0.01	or	10^{-2}	centi	c
0.001	or	10^{-3}	milli	m
0.000001	or	10^{-6}	micro	μ (Greek letter mu)
0.000000001	or	10^{-9}	nano	n
0.000000000001	or	10^{-12}	pico	p
0.000000000000001	or	10^{-15}	femto	f
0.000000000000000001	or	10^{-18}	atto	a

Further, the units can be treated just like numbers. Centimeters times centimeters gives square centimeters:

$$cm \times cm = cm^2$$

Kilometers divided by kilometers equals 1:

$$km/km = 1$$

If meters are multiplied by centimeters per meter, the meters divide to become 1 and the product will be centimeters:

$$(m)(cm/m) = cm$$

To divided by a fraction, we invert the denominator and multiply the result. The same is true with dimensions. If kilometers per hour is divided by minutes per hour, the hours divide out and the result becomes kilometers per minute:

$$(km/h)/(min/h) = (km/h)(h/min) = km/min$$

Only numbers having identical units can be added or subtracted. Grams can only be added to grams, meters to meters, or feet to feet. Ten pounds of copper plus 15 pounds of copper produces 25 pounds of copper:

$$(10 \text{ lb copper}) + (15 \text{ lb copper}) = 25 \text{ lb copper}$$

Grams cannot be added to kilograms without first converting the grams to kilograms or the kilograms to grams. You cannot add 3 pounds of apples to 2 kilograms of oranges and get anything but a lot of fruit salad. Always write down the units with all numbers as you work a problem. The units will help keep you from making mistakes.

If 40 milliliters is to be added to 2 liters, the final volume could be figured by converting the milliliters to liters, then adding the 2 liters:

$$\frac{40 \text{ mL}}{1000 \text{ mL/L}} + 2 \text{ L} = 2.040 \text{ L}$$

The volume could also be expressed in milliliters by converting the liters to milliliters, then adding the 40 milliliters:

$$(2 \text{ L})(1000 \text{ mL/L}) + 40 \text{ mL} = 2040 \text{ mL}$$

To illustrate the use of dimensions in solving problems, consider the problem of converting 7 kilometers to meters. Just look at the units. If the kilometers are multiplied by the number of meters per kilometer, the product would be meters:

$$(\text{km})(\text{m/km}) = \text{m}$$

We then only need to put in the associated numbers:

$$(7 \text{ km})(1000 \text{ m/km}) = 7000 \text{ m}$$

To convert 4.3 grams per cubic centimeter to kilograms per cubic meter, the units would be

$$(\text{g/cm}^3)(\text{cm/m})^3/(\text{g/kg}) = (\text{g/cm}^3)(\text{cm}^3/\text{m}^3)(\text{kg/g}) = \text{kg/m}^3$$

Put the numbers in the problem and it becomes

$$(4.3 \text{ g/cm}^3)(100 \text{ cm/m})^3/(1000 \text{ g/kg}) = 4300 \text{ kg/m}^3$$

If a person really understands a problem, formulas are unnecessary, but they are often a help in solving problems. The most common formulas used in materials testing involve areas of circles, rectangles, and squares and the volumes of cylinders, cubes, and rectangular solids. Some of these follow.

Area (A):

Square	$A = s^2$	where s = length of the side of the square
Rectangle	$A = L \times W$	where L = length of the rectangle W = width of the rectangle
Circle	$A = \pi r^2$	where r = radius of the circle π = 3.1416 (approximately)
or	$A = \dfrac{\pi d^2}{4}$	where d = diameter of the circle

Volume:

Cube	$V = s^3$	where s = length of the edge of the cube
Rectangular solid	$V = L \times W \times D$	where L = length of the solid W = width of the solid D = depth of the solid
Right circular cylinder	$V = \pi r^2 h$	where r = radius of the base h = height of the cylinder

Some examples of area calculations are as follows:

The area of a square 15 cm on a side is

$$A = (15 \text{ cm})^2 = 225 \text{ cm}^2$$

The area of a rectangle that is 20 cm long and 18 cm wide is

$$A = (20 \text{ cm})(18 \text{ cm}) = 360 \text{ cm}^2$$

The area of a circle having a diameter of 5.20 cm is

$$A = \frac{(3.1416)(5.20 \text{ cm})^2}{4} = 21.3 \text{ cm}^2$$

Some examples of volume calculations follow.

The volume of a cube 17 mm on an edge is

$$V = (17 \text{ mm})^3 = 4913 \text{ mm}^3$$

The volume of a rectangular swimming pool measuring 30 ft wide, 50 ft long, and 4 ft deep is

$$V = (30 \text{ ft})(50 \text{ ft})(4 \text{ ft}) = 6000 \text{ ft}^3$$

The volume of an oil can 10 cm in diameter and 18 cm high is (if the diameter is 10 cm, the radius is 10 cm/2 or 5 cm)

$$V = (3.1416)(5 \text{ cm})^2(18 \text{ cm}) = 1414 \text{ cm}^3$$

PROBLEM SET 1-1

1. Convert 6 m to centimeters.
2. Convert 3.72 km to meters.
3. Change 39,750 m to kilometers.
4. How many decigrams in 7.8 kg?
5. How many milligrams in 4.32 kg?
6. 0.893 g equals how many micrograms?
7. 32 liters is equivalent to how many milliliters?
8. Which is larger, 4.7 dag or 5936 cg?
9. Which is longer, 37 cm or 95 mm?
10. How many square meters are there in 3.7 km²?
11. What is the area of a rectangle in square meters that measures 40 m by 72 m?
12. A rectangular (cross section) steel bar measures 2 cm by 5 cm and is 10 m long. What is its volume in cubic centimeters?
13. A cylindrical tank is 10 cm in diameter and 40 cm high. What is its volume in cubic centimeters?
14. How many picoseconds are in 32.000000000000 min? (Express the answer in scientific notation.)
15. How many centimeters per second would 100 km/h be?
16. A snail can speed along at a rate of about 10 cm/min. What is this speed in kilometers per hour?
17. If a person can run 200 m in 22 s, what would be the average speed in kilometers per hour?
18. If 42 mL is added to 32 cL, what will be the volume in liters?
19. A cube is 32 mm on a side.
    (a) What is the volume of the cube in cubic centimeters?
    (b) What is the surface area of the cube in square centimeters?

20. A cylindrical tank that is 2 m tall with a radius of 1.5 m is to be filled with water at the rate of 10 L/min. How long will it take to fill the tank? (*Hint:* 1 cubic centimeter of water is 1 milliliter.)

The engineering technician must also be able to convert English units into metric equivalents, and vice versa. This can usually be accomplished by remembering only a few conversion factors, the metric multipliers, and the English system multipliers.

The metric-to-English conversion factors most essential are the following:

Length:

$$2.54 \text{ centimeters} = 1 \text{ inch}$$

$$39.37 \text{ inches} = 1 \text{ meter}$$

Mass:

$$2.2 \text{ pounds} = 1 \text{ kilogram}$$

$$454 \text{ grams} = 1 \text{ pound}$$

Volume:

$$1 \text{ quart} = 0.946 \text{ liter}$$

$$1 \text{ gallon} = 3.784 \text{ liters}$$

Force:

$$1 \text{ pound} = 4.448 \text{ newtons}$$

Also remember that 1 milliliter is approximately equal to 1 cubic centimeter. The English system equivalents must also be remembered.

12 inches	= 1 foot
3 feet	= 1 yard
5280 feet	= 1 statute mile
16 ounces	= 1 pint = 1 pound of water (approximately)
2 pints	= 1 quart
4 quarts	= 1 U.S. gallon
231 cubic inches	= 1 gallon

EXAMPLE 1-1:

Consider the following examples. Convert 62 ft to meters.

Solution:

$$\frac{(62 \text{ ft}) (12 \text{ in./ft}) (2.54 \text{ cm/in.})}{100 \text{ cm/m}} = 18.9 \text{ m}$$

EXAMPLE 1-2:

Convert 55 mi/hr to meters per second.

Solution:

$$\frac{(55 \text{ mi/hr})\ (5280 \text{ ft/mi})\ (12 \text{ in./ft})\ (2.54 \text{ cm/in.})}{(60 \text{ min/hr})\ (60 \text{ sec/min})\ (100 \text{ cm/m})} = 24.6 \text{ m/s}$$

EXAMPLE 1-3:

What cubic-inch displacement does a 2-L engine have?

Solution:

$$\frac{(2 \text{ L})\ (1000 \text{ cm}^3/\text{L})}{(2.54 \text{ cm/in.})^3} = 122 \text{ in}^3$$

PROBLEM SET 1-2

1. Convert 9300 g to pounds.
2. A 350-in³ engine would be how many cubic centimeters?
3. A 1.3-L motorcycle engine would be how many cubic inches?
4. A person weighing 150 lb would weigh how much in kilograms?
5. (a) If a person wanted to reduce his weight to 160 lb while in Europe, and weighed in at 75.3 kg, how many kilograms would he have to lose?
   (b) What would be that weight loss in pounds?
6. A 16-oz bottle would hold how many liters?
7. A force of 200 lb is applied to a lever. What is that force in newtons?
8. One cubic foot of fresh water weighs 62.4 lb. What is the weight of 1 gal of water?
9. If 1 gal of oil weighs 7.9 lb, what will 1 L of that oil weigh?
10. A rectangular tank measures 25 in. long by 15 in. wide and is 10 in. deep (inside measurements). Determine the capacity of the tank in
    (a) cubic inches;
    (b) gallons;
    (c) liters.
    (d) If water weighs 62.4 lb/ft³, what will be the weight of the water in the tank?
11. A king-size water-bed mattress is 80 in. long, 70 in. wide, and 8 in. deep.
    (a) What is the volume of the water bed in cubic feet?
    (b) What is the weight of water in the mattress? (Water weighs 62.4 lb/ft³.)
12. Convert a speed limit of 55 mi/hr to kilometers per hour.
13. A speed limit of 100 km/h would be how many miles per hour?
14. A train traveling at 90 mi/hr would be going how many feet per second?
15. The speed of light is 186,000 mi/sec. Some English horse-racing fans would like to know the speed of light in "furlongs per fortnight." (There are 8 furlongs per mile and 14 days and nights make a fortnight.)

16. The SR-71 can fly at a speed of 2195 mi/hr. The speed of a 30-06 rifle bullet is 3000 ft/sec. Which is faster?
17. At a speed of 2000 ft/sec, how long would it take a jet aircraft to fly 1 mi?
18. At a speed of 800 m/s, how many miles would a jet aircraft fly in 1 min?
19. The speed of sound in air at sea level is about 1100 ft/sec. What is this speed in kilometers per second?
20. What is the speed of sound in miles per hour?

Pressure is a unit often used in engineering calculations. In materials testing, it has the same units as stress. Pressure and stress are defined as a force per unit area, or

$$P = \frac{F}{A}$$

Force is further defined as mass times acceleration, or

$$F = m \times a$$

On the earth, a body at rest always has the acceleration of gravity applied to it. Therefore, weight is a force. In the English system of units, the force is measured in pounds. Since mass is also measured in pounds, care must be taken to differentiate between the two. A pound force is often denoted lb +, sometimes called the poundal. When weights or forces are given in the English system, they are obviously in pounds force.

The metric counterpart of the pound force is the newton (N). A newton is a kilogram-meter per second squared. To derive newtons, kilograms are multiplied by the acceleration of gravity—9.8 meters per second squared—or

$$newtons = kilograms \times 9.8 \ m/s^2$$

Also remember that 2.2 pounds equals 1 kilogram. To illustrate, if a load of 24 lb is applied to a material, the force in newtons would be

$$\frac{24 \ lb}{2.2 \ lb/kg} = 10.9 \ kg$$

$$(10.9 \ kg) (9.8 \ m/s^2) = 107 \ kg\text{-}m/s^2 = 107 \ N$$

Pressure is often calculated in pascal (Pa). A pascal is a newton per square meter. If the 107-N force of the previous example were applied to an area of 100 cm², the pressure in that area would be

$$P = \frac{107 \ N}{100 \ cm^2} = 1.07 \ N/cm^2$$

or

$$P = (1.07 \ N/cm^2) (100 \ cm/m)^2 = 10,700 \ N/m^2 = 10,700 \ Pa$$

The pascal is a fairly small unit, so the megapascal is usually used. One megapascal is 1 million pascal. Consider the following problem.

EXAMPLE 1-4:

A 100-kg football player stands on a 3-cm-diameter solid metal cylinder. Determine the pressure on the cylinder.

Solution:

$$\text{Pressure} = \frac{\text{force}}{\text{area}}$$

$$P = \frac{(100 \text{ kg}) \ (9.8 \text{ m/s}^2) \ (100 \text{ cm/m})^2}{(3.1416) \ (1.5 \text{ cm})^2} = 1{,}390{,}000 \text{ Pa}$$

or 1.39 MPa.

The technician should be able to convert pounds per square inch to megapascal. Consider the following problem.

EXAMPLE 1-5:

Convert 25,000 lb/in² to MPa.

Solution:

$$\frac{(25{,}000 \text{ lb/in}^2) \ (9.8 \text{ m/s}^2)}{(2.2 \text{ lb/kg}) \ (2.54 \text{ cm/in.})^2 \ (0.01 \text{ m/cm})^2} = 1.7 \times 10^8 \text{ Pa}$$

or 170 MPa.

Note that the calculator gave an answer of 1.7261398×10^8. Since the original values were given to only two significant figures, it makes no sense to calculate an answer to more than two significant figures. It would be ridiculous to calculate an answer to a millionth of an inch, for example, if the original measurement were measured only to the nearest inch. The number of significant figures in any solution must never exceed the number of significant figures in the least accurate measurement of the problem. Match the habit of good technicians and engineers and round off your final answers to the proper number of significant figures.

PROBLEM SET 1-3

1. Convert a load of 52 kg to newtons.
2. A load of 7500 lb would be how many newtons?
3. Convert 500 N to pounds.
4. A load of 1000 kg is pulling on a cylindrical bar that is 12.8 mm in diameter. What is the pressure on the bar in pascal?
5. Convert a pressure of 14.7 lb/in² to newtons per square centimeter.
6. Which is the larger pressure, 8500 lb/in² or 100 MPa?
7. A 2000-lb wrecking ball is set on a steel tee 2 in. in diameter. What is the pressure on the tee
   (a) in pounds per square inch;
   (b) in megapascal?
8. Concrete is often designed to withstand a pressure of 3000 lb/in². What would this pressure be in pascal?

9. A flood gate has dimensions of 5 ft by 8 ft. The pressure on this gate is 30 lb/in². What is the total force on the gate
    (a) in pounds;
    (b) in newtons?

10. A 3-m by 5-m billboard sign is designed to withstand a wind pressure of 100,000 Pa. If a wind force of 750 lb were applied to the face of the sign, would the sign stay standing?

11. The *Guinness Book of Records* lists the highest sustained pressure in a laboratory as 12,300 tons/in². What is this pressure in megapascal?

12. The pressure at the center of the earth is reported to be 23,600 tons/in². What is this pressure
    (a) in pounds per square inch;
    (b) in megapascal?

13. A tire carries a gauge pressure of 32 lb/in². What is this pressure in pascal?

14. A theater organ needs a pressure of about 15 lb/in² of air pressure to work properly. If a four-rank organ has 244 pipes averaging a cross-sectional area of 0.75 in² each at the throat, what is the air pressure of a pipe in pascal?

15. The fire department states that they need 180 lb/in² pressure on their water mains. What is that pressure in pascal?

16. The pressure of air at sea level is 14.7 lb/in². What is the pressure of air at sea level in pascal?

17. A rectangular tank is 12 in. by 12 in. on the bottom and 6 ft high. If the tank were filled with water, what would be the pressure on the bottom of the tank in
    (a) pounds per square inch;
    (b) pascal?

18. A total force of 90 N is acting on a damper that measures 20 cm by 30 cm. What is the pressure on the damper
    (a) in pascal;
    (b) in pounds per square inch?

19. A pressure of 3,200,000 Pa is exerted in an imported air tank. What is that pressure in pounds per square inch?

20. The formula for the surface area of a sphere is $S = 4\pi r^2$. If a basketball has a diameter of 12 in. and is inflated to a pressure of 13 lb/in², what is the total force on the inner surface of the basketball?

PROPERTIES OF MATERIALS

Every material has several types of properties. There are *physical properties* that include the material's *mechanical properties*, *thermal properties*, *electrical properties*, *magnetic properties*, and *optical properties*. *Chemical properties* of a material involve how that material will react with other elements, acids, bases, solvents, and other compounds. Figure 1-8 summarizes some of the various properties of materials. Not all of these properties are discussed in this book. References to them can be found in physics and chemistry books.

Density is a property of any material. Density is defined as the mass per unit volume:

$$\text{density} = \frac{\text{mass}}{\text{volume}}$$

To calculate the density of a solid, it must be weighed and then its volume must be determined. If it is of a measurable shape, such as a cube or rectangular solid, its volume can be determined by use of a tape measure, micrometer, or other scale. If the solid is an irregular shape, its volume can be determined by immersing it in water and measuring the amount of water that it displaces. Since 1 milliliter of water (1 cubic centimeter) weighs 1 gram, the weight of water displaced can be converted to the volume of water, which is also the volume of the solid.

Thinker 1-11:

How could you find the volume of a solid such as a salt crystal or sugar crystal which dissolves in water?

Consider the following problem.

EXAMPLE 1-6:

A cube of metal 10 cm on an edge weighs 2.7 kg. What would be its density in grams per cubic centimeter?

Solution:

$$\text{Density} = \frac{\text{mass}}{\text{volume}}$$

$$= \frac{(2.7 \text{ kg}) (1000 \text{ g/kg})}{(10 \text{ cm})^3}$$

$$= 2.7 \text{ g/cm}^3$$

EXAMPLE 1-7:

In the English system, find the density of a rectangular solid piece of metal that measures 2 in. by 5 in. by 10 in. and weighs 97.5 lb.

Solution:

$$\text{Density} = \frac{97.5 \text{ lb}}{(2 \text{ in.}) (5 \text{ in.}) (10 \text{ in.})}$$

$$= 0.975 \text{ lb/in}^3$$

The metric system is based on forces, distances, temperatures, and other units occurring in nature. The density of water in the metric system is 1g/mL or 1 g/cm^3. In the English system pure water weighs 62.4 lb/ft^3. The term *specific gravity* (abbreviated SG) means the number of times the density of a material is heavier than the density of water. To get the specific gravity of a material, divide its density by the density of water. In the metric system a metal having a density of 10.7 g/cm^3 would have a specific gravity of 10.7:

$$\text{specific gravity} = \frac{10.7 \text{ g/cm}^3}{1 \text{ g/cm}^3} = 10.7$$

Density always has units. Specific gravity is unitless. It is incorrect to say (though often heard) "the density of the metal was 10.7." The specific gravity of

FIGURE 1-8   Types of properties of materials.

Physical Properties

Mechanical Properties

Tensile strength
Compressive strength
Flexure strength
Shear strength
Torsion strength
Impact strength
Fatigue resistance
Elasticity
Plasticity
   Ductility
   Malleability
Hardness
Viscosity

Thermal Properties

Thermal expansion
Heat capacity
Heat of fusion
Heat of vaporization
Melting point
Boiling point
Flash point
Flame point
Heat of combustion
Specific heat

Electrical Properties

Electrical resistance and resistivity
Electrical conductance and conductivity

Magnetic Properties

Dia-, para-, and ferromagnetism

Optical Properties

Optical transparency
Index of refraction
Turbidity
Optical absorption
Color

Chemical Properties

Electronegativity
Reaction rates
Corrosion resistance
Electronic structure
Solubility

the metal would be 10.7, but the density would be 10.7 *grams per cubic centimeter*. In the English system of measurement, the density would be given in pounds per cubic foot, pounds per gallon, pounds per cubic inch, or ounces per cubic inch. Steel that has a density of 490 lb/ft^3 would have a specific gravity of

$$\text{specific gravity} = \frac{490 \text{ lb/ft}^3}{62.4 \text{ lb/ft}^3} = 7.85$$

All that is necessary to convert the density of a material from the metric system to the English system is to multiply the specific gravity of the material by the density of water in the English system (62.4 lb/ft^3). The specific gravity of a material is the same in any system of measurement.

Thinker 1-12:

If a gas station attendant told you that the antifreeze in your car's radiator had a density of 3.1 and that you ought to get it changed, would you believe him? State your reasons.

PROBLEM SET 1-4

1. A cube of lead 1 in. on a side weighs 0.4095 lb. What is its density in
    (a) pounds per cubic inch;

(b) pounds per cubic foot;

(c) grams per cubic centimeter?

2. A sample of a metal displaced 12 mL of water when submerged. It weighed 88 g. What was its density in grams per cubic centimeter?

3. A piece of wood measuring 2 in. by 4 in. by 6 in. weighed 1.4 lb. What was its density in pounds per cubic foot?

4. What was the specific gravity of the wood in Problem 3?

5. A metallurgist was brought a nugget and asked if it was gold. He found that it displaced 1.3 mL of water and weighed 25.116 g. If the specific gravity of gold is 19.32, was the nugget gold?

6. The specific gravity of iron wood is 1.4. What is its density in
   (a) the metric system;
   (b) the English system?

7. A gallon of gasoline weighs 7.6 lb. What is its density in
   (a) pounds per cubic foot;
   (b) kilograms per liter?

8. Silver has a specific gravity of 10.5. If a person tried to sell you a cube of metal 0.75 in. on a side that weighed 60.1 g, would you buy the cube as silver?

9. Steel has a density of 490 lb/ft³. What is its specific gravity?

10. A beam of wood measured 6 in. by 10 in. by 16 ft long. If the wood had a specific gravity of 0.49, how much did the beam weigh?

11. A metal bar measured 32 cm by 0.5 cm by 1.5 cm and weighed 132 g. What was the density in
   (a) the metric system;
   (b) the English system?
   (c) What was its specific gravity?

12. Mercury has a specific gravity of 13.6. How much would a gallon of mercury weigh?

13. A cylindrical bar of metal 2 in. in diameter and 10 in. long weighed 8.8 lb. It was either iron with a specific gravity of 7.8, nickel with a specific gravity of 8.9, or zinc with a specific gravity of 7.1. Which do you suspect it was?

14. A rectangular zinc bar measuring 12 cm by 6 cm by 1 m long would weigh how much if its specific gravity were 7.1?

15. Magnesium is one of the least dense metals available at 1.738 g/cm³. How much would a cubic foot of magnesium weigh in the English system?

16. A rectangular sheet of aluminum with a surface area of 2 ft by 3 ft weighed 10.5 lb. If the specific gravity of the aluminum was 2.7, how thick was the sheet of aluminum?

17. A cube of aluminum having a specific gravity of 2.7 weighed 0.585 lb. What was the length of the edge of the cube?

18. The density of the iron core of the center of the earth is estimated to be 13.09 g/cm³. What would this density be in pounds per cubic foot?

19. The specific gravity of mercury is 13.546. What is its density in
   (a) the metric system;
   (b) the English system?

20. A friend who weighs 150 lb bets you that he can lift a bar of steel of square cross section 6 in. by 6 in. and 3 ft long. Would you cover the bet? Steel has a specific gravity of 7.8.

The *thermal properties* of a material are those related to heat. Temperature, coefficient of thermal expansion, heat capacity, specific heat, and heating value are all thermal properties of a material. In the English system temperature is measured in degrees Fahrenheit; the metric system uses the centigrade system and measures temperature in degrees Celsius. On the Fahrenheit scale, water melts at 32 degrees and boils at 212 degrees. The corresponding Celsius temperatures for the melting and boiling points of water are 0 and 100 degrees, respectively. Therefore a 100-degree temperature difference on the Celsius scale corresponds to a 180-degree change on the Fahrenheit thermometer (Figure 1-9). Stated another way, 1 degree Celsius is equal to 1.8 degrees Fahrenheit. Zero degrees Celsius corresponds to 32 degrees Fahrenheit. In converting from Celsius to Fahrenheit, the Celsius temperature is multiplied by 1.8, then 32 degrees are added. In converting from Fahrenheit, 32 degrees are subtracted from the temperature before the result is divided by 1.8. The formulas for the conversions from Celsius to Fahrenheit, and vice versa, are

$$°C = \frac{°F - 32}{1.8}$$

and

$$°F = (1.8 \times °C) + 32$$

where °C is the Celsius thermometer reading and °F is the Fahrenheit thermometer reading.

It has been determined by physical chemists that the coldest that anything could theoretically get is $-273.15°C$, which is known as *absolute zero*. (We will use $-273°C$ for our calculations.) The temperature scale that uses absolute zero as the starting point is called the *Kelvin scale*. Zero degrees Celsius is 273 kelvin. Absolute zero is $-459.67° F$. (We will use $-460°F$ here.) The temperature scale based on Fahrenheit degrees that uses absolute zero as the starting point is the *Rankine scale*. Zero degrees Fahrenheit is 460 degrees Rankine. Some calculations require the use of the absolute temperatures. To convert a Celsius reading to kelvin, just add 273 to it. To convert a Fahrenheit reading to Rankine, add 460. To convert a Fahrenheit reading to kelvin, first convert the Fahrenheit to Celsius, then add 273.

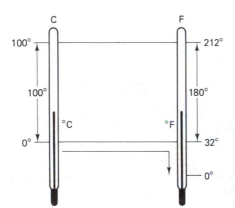

FIGURE 1-9 Converting Celsius to Fahrenheit.

PROBLEM SET 1-5

1. Convert 68°F to degrees Celsius.
2. Convert 1000°F to degrees Celsius.

  3. Convert $-40°F$ to degrees Celsius.

  4. Convert $22°C$ to degrees Fahrenheit.

  5. Convert $327°C$ to degrees Fahrenheit.

  6. Normal body temperature for a person is established at $98.6°F$. What is this temperature in degrees Celsius?

  7. Pure magnesium melts at $651°C$. What is its melting point in degrees Fahrenheit?

  8. Liquid nitrogen boils at $-195.8°C$. What is this temperature in
    (a) degrees Fahrenheit;
    (b) kelvin?

  9. A temperature of $120°F$ is about as hot as one can comfortably hold in one's hand. Could a person be comfortable holding a metal heated to $45°C$?

10. A critical temperature in the heat treatment of steel is $1333°F$. What is this temperature in degrees Celsius?

11. The temperature at the center of the earth is estimated to be about $4500°C$. What is this temperature in
    (a) degrees Fahrenheit;
    (b) kelvin;
    (c) degrees Rankine?

12. An ideal temperature of bath water is thought to be $104°F$. What is this temperature in degrees Celsius?

13. The melting point of a cast iron is $1130°C$. What is this temperature in degrees Rankine?

14. Superconductors are now able to operate at $90K$. What is this temperature in degrees Celsius?

15. What is $90K$ in degrees Fahrenheit?

16. In Europe, the swimming pool water was listed at $12°C$. Would you want to go swimming in that pool? (Find the temperature in $°F$.)

17. The melting point of gallium is $29.78°C$. Would gallium melt in your hand? The temperature of a human hand is about $98.6°F$.

18. Lipowitz metal is a mixture of cadmium, tin, lead, and bismuth. Its melting point is $60°C$. Could a spoon made of this metal melt in a cup of coffee having a temperature of $145°F$?

19. Mercury melts at $-38.87°C$. What is the temperature in
    (a) Kelvin;
    (b) degrees Fahrenheit;
    (c) degrees Rankine?

20. The temperature of the surface of the sun is $5525°C$. What is this temperature in
    (a) degrees Fahrenheit;
    (b) Kelvin?

Temperature is not heat. Temperature is caused by the amount of heat in a material. *Heat* is the internal energy of the material caused by the movement of the atoms. At absolute zero, there is no movement of atoms, therefore no heat. The temperature is a measure of the number of collisions per unit of time of the atoms. The more collisions per second, the higher the temperature. A gas gets hotter when compressed because the molecules are pushed closer together, which results in more collisions per unit of time.

Heat is measured in *calories* in the metric system and in *Btu (British thermal units)* in the English system. A calorie is defined as the amount of heat required to raise 1 gram of water 1 degree Celsius (to be precise, that degree Celsius is specified as being from 14°C to 15°C). The Btu is defined as the heat required to raise 1 pound of water 1 degree Fahrenheit. The conversion factor is

$$1 \text{ Btu} = 252 \text{ cal}$$

A larger unit of heat in the metric system is the *kilocalorie* or "large calorie," which is sometimes abbreviated Cal (using a capital C); a kilocalorie is 1000 calories. One hundred thousand Btu is a *therm*. Natural gas is supposed to carry 1000 Btu per cubic foot, but many states have laws requiring that gas companies bill their customers by the therm rather than by the cubic foot of gas used.

The amount of heat required to raise a unit of any material 1 degree of temperature is called its *heat capacity*. In the metric system, heat capacity would be expressed in calories per gram per degree Celsius (cal/g/°C). In the English system it would be Btu per pound per degree Fahrenheit (Btu/lb/°F). The heat capacity of water in the metric system is 1 cal/g/°C, while in the English system it is 1 Btu/lb/°F. The *specific heat* of any material is its heat capacity divided by the heat capacity of water. Like specific gravity, specific heat has no units. The specific heat of steel is about 0.1, but the heat capacity of steel is 0.1 cal/g/°C or 0.1 Btu/lb/°F.

EXAMPLE 1-8:

If a metal had a specific heat of 0.1, how many calories would it take to heat 30 g of it from 20°C to 95°C?

Solution: If the specific heat is 0.1, the heat capacity is

$$0.1(1 \text{ cal/g/°C}) = 0.1 \text{ cal/g/°C}$$

The temperature difference is

$$95 - 20 = 75°C$$

The heat required would be

$$\text{heat} = (30 \text{ g}) \ (75°C) \ (0.1 \text{ cal/g/°C}) = 225 \text{ cal}$$

EXAMPLE 1-9:

How many therms would it take to heat a swimming pool holding 10,000 gal of water from 70°F to 82°F?

Solution: Water has a heat capacity of 1 Btu/lb/°F.

$$\text{Therms} = \frac{(10,000 \text{ gal}) \ (8.33 \text{ lb/gal}) \ (1 \text{ Btu/lb/°F}) \ (12°F)}{100,000 \text{ Btu/therm}} = 10 \text{ therm}$$

Most metals expand when they are heated. The amount a metal will expand per unit of original length per unit of temperature is called its *coefficient of thermal expansion*. In the metric system the units for the coefficient of thermal expansion would be centimeters per centimeter per degree Celsius. In the English system it would be inches per inch per degree Fahrenheit.

FIGURE 1-10   Expansion joint in an elevated highway.

If a copper bar 15 in. long had a coefficient of thermal expansion of 9.2×10^{-6} in./in./°F, how much would it lengthen if heated 20°F?

$$\text{Change in length} = \Delta L = (15 \text{ in.}) (9.2 \times 10^{-6} \text{ in./in./°F}) (20°F)$$

$$= 0.0028 \text{ in.}$$

In the construction of long bridges and large buildings, room must be left for the thermal expansion of the materials. Expansion joints such as the one in the elevated concrete highway shown in Figure 1-10 must be a part of the design of the structure.

Thinker 1-13:

Outline an experiment whereby the coefficient of thermal expansion for a rod of metal could be found. Draw a sketch of your apparatus for this experiment.

The *thermal conductivity* of a material is the amount of heat it will transmit

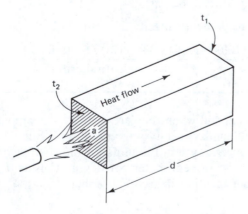

FIGURE 1-11   Thermal conductivity.

over a given area through a given thickness per degree difference between the hot and cold sides per unit of time. The units of thermal conductivity in the metric system are calorie-centimeters per second per square centimeter per degree Celsius. Translating this to the English system, it becomes Btu-inch (or foot) per second (or hour) per square foot per degree Fahrenheit. The formula for thermal conductivity is

$$Q = \frac{K(t_2 - t_1)aT}{d}$$

where Q = heat conducted (Btu)
K = coefficient of thermal conductivity
t_2 = temperature of the hot face
t_1 = temperature of the cooler face
a = area through which the heat is transmitted
T = time of the heat transmission
d = thickness of the material through which the heat is going

A look at Figure 1-11 explains this principle.

EXAMPLE 1-10:

If a furnace was heated to 1600°F and had a 4-in.-thick firebrick wall with a thermal conductivity of 50 Btu/ft²/in./°F/hr, and the outside temperature of the wall was kept at 100°F, how much heat would be lost through a 1-ft² area in a period of 5 hours through the wall?

Solution:

$$Q = \frac{(0.21 \text{ Btu-in./ft}^2/°F/hr)\ (1500°F)\ (5\ hr)\ (1\ ft^2)}{4\ in.} = 79 \text{ Btu}$$

The *heat of combustion* of a fuel or other material is the amount of heat per unit of weight that the material will emit when it is burned. The units are in calories per gram or Btu per pound. A slice of bread rated at 200 Cal will give off that amount of energy when oxidized whether burned in the atmosphere or oxidized in the body.

Thinker 1-14:

a. How could you determine the approximate heat of combustion of the wax in a candle?
b. Outline an experiment of how the heat of combustion of a piece of bread could be determined. Sketch your apparatus.

PROBLEM SET 1-6

1. How many calories would be required to heat a cup of water having 114 mL from a temperature of 20°C to 40°C?
2. How many Btu is needed to heat 50 gal of water from 70°F to 140°F? One gallon of water weighs 8.33 lb.

3. The coefficient of thermal expansion of aluminum is 13×10^{-6} in./in./°F. How much would a 20-in. bar expand with a temperature change from 70°F to 150°F?

4. What would be the coefficient of thermal expansion of aluminum in the metric system? (Use the information in Problem 3.)

5. A steel bridge is 50 ft long. If its coefficient of thermal expansion is 8.4×10^{-6} in./in./°F, how much will it elongate when the temperature goes up 20°F?

6. If the sun shines on 1 ft² of surface during the summer, approximately 230 Btu will be absorbed per hour. If a container holding 8 gal of water and having a surface area of 2 ft² sits in the sun for 2 hours, what will be the rise in temperature?

7. A furnace having a sidewall 3 in. thick is made of a firebrick having a thermal conductivity of 0.21 Btu-in./ft²/°F/hr. It has an inside temperature of 1000°F, an outside temperature of 100°F, and measures 12 in. by 18 in. How much heat will it lose per hour?

8. A candle that can provide 500 cal/min is used to heat up a cup of water for instant coffee. How long would it take to heat the 4 oz of water from 70°F to 120°F?

9. A certain metal has a specific heat of 0.11. How many calories would it take to heat a 250-g bar of this metal from 50° to 90°C?

10. Five pounds of the metal having a specific heat of 0.11 is heated from 70°F to 110°F. How many Btu is required?

11. Heat is removed from a metal having a specific heat of 0.10 at the rate of 300 cal/s. How long would it take to cool a 200-g bar from 700°C to 20°C?

12. If a gas water heater was able to produce 25,000 Btu/hr, how long would it take to heat 30 gal of water from 70°F to 140°F?

13. A firebrick for a furnace wall had a coefficient of thermal conductivity of 0.21 Btu-in./ft²/°F/hr. It was desired to lose no more than 100 Btu/hr. If the inside temperature of the wall was to be 1500°F, the outside temperature 100°F, and the area of the wall 20 in. by 24 in., how thick must the wall be?

14. If 2.0 ounces of fuel produced 2300 Btu, what is its heat of combustion in Btu per pound?

15. If it took 10 g of an oil to produce 50,000 cal, what is its heat of combustion in calories per gram?

16. If it took 0.2 oz of fuel to raise 2 lb of water 60°F, what is the heat of combustion of the fuel?

17. If 1.5 g of fuel was burned and it heated 500 g of water 7°C, what was the heat of combustion of the fuel?

18. The specific heat of a metal is 0.13. What is its heat capacity in
    (a) the metric system;
    (b) the English system?

19. A bar of metal was heated uniformly over its entire length from 20°C to 80°C. It expanded from a length of 30.00 cm to 30.015 cm. What is its coefficient of thermal expansion?

20. A 100-m-long bridge was made of steel having a coefficient of thermal expansion of 4.6×10^{-6} cm/cm/°C. How much space should it be given for expansion for a temperature range of 20 to 40°C?

Electrical properties of electrical conductivity, resistance, and resistivity are necessary to the design of motors, heating elements, electronic equipment, and electrical transmission lines.

Electricity is the flow of electrons from one atom to another along a conductor. To make these electrons move along the wire, a force is necessary. The unit of electrical force (or electromotive force) is the *volt*. A quantity of electrons that will deposit 0.001118 g of silver in an electroplating cell (electroplating cells are discussed in Chapter 4) is called a *coulomb*. This amounts to 6.24×10^{18} electrons. A coulomb of electrons flowing past a given point per second is an *ampere*. An ampere is a flow rate of electricity. One ampere driven by a force of 1 volt is a *watt*. The watt is a unit of electrical power. There is resistance to the flow of electrons along a wire. This resistance is measured in *ohms*. An ohm is the resistance to 1 ampere driven by a force of 1 volt. The energy used by electric appliances is measured in *watt-hours*. Kilowatt-hours are the units of electrical energy by which power companies charge their customers. The formulas for electricity are given in Figure 1-12. These formulas are actually for direct current, but work very well for house current also. Consider the following problem.

EXAMPLE 1-11:

A 100-W electric light bulb operates on a 110-V circuit. How many amperes does it draw?

Solution:

$$P = EI$$

$$I = \frac{P}{E}$$

$$= \frac{100 \text{ W}}{110 \text{ V}} = 0.9 \text{ A}$$

EXAMPLE 1-12:

What is the resistance of a heating element operating on a 110-V line drawing 10 A?

Solution:

$$R = \frac{E}{I}$$

$$= \frac{110 \text{ V}}{10 \text{ A}} = 11 \text{ }\Omega$$

The amount of electric current (amperes) a wire can conduct is determined by its resistivity and its thickness. The larger the wire, the more amperes it can conduct. If more current is forced through the wire than it can take, the resistance in the wire causes it to get hot. For this reason, electric circuit breakers and fuses are placed in the circuits to protect against possible fire hazards. Fuses limit the amount of amperage a circuit can draw.

The voltage is limited by the insulation around the wire. Without sufficient insulation, two wires touching can "short circuit," which can cause damage to the apparatus.

Voltage	$E = IR$
Power	$P = EI$ or $P = I^2R$
Resistance	$R = E/I$
Energy	$e = Pt$

where E = voltage (volts)
I = Amperage (amperes)
P = the power (watts)
R = the resistance (ohms)
e = the energy (watt-hours)
t = time (hours)

FIGURE 1-12   Electrical formulas.

PROBLEM SET 1-7

1. How many amperes does an 1100-W toaster operating on a 110-V circuit draw?

2. A clothes dryer operates on a 220-V line and is rated at 10 A. How many watts does it use?

3. A household circuit has a 20-A circuit breaker. Could that circuit take three 1100-W heating elements operating on 110 V?

4. A 100-W light bulb is left on continuously for 30 days. How many kilowatt-hours of energy does it use in that time?

5. What is the resistance of a 100-W light bulb operating on a 110-V line?

6. If electricity costs 13 cents per kilowatt-hour, what would be the electrical cost of drying a load of clothes in a dryer operating on a 220-V line drawing 10 A for 45 minutes?

7. A freezer is "on" approximately 15 minutes out of every hour. If it works on a 110-V line, draws 6 A, and the electricity costs 10 cents per kilowatt-hour, how much would it cost to operate that freezer for 30 days?

8. A water pump for a fish pool operates continuously for 30 days. It draws 0.5 A on a 110-V line. The price of the electricity is 10 cents per kilowatt-hour. What is its monthly (30-day) cost of operation?

9. An automobile light draws $\frac{1}{2}$ A on a 12-V circuit. What is the resistance of the light-bulb filament?

10. You forget to turn out your 200-W reading lamp prior to leaving for a party and are gone 4 hours. At a price of 12 cents per kilowatt-hour, what did this mistake cost you?

11. There are 746 W in 1 horsepower. A 1.0-hp motor is operated on a 220-V line. What is the minimum-size circuit breaker (amperes) which should be in that line to let the motor operate at its design horsepower?

12. A 1-hp pool pump is operated 4 hours a day on a 110-V line. At the rate of 11 cents per kilowatt-hour, how much would it cost to operate that pump for 30 days?

13. A cook decided to plug in a 1600-W skillet, an 1100-W coffee pot, and a 1300-W toaster on the same 30-A 110-V circuit. Would the circuit breaker take this load?

14. A little night light rated at 5 W is operated on a 110-V line and is left on continuously. How much does this add to the monthly electric bill (30 days) at the rate of 10 cents per kilowatt-hour?

15. An air conditioner that draws 20 A on a 220-V line runs 15 minutes out

of every hour. At the rate of 11 cents per kilowatt-hour, what is the monthly electric bill (30 days)?

16. There are 2547 Btu per hour in 1 horsepower. One horsepower also equals 746 W. How many Btu per hour equals 1 watt?

17. A 100-W light bulb puts out 96% of the energy it uses as heat. How many Btu per hour does this light bulb emit? (Use the information in Problem 16.)

18. A heating element has a resistance of 80 Ω and draws 5 A. What is its power consumption in watts?

19. A 1300-W heating element operates on a 110-V circuit. What is the resistance on this heating element?

20. A 6-V circuit has a resistance of 18 Ω.
    (a) How many amperes does it draw?
    (b) How many watts does it use?

Thinker 1-15:

If a motor can be set to operate either on 110 V or 220 V and would deliver the same power at either setting, would there be any advantage to operating the motor at 220 V?

Tests of Mechanical Properties

TENSILE TEST

One of the most common tests of metals, wood, plastics, and other materials is the *tensile test*. ASTM, SAE, ASM, ANSI, and other testing organizations generally agree on the procedures to be used in this test. Several companies market machines for this test. The tensile test is usually conducted using a standardized specimen, but it is not limited to these. All that is required is that the cross-sectional area of the test sample and its original length be measured.

Prior to discussing the procedures by which this test is performed, some basic terms associated with the test must be defined. Before the test specimen is placed in the testing machine, its cross-sectional area and its gauge length must be measured. If the specimen is cylindrical in shape, its cross-sectional area can be found by measuring the diameter with a micrometer. If it is other than a cylindrical shape, calipers or a micrometer can be used to determine its dimensions. The gauge length or original test length of the bar is often determined by a two-point gauge punch. The specimen is then mounted in the testing machine and a *tensile load* (usually denoted by the letter P in formulas) is applied to it. The dials on the testing machine indicate this load in pounds or kilograms. A load of 1000 lb applied to a $\frac{1}{2}$-in^2 cross section would be equivalent to a load of 4000 lb applied to an area of 1 in^2. Similarly, if the load of 1000 lb was applied to a circular cross-sectional sample with a diameter of 0.505 in., the cross-sectional area would be 0.2 in^2 and the stress would be 5000 lb/in^2. Figure 2-1 depicts this concept.

The load equivalent to that which would be on a 1-in^2 cross section of the material is called the *stress*. Stress (conventionally denoted by the letter S) is therefore defined as the load per unit of cross-sectional area and is measured in pounds per square inch, or pascal. The equation for stress is

$$\text{stress} = \frac{\text{load}}{\text{unit cross-sectional area}}$$

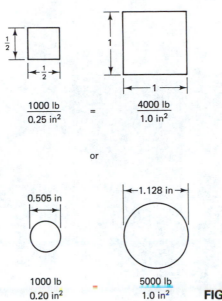

FIGURE 2-1   Equivalent loads.

If the test bar has a rectangular cross section, the formula for the stress becomes

$$S = \frac{P}{W \times D}$$

where S = stress
P = load
W = width of the bar
D = depth of the bar

If the test sample is cylindrical in shape, the stress becomes

$$S = \frac{P}{\pi d^2/4}$$

where d is the diameter of the bar.

It stands to reason that if a load is applied to the bar, it will get longer. This stretch is called the *elongation* and is measured in inches or millimeters. The elongation that each unit of original length stretches is known as the *strain* (denoted by the letter e). The elongation is measured by an *extensometer* (also called a strain gauge) similar to the one shown in Figure 2-2.

Strain can be defined as the elongation divided by the gauge length:

$$\text{strain} = \frac{\text{elongation}}{\text{gauge length}}$$

As a load (shown as P in Figure 2-3) is applied, the extensometer will show the bar getting longer. Let the original or gauge length be designated as L_0; then the length at any load (P) will be L. The elongation is the change in length. The strain (designated e) becomes the change in length divided by the original length:

$$e = \frac{L - L_0}{L_0}$$

FIGURE 2-2   Extensometer.

EXAMPLE 2-1:

Assume that a test bar was circular in cross section and had an original diameter of 0.75 in. Further, this sample was fitted with an extensometer which set at "zero" when it spanned an original length of 6 in. The original cross-section area would then be

$$\text{area} = \frac{\pi d^2}{4}$$

$$= \frac{3.1416 \times 0.75^2}{4} = 0.44 \ \text{in}^2$$

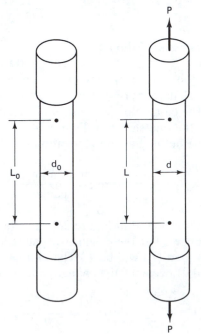

FIGURE 2-3   Test sample notations.

At the point where the tensile test machine showed a load of 1000 lb, the stress would be

$$\text{stress} = \frac{1000 \text{ lb}}{0.44} = 2300 \text{ psi}$$

In the metric system, the load would be recorded in kilograms and converted to newtons. The cross-sectional area would be expressed in square meters. Of course, a newton per square meter is a pascal.

An example of stress using the metric system would be as follows.

EXAMPLE 2-2:

Suppose that a cylindrical test sample was 1.00 cm in diameter and had a load of 5000 kg applied to it. The stress would be

$$\text{load} = (5000 \text{ kg})(9.8 \text{ m/s}^2) = 49,000 \text{ N}$$

$$\text{area} = \frac{3.1416(1 \text{ cm})^2/(100 \text{ cm/m})^2}{4} = 7.85 \times 10^{-5} \text{ m}^2$$

$$\text{stress} = \frac{49,000 \text{ N}}{(7.85 \times 10^{-5}) \text{ m}^2} = 6.24 \times 10^8 \text{ Pa} = 624 \text{ MPa}$$

If at the point where the load was 1000 lb, the extensometer had a reading of 0.024 in., the strain would be

$$\text{strain} = \frac{0.024 - 0}{6} = 0.004 \text{ in./in.}$$

It should be noted that the strain is recorded in inches of elongation per inch of original gauge length. In the metric system the strain could be recorded in meters of elongation per meter of original length.

PROBLEM SET 2-1

1. What is the stress in a 0.5-in.-diameter bar if it is subjected to a 3000-lb load?
2. What is the stress in a 0.75-in.-diameter bar when it is under a load of 10,000 lb?
3. What is the strain on a bar if a gauge length of 2.000 in. is elongated by 0.006 in.?
4. What is the strain in a test specimen if an 8.000-in. gauge length is elongated to 8.024 in.?
5. What is the stress in a 0.505-in.-diameter sample of metal that is under a load of 30,000 lb?
6. What is the stress in a bar that has a gauge length of 8.000 in. and a diameter of 0.75 in., is under a load of 25,000 lb, and is elongated by 0.320 in.?
7. What is the strain on the bar in Problem 6?
8. What is the stress in a 3.0-cm-diameter bar under a tensile load of 10,000 kg?
9. What is the stress in a 1.20-cm-diameter bar that has an applied tensile load of 1500 kg?

10. What is the strain in a bar with a 12-cm gauge length when it is elongated by 0.040 cm?

11. A bar with an 8-cm gauge length is elongated by 1.6 mm. What is the strain at that point?

12. A solid square bar of steel in a bridge measures 2 in. on a side in cross section and has a load of 2 tons applied. What would be the stress in that bar in lb/in²?

13. What would be the stress on the bar in Problem 12 in MPa?

14. If a 2-in.-diameter cylindrical steel bar was designed to withstand a tensile stress of 50,000 lb/in², would a load of 100 tons be too much for that bar?

15. A 5- by 8-cm (cross section) steel bar was designed to support a load of 50 MPa.
    (a) What would be the force in newtons equal to the design strength?
    (b) What load in kilograms would that be?

16. A steel bar must not be strained over 0.01 in./in. If the bar is 12 ft long, what would be the total elongation of the bar at maximum strain?

17. The stress on a bar is not to be over 80,000 lb/in². What must be the diameter of the bar to keep a load of 10 tons at that stress?

18. The stress on a bar is limited to 65,000 lb/in². What is the maximum load that could be put on a cylindrical bar having a diameter of 0.875 in.?

19. The strain on a bar in tension is 0.008 in./in. The original length of the bar was 6.000 in. What is the length of the bar having that strain?

20. The strain on a bar in tension is limited to 0.016 in./in. What was the original gauge length of the bar if it had a strain of 0.016 in./in. at a load that elongated the bar to a length of 8.128 in.?

The tensile testing machines are available in several sizes. Figure 2-4 shows a 60,000 lb machine, while Figure 2-5 shows one of the largest machines made. It is a one million pound machine that easily breaks 2 in. by 3 in. bars of steel in tension.

FIGURE 2-4   Tensile test machine.

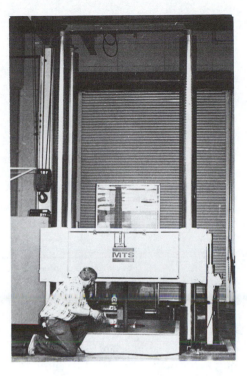

FIGURE 2-5 One million pound test machine.

In conducting a standard tensile test, a specimen that has been measured for its cross-sectional area and gauge length is placed in the testing machine and the extensonometer attached as shown in Figure 2-2. Simultaneous readings of load and elongation are taken at uniform intervals of load. The data are either recorded on a form such as shown in Figure 2-6, or fed directly into a computer. From the load and elongations readings, stress and strain can be calculated for each set of readings. A *stress–strain curve* can then be plotted using stress in lb/in² or MPa as the ordinate (vertical axis) and strain in in/in. or m/m as the abscissa (horizontal axis). A stress–strain curve for a typical ductile material is shown in Figure 2-7. (See the appendix for the proper method of constructing a graph.)

From the graph in Figure 2-7, many properties of the material can be determined. Note that the left portion of the curve is a straight line. In this linear region, the material will return to its original length if the load is removed. This is known as the *elastic region* of the curve. At some point on the curve, it will cease to be

Material: _____	Date: _____		
Diameter: _____	Operator(s): _____		
Gauge length: _____			
Reading No.	Load	Elongation	Stress	Strain
1				
2				
3				
.				
.				
.				

FIGURE 2-6 Tensile test report sheet.

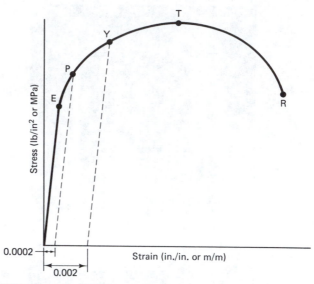

FIGURE 2-7   Typical stress–strain curve for a ductile material.

a straight line and starts to curve. The theoretical stress where the curve starts to bend is called the *elastic limit* (Point E in Figure 2-7) of the material. For many materials this exact point is difficult to detect. It is about the equivalent of measuring the height of a cloud: we can tell when we are in the cloud or out of it, but not exactly where it starts. Therefore, some artificial standard must be applied to fix a point at which the curve can be said to be nonlinear. One standard technique is to draw a line parallel to the straight-line portion of the graph which is offset by a distance of 0.0002 in./in. or 0.0002 m/m on the strain axis. The point where this offset line intersects the curve (point *P* in Figure 2-7) is known as the *proportional limit* of the material. After the elastic limit has been exceeded, the bar will have been permanently "stretched" and will no longer return to its original length when the load is removed. The area past the elastic limit is known as the *plastic region* of the curve.

Engineers use still another convention. If a line offset by a distance of 0.002 in./in. or m/m is drawn parallel to the elastic portion of the curve, the intersection of this line and the curve is known as the *yield point* (point *Y* on the graph) of the material. The yield stress is the one that engineers use as the maximum strength of the material for calculating the size of beams, thicknesses of gears, and other design work. Further, *safety factors* are used to reduce the design limit even more.

Safety factors have been called "ignorance factors." All engineering design involves a certain amount of assumption or guess work. In designing a bridge, gear train, or any other project, one never knows exactly what maximum loads it will eventually have to withstand. Further, one can never know exactly what the yield strengths of any given piece of material will be. Just because the test sample had a given strength, does that mean that all pieces made from that material have the same strength? (Not necessarily.) Compound these uncertainties with the uncertainty of workers (welders, riveters, etc.) always doing a perfect job and one can understand the need for safety factors.

The safety factor is the number of times that the design strength will go into the yield strength. If the yield strength of a steel were 50,000 lb/in², a safety factor of 2 would allow the engineer to design the part at 25,000 lb/in². Most modern machinery is designed with safety factors of between 2 and 3. There are exceptions, however, where safety factors outside this range are used.

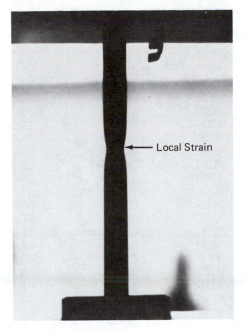

←— Local Strain

FIGURE 2-8   Local strain.

The highest stress that a material will take before it starts to fail is the *tensile strength* or *ultimate strength* of the material. The tensile strength is indicated by point *T* on the stress–strain curve in Figure 2-7. After this stress has been reached, the cross-sectional area of a ductile test specimen will start to decrease rapidly in what is known as necking or *local strain* (see Figure 2-8). After the tensile strength has been reached the loads indicated by the testing machine will begin to drop quickly. Finally, the bar will break. The stress at the breaking point is the *rupture strength* or *breaking strength* of the material (point *R* on Figure 2-7).

Another parameter that can be obtained from the stress–strain curve is the *modulus of elasticity* or *Young's modulus* of the material. The modulus of elasticity (conventionally labeled *E*) is defined as the stress divided by the strain in the elastic region of the stress–strain curve. Basically, it is the slope of the straight-line portion of the curve.

EXAMPLE 2-3:

Consider a sample problem using data taken from a tensile test of cold-worked copper (Figure 2-9). The modulus of elasticity can be calculated as follows:

$$\text{modulus of elasticity} = E = \frac{\text{stress}}{\text{strain}} \text{ (in the elastic region)}$$

$$\text{area} = A = \frac{\pi(0.505)^2}{4} = 0.200 \text{ in}^2$$

$$E = \frac{(2000 \text{ lb})/(0.200 \text{ in}^2)}{(0.0008 \text{ in.})/(2.000 \text{ in.})}$$

$$= \frac{10,000 \text{ lb/in}^2}{0.0004 \text{ in./in.}}$$

$$= 25,000,000 \text{ lb/in}^2$$

Material: __Copper (Sample Problem)__ Date: _____8-29-_____

Diameter: _____0.505 in._____ Operator(s): _____W.O.F._____

Gauge length: _____2.000 in._____

Reading No.	Load (lb)	Elongation (in.)	Stress (lb/in²)	Strain (in./in.)
1	0	0	0	0
2	1000	0.0004	5,000	0.0002
3	2000	0.0008	10,000	0.0004
4	3000	0.0012	15,000	0.0006
5	4000	0.0016	20,000	0.0008
6	5000	0.0026	25,000	0.0013
7	6000	0.0035	30,000	0.0017
8	6420	0.0040	32,100	0.0020
9	7060	0.0050	35,300	0.0025
10	7480	0.0060	37,400	0.0030
11	7800	0.0070	39,000	0.0035
12	7860	0.0080	39,300	0.0040
13	8000	0.0090	40,000	0.0045
14	8050	0.0100	40,200	0.0050
15	8150	0.0150	40,700	0.0075
16	8150	0.0200	40,700	0.0100
17	8040	0.04	40,200	0.02

Note: Extensonometer removed at this point and readings were made by dividers, thus the loss of some precision.

18	7940	0.06	39,700	0.03
19	7840	0.08	39,200	0.04
20	7710	0.10	38,500	0.05
21	7580	0.12	37,400	0.06
22	7430	0.14	37,200	0.07
23	7250	0.16	36,200	0.08
24	7060	0.18	35,200	0.09
25	6850	0.20	34,200	0.10
26	6600	0.22	33,000	0.11
27	6330	0.24	31,600	0.12
28	6020	0.26	30,100	0.13
29	5660	0.28	28,300	0.14
30	5270	0.30	26,300	0.15
31	3700	0.36	18,500	0.18
32	3000	0.38	15,000	0.19
33	2500	0.40	12,500	0.20
34	2000	0.42	10,000	0.21
35	1000	Break (0.44)	5,000	0.22

FIGURE 2-9   Tensile test report sheet for Example 2-3.

In the metric system this could be converted to read

$$A = (0.200 \text{ in}^2)(2.54 \text{ cm/in.})^2(0.01 \text{ m/cm})^2 = 1.29 \times 10^{-4} \text{ m}^2$$

$$E = \frac{\left(\dfrac{2000 \text{ lb}}{2.2 \dfrac{\text{lb}}{\text{kg}}}\right)\left(\dfrac{9.8 \dfrac{\text{m}}{\text{s}^2}}{1.29 \times 10^{-4} \text{ m}^2}\right)}{\dfrac{(0.0008 \text{ in.})\left(2.54 \dfrac{\text{cm}}{\text{in.}}\right)}{(2 \text{ in.})\left(2.54 \dfrac{\text{cm}}{\text{in.}}\right)}}$$

$$= 1.73 \times 10^{11} \text{ Pa}$$

or

$$E = \frac{1.73 \times 10^{11} \text{ Pa}}{1 \times 10^6 \text{ Pa/MPa}}$$

$$= 1.73 \times 10^5 \text{ MPa}$$

The modulus of elasticity is especially valuable since it allows a designer to calculate how much a bar of given length will stretch under given loads. Consider the following example.

EXAMPLE 2-4:

A 1-in.-diameter steel bar 10 ft long is subjected to a load of 50,000 lb. If the modulus of elasticity for the steel is 30,000,000 lb/in^2 what would be its elongation?

Solution:

$$E = \frac{\text{stress}}{\text{strain}} = \frac{\text{load/area}}{\text{elongation/original length}}$$

$$= \frac{50,000 \text{ lb}/[\pi(0.5 \text{ in.})^2]}{\text{elongation}/[(10 \text{ ft})(12 \text{ in./ft})]} = 30,000,000 \text{ lb/in}^2$$

$$30,000,000 \text{ lb/in}^2 = \frac{(50,000 \text{ lb})(120 \text{ in.})}{(0.785 \text{ in}^2)(\text{elongation})}$$

or

$$\text{elongation} = \frac{(50,000 \text{ lb})(120 \text{ in.})}{(0.785 \text{ in}^2)(30,000,000 \text{ lb/in}^2)}$$

$$= 0.255 \text{ in.}$$

A similar problem in the metric system would be as follows.

EXAMPLE 2-5:

A 2.0-cm-diameter bar 3 m in length is subjected to a load of 2000 kg. If its modulus of elasticity were 205,000 MPa, what would be its elongation?

Solution:

$$205,000 \text{ MPa} = 2.05 \times 10^{11} \text{ Pa}$$

Since a pascal can be written as "a newton per square meter" and a newton is a kilogram-meter per second squared, therefore

$$2.05 \times 10^{11} \text{ Pa} = 2.05 \times 10^{11} \text{ kg-m/m}^2\text{-s}^2$$

Converting this to kilograms per square meter yields

$$\frac{2.05 \times 10^{11} \text{ kg-m/m}^2\text{-s}^2}{9.8 \text{ m/s}^2} = 2.09 \times 10^{10} \text{ kg/m}^2$$

Using the formula for the modulus of elasticity gives

$$2.09 \times 10^{10} \text{ kg/m}^2 = \frac{2000 \text{ kg}/[\pi(0.02 \text{ m})^2/4]}{\text{elongation}/3 \text{ m}}$$

Then

$$2.09 \times 10^{10} \text{ kg/m}^2 = \frac{(2000 \text{ kg})(3 \text{ m})}{\pi(0.0001 \text{ m}^2)(\text{elongation})}$$

or

$$\text{elongation} = \frac{(2000 \text{ kg})(3 \text{ m})}{\pi(0.0001 \text{ m}^2)(2.09 \times 10^{10} \text{ kg/m}^2)}$$

$$= 9.1 \times 10^{-4} \text{ m}$$

or

$$\text{elongation} = 0.91 \text{ mm}$$

The *ultimate percent elongation* is that elongation which the material has at its rupture point. From the graph in Figure 2-10 the ultimate percent elongation would be

$$\% \text{ elongation} = \frac{(L_f - L_0)(100)}{L_0}$$

$$= \frac{(2.44 - 2.000)(100)}{2.000} = 22\%$$

Note that the elongation at the rupture point should be extrapolated from the stress–strain curve. It is not the last elongation reading taken before rupture.

As a material is stretched, the cross-sectional area of the material gets smaller. The negative of the ratio of the strain in the cross-sectional direction to the strain in the direction of the tension is called *Poisson's ratio* (pronounced "pwah-sons") of the material. The lowercase Greek letter eta (η) is generally used as the symbol for Poisson's ratio. This is measured only in the elastic region of the stress–strain curve. For example, if a 2.000-in. gauge length of a steel bar were elongated to a length of 2.010 in., at which time the diameter changed from 0.5050 in. to

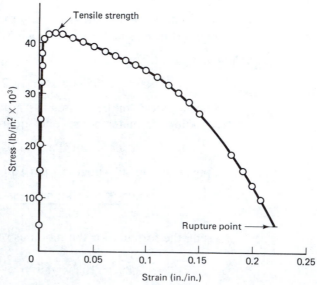

FIGURE 2-10   Stress–strain curve for sample problem.

0.5042 in., Poisson's ratio for that material would be

$$\text{Poisson's ratio} = \eta = \frac{-e_{\text{cross section}}}{e_{\text{longitudinal}}}$$

$$= \frac{-(d - d_0)/d_0}{(L - L_0)/L_0}$$

$$= \frac{-(0.5042 - 0.505)/0.505}{(2.010 - 2.000)/2.000}$$

$$= 0.316$$

Poisson's ratio for a perfectly incompressible material is 0.500. Poisson's ratio for most materials ranges from 0.25 to 0.5. The accepted value for Poisson's ratio for steel is 0.303. Poisson's ratio for pure solid rubber is 0.5. Until recently there was no material that had a Poisson's ratio greater than 0.5 since that would indicate that the volume of the material actually increased as the material was stretched. Now a few composites have been reported to have a Poisson's ratio in excess of 0.5. The complete curve for the data shown in the sample problem above is shown in Figure 2-10.

It is sometimes desirable to construct an expanded stress–strain curve showing only the straight-line portion of the data. This is an aid in determining the proportional limit, the yield strength, and the modulus of elasticity of the material. Figure 2-11 shows the expanded stress–strain curve for the data above.

A summary of the data from the test of copper shown above would be as follows:

Proportional limit:	30,000 lb/in²
Yield stress:	36,000 lb/in²
Tensile strength:	40,700 lb/in²
Rupture strength:	5,000 lb/in²

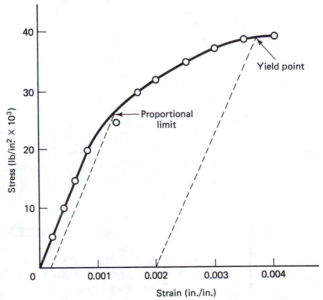

FIGURE 2-11   Expanded stress–strain curve for sample problem.

Average modulus of
    elasticity: 25 × 10 lb/in²
Ultimate percent
    elongation: 22%

While the curve shown in Figure 2-7 is typical for most ductile nonferrous materials, there are some exceptions. Steel has a distinctive stress–strain curve, as shown in Figure 2-12.

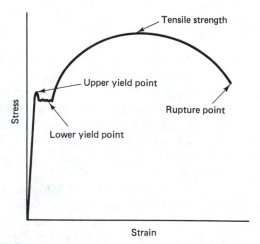

FIGURE 2-12   Stress–strain curve for steel.

Cast iron and other brittle materials always break while the load is still rising. Further, many brittle materials may not have a clearly linear portion of the graph. Figure 2-13 shows a typical cast-iron curve.

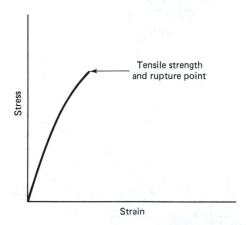

FIGURE 2-13   Stress–strain curve for cast iron.

The yield point for these nonlinear brittle materials is generally taken to be one-fourth of the average of breaking stresses of the samples tested. The modulus of elasticity is the slope of the line at very low stresses.

PROBLEM SET 2-2

1. A point in the linear portion of a stress–strain graph has a stress of 5000 lb/in² and a strain of 0.004 in./in. What would be the modulus of elasticity of the material in lb/in²?

2. At a point on the linear portion of a stress–strain curve, the stress was

175 MPa and the strain was 0.001 m/m. What was the modulus of elasticity of the material in MPa?

3. Given the following data for the tensile test of a material:

| Diameter: | 0.505 in. |
| Gauge length: | 2.000 in. |

Reading No.	Load (lb)	Elongation (in.)
1	0	0
2	1000	0.0003
3	2000	0.0006
4	3000	0.0009
5	4000	0.0012
6	5000	0.0015
7	6000	0.0018
8	7000	0.0022
9	8000	0.0026
10	8400	0.0032
11	8600	0.0038
12	8800	0.0044
13	9000	0.0052
14	9200	0.0062
15	9800	0.0100
16	9700	0.0150
17	9500	0.0210
18	9300	0.0300
19	9100	Break

(a) What is the stress at a load of 6000 lb?
(b) What is the strain at a load of 8000 lb?
(c) Plot the complete stress–strain curve for these data.
(d) What is the tensile strength of the material?
(e) What is the rupture strength of the material?
(f) What is the ultimate percent elongation of the material?
(g) What is the modulus of elasticity for the material?

4. Consider a square metal bar 7 ft long with a cross-sectional area of 0.75 in². The modulus of elasticity is 30×10^6 lb/in². How much would the bar elongate under a tensile load of 20,000 lb?

5. A 5-cm-diameter solid cylindrical bar 3.2 m long is placed under a tensile load of 3000 kg. Its modulus of elasticity is 200,000 MPa. How much would it elongate?

6. A solid square-cross-section copper bar 9 ft long with a modulus of elasticity of 16×10^6 lb/in², under a tensile load of 5000 lb, should elongate only 0.03 in. What should be the cross-sectional dimensions of the bar?

7. If the bar in Problem 6 had a circular cross section, what would be its diameter?

8. A rectangular bar 1.5 cm by 3 cm in cross section and 2.5 m long has a modulus of elasticity of 200,000 MPa. How much would it elongate under a 1000-kg load?

9. A 0.7500-in.-diameter 8.0000-in. long cylindrical metal bar had its diameter reduced to 0.7440 in. under a load that elongated it 0.1500 in. What is the Poisson's ratio of the material?

10. A cylindrical metal solid bar 0.5250 in. in diameter and 10.0000 in. long has a Poisson's ratio of 0.300. Under a load that elongates the bar 0.2000 in., what would be the new diameter of the bar?

11. A circular-cross-section metal bar 2.000 cm in diameter had a reduction in diameter of 0.003 cm while elongating from 10.0000 cm to 10.0580 cm. What is the Poisson's ratio of the metal?

12. The modulus of elasticity of Kevlar (a plastic used in composites) is 138 gPa. What is the modulus of elasticity in pounds per square inch?

13. The modulus of elasticity of a graphite reinforcement fiber is listed as 37×10^6 lb/in². What is the modulus of elasticity in the metric system?

14. A fiber of Kevlar is 0.04 mm in diameter. It has a load of 500 g applied. How much would a 50-cm-long piece of Kevlar be expected to elongate if its modulus of elasticity is 138 gPa?

15. A 5-ft-long bar of steel having a modulus of elasticity of 30×10^6 lb/in² is elongated 0.240 in. under a load of 8000 lb. What is the cross-sectional area of the bar?

16. A bar of copper has a modulus of elasticity of 16×10^6 lb/in². It is circular in cross section and a 10-in. length elongates 0.0200 in. under a load of 5 tons. What is the diameter of the bar?

17. A 3-m-long metal brace for a bridge is stretched 0.600 cm under a load of 2800 kg. If the diameter of the bar were 1.5 cm, what would be the modulus of elasticity of the material? (Assume that it is still in the elastic region of the curve.)

18. A structure weighing 17 tons is to be hung on a steel cable having a cross-sectional area of 0.75 in². If the yield point of the steel is 65,000 lb/in², would the cable be able to hold the load?

19. A steel cable 0.625 in. in diameter is to be used to hoist a block of concrete weighing 10 tons. If the modulus of elasticity is 30×10^6 lb/in² and the cable was 10 ft long, how much would it stretch under the load?

20. A graphite rope having modulus of elasticity of 200 gPa must lift a load of 500 kg. If the rope has a diameter of 12 mm and a length of 3 m, how much will it elongate under that load?

COMPRESSION TESTS

Compression is the result of forces pushing toward each other.

> *Thinker 2-1:*
>
> Analyze a bicycle and state which parts are in compression.

Compression tests are run in much the same manner as the tension test. The test specimen is placed in the testing machine and the compressometer is attached (see Figure 2-14).

After the cross-sectional area and the gauge length are determined, a load is applied at a constant rate and the load and deformation are recorded simultaneously for points at given intervals. From these data, a compressive stress–strain curve can be drawn. Figure 2-15 shows a typical compressive stress–strain curve for wood tested parallel to the grain.

In testing a specimen in compression, care must be taken to keep the com-

FIGURE 2-14   Compression test setup.

pressive forces exactly in the same line. If one of these pushing forces tilts or gets off-center, the specimen will start to bend and will break in flexure, not compression. To assure that the specimen is truly in compression, the ends of the specimen must be flat and parallel and the *length of the sample should not be over twice the smallest cross-section dimension* (see Figure 2-16). In many instances, only the ultimate compressive strength of the material is needed. In this case, the compression gauge is not used and only the highest load at which the specimen withstands before breaking is recorded.

Compressive strength is always reported in pounds per square inch or pascal. The load read from the test machine must be divided by the cross-sectional area. For example, if a 2 in. by 2 in. by 6 in. Douglas fir wood specimen fails under a load of 8000 lb, its compressive strength will be reported as 2000 lb/in^2:

$$\text{compressive strength} = \frac{8000 \text{ lb}}{(2 \text{ in.})^2}$$

$$= 2000 \text{ lb/in}^2$$

Concrete is one material nearly always tested in compression. The standard commercial test specimen is 6 in. in diameter and 12 in. long. The concrete is tested to failure using a steady loading rate (see Figure 2-17). The highest load is then

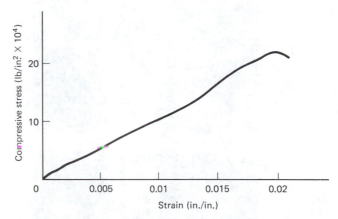

FIGURE 2-15   Typical compressive stress–strain curve for wood.

Bad

L = 2d

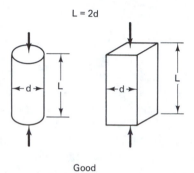

Good

FIGURE 2-16   Compressive test specimen.

recorded and the compressive strength of the concrete is calculated in pounds per square inch. As an example, if a standard 6- by 12-in. concrete sample withstood a load of 120,000 lb before it failed, the compressive strength of the concrete would be calculated as follows:

$$\text{area} = \pi r^2 = 3.1416 \times 3^2 = 28.3 \text{ in}^2$$

$$\text{compressive strength} = \frac{120,000 \text{ lb}}{28.3 \text{ in}^2} = 4240 \text{ lb/in}^2$$

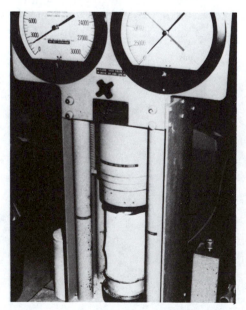

FIGURE 2-17   Concrete in a compression test.

Wood, plastics, metals, brick, ceramics, and other materials are tested in a similar fashion.

PROBLEM SET 2-3

1. A 2- by 2-in. square-cross-section piece of wood failed in compression at a load of 6250 lb. What was its compressive strength?

2. A standard 6-in.-diameter by 12-in.-long concrete test sample failed in compression under a load of 90,000 lb. What was the compressive strength of the concrete?

3. An architect specified that the concrete used in a certain building had to have a compressive strength of 3000 lb/in². Three standard 6-in.-diameter by 12-in.-long test samples broke under compression loads of 87,000, 89,500, and 92,000 lb, respectively. Is the concrete acceptable for use in the building based on the average compressive strengths of the samples?

4. The heel of a woman's high-heeled shoe has a square cross section of 0.25 in. by 0.25 in. What would be the compressive stress on the heel if a woman weighing 125 lb puts her entire weight on that heel?

5. A table (with four legs) has a 94-lb sack of portland cement placed in the middle of it. The legs are made of wood and are 3 in. by 3 in. in cross section. What is
    (a) the load and
    (b) the stress on each leg?

6. A concrete sample was found to have a compressive strength of 3000 lb/in². What should be the dimensions of a square (cross-section) pedestal to hold up a bronze statue weighing 150,000 lb?

7. If the pedestal in Problem 6 were circular in cross section instead of square, what should be its diameter?

8. A 5- by 5-cm square cross section of wood failed at a compressive load of 3100 kg. What is the compressive strength of the wood in MPa?

9. If a circular-cross-section piece of wood 10 cm in diameter failed under a compressive load of 310 kg, what would be the compressive strength of the wood in MPa?

10. Would it be safe to place a machine weighing 4800 lb on a three-legged bench which had cylindrical wooden legs 2.5 in. in diameter if the wood had a compressive strength of 920 lb/in²? Assume that the load is evenly distributed over the three legs and that the table will not tip over.

11. A square (cross-section) concrete piling was to hold up a load of 45,000 kg. The concrete's compressive strength was 22 MPa. What was the cross-section dimensions of the concrete in centimeters?

12. A column 6 cm by 6 cm (square) in cross section had a compressive strength of 27 MPa. What was the maximum load that it could support?

13. A load of 7100 kg was to be supported by columns each 2.5 cm by 2.0 cm (rectangular) in cross section. If the compressive strength of the columns was 20 MPa, how many columns would be required to support the load?

14. A load of 13,000 lb has to be supported by circular-cross-section columns each with a compressive strength of 1500 lb/in². How many legs would be required if the diameter of each leg was 1.5 in.?

15. A reinforced concrete column with a circular cross section 9 in. in diameter was designed to support 30 tons. What must be the compressive strength of the column in lb/in²?

16. A concrete pier was designed to take a maximum of 2800 lb/in². The square pier measured 6 in. on a side. Would this pier safely support a load of 20 tons?

17. A concrete pier was designed to take a maximum of 200 MPa. The square pier measured 15 cm on a side. Would this pier safely support 10,000 kg?

18. A particular batch of concrete has a compressive strength of 3200 lb/in². It is to support a load of 50 tons. What must be the diameter of a circular column of the concrete?

19. It is desired to break a test block of wood 8 in. square having a compressive strength of 1200 lb/in². Would a tester that could apply a load of 60,000 lb be able to break the block?

20. What is the maximum-diameter cylinder of concrete having a compressive strength of 7000 lb/in² which could be broken on a test machine that could exert a 120,000-lb load?

SHEAR TESTS

Shear is the result of parallel but slightly off-axis forces being applied to a body. Shear can be applied either in compression or tension (see Figure 2-18). Rivets, bolts, screw threads, and cotter pins are but a few examples of parts that have shear forces applied to them.

Thinker 2-2:

Are there any parts in shear in the chair on which you are sitting?

Shear tests of the materials are again run in the universal testing machine using a standard shear testing tool (see Figure 2-19). In setting up a shear test, care must be taken that the forces applied are only slightly off-center. If the top part of a shear tool is allowed to overlap the bottom part of the tool, part of the specimen will be in compression. If there is too much offset distance between the top force and the bottom force, the specimen will tend to go into flexure. Further, the forces must be kept perpendicular to the test specimen or the test sample could be put in a tension mode. Any of these mistakes would cause the test to be faulty (see Figure 2-20). Only the ultimate load required to fail the material is usually needed. This load is then divided by the entire original cross section of the *sheared surfaces* to get the shear strength of the material in pounds per square inch or pascal.

Shear can be tested by single shear or double shear, depending on the number

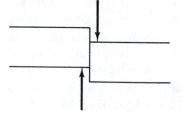

FIGURE 2-18   Shear.

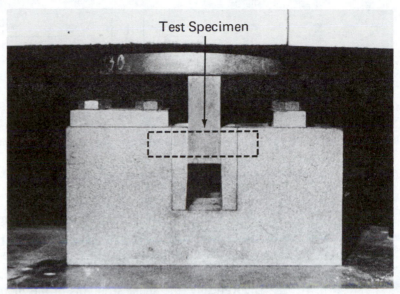

FIGURE 2-19   Shear test tool.

of surfaces cut (see Figure 2-21). If a $\frac{1}{2}$-in.-diameter bar required a load of 8000 lb in single shear to make it fail, the shear strength of the bar would be calculated as follows:

$$\text{area} = 3.1416(0.25 \text{ in.})^2 = 0.196 \text{ in}^2$$

$$\text{shear strength} = \frac{8000 \text{ lb}}{0.196 \text{ in}^2}$$

$$= 40,800 \text{ lb/in}^2$$

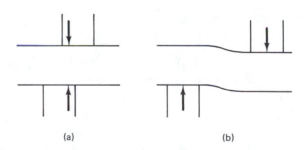

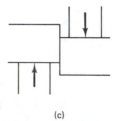

FIGURE 2-20   Shear alignment: (a) wrong (compression); (b) wrong (flexure); (c) correct (shear).

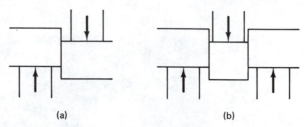

(a)   (b)

FIGURE 2-21   (a) Single and (b) double shear.

If the same bar, tested in double shear, took 16,000 lb to cause it to fail, the calculations would be

$$\text{area} = 3.1416 \times (0.25 \text{ in.})^2 \times 2 \text{ surfaces} = 0.393 \text{ in}^2$$

$$\text{shear strength} = \frac{16{,}000 \text{ lb}}{0.393 \text{ in}^2} = 40{,}700 \text{ lb/in}^2$$

Theoretically, the shear strengths of the metal as determined by the single-shear and double-shear techniques should be the same, but there is usually a slight difference due to the test bar bending or going into flexure when tested in single shear. Also, some difference will occur between the tests due to errors in reading the dials on the test machine. The results of the single- and double-shear tests in the example shown here are due to rounding off and are the same to two significant figures.

In tension testing there is a modulus of elasticity or Young's modulus (E), defined as the stress per unit strain in the elastic region of the material. There is also a *shear modulus* (denoted G by convention), which is defined as the shear stress divided by the shear strain in the elastic region. The shear stress is usually denoted by the lowercase Greek letter tau (τ) and the shear strain by the lowercase Greek letter gamma (γ). The units of shear stress are either pounds per square inch (lb/in²) or pascal (Pa). The equation for the shear modulus is

$$G = \frac{\tau}{\gamma}$$

The shear strain is defined in terms of the angle (α, alpha) which a shear stress produces on the atoms (see Figure 2-22). The formula for shear strain becomes

$$\gamma = \tan \alpha$$

The shear modulus is related to the modulus of elasticity (Young's modulus) by the equation

$$E = 2G(1 + \eta)$$

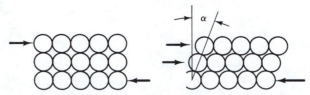

FIGURE 2-22   Shear strain angle.

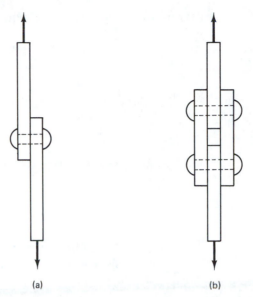

FIGURE 2-23 (a) Single and (b) double shear of rivets.

where η is Poisson's ratio. Since Poisson's ratio usually falls between 0.25 and 0.5, the shear modulus is generally about one-third to two-fifths the modulus of elasticity. The shear modulus of steel (Poisson's ratio 0.303) is 11×10^6 lb/in^2.

The test of rivets is one practical example of shear test applications. Samples of steel or other material are riveted together in either single or double shear and then placed in the tensile test machine. The pulling apart of the steel places the rivets in shear (see Figure 2-23). The rivets will be sliced by the steel plates. The diameter of the rivets is measured and the maximum load it takes to cut the rivets is recorded. The shear strength of the rivets is calculated either as the strength of the metal in pounds per square inch or pascal, or as the strength per rivet. For example, suppose that two $\frac{1}{4}$-in.-diameter rivets, tested in single shear, required a force of 4000 lb to break. The shear strength of the metal in the rivets would be

$$\text{area per rivet} = \pi r^2 = 3.1416 \left(\frac{0.25}{2}\right)^2 = 0.049 \text{ in}^2$$

$$\text{shear strength} = \frac{4000 \text{ lb}}{(0.049 \text{ in}^2)(2)} = 41,000 \text{ lb/in}^2$$

The shear strength per rivet would be

$$\text{strength} = \frac{4000 \text{ lb}}{2 \text{ rivets}} = 2000 \text{ lb/rivet}$$

Using the data above, if a load of 10 tons had to be supported by these rivets, it would require a cross-sectional area of the rivets of

$$\frac{(10 \text{ tons})(2000 \text{ lb/ton})}{41,000 \text{ lb/in}^2} = 0.49 \text{ in}^2$$

or 0.49 in² of metal on one rivet with a diameter of

$$0.49 = \frac{\pi d^2}{4} = 0.7854 d^2$$

$$d^2 = \frac{0.49}{0.7854}$$

$$d = 0.8 \text{ in.}$$

Figured another way, the 10 tons could be supported by ten $\frac{1}{4}$-in.-diameter rivets:

$$\frac{(10 \text{ tons})(2000 \text{ lb/ton})}{2000 \text{ lb/rivet}} = 10 \text{ rivets}$$

If the same $\frac{1}{4}$-in.-diameter rivets were tested in double shear and required 8000 lb to shear them, the calculations would be

$$\text{area of cut rivet} = \pi \left(\frac{0.25 \text{ in.}}{2} \right)^2 \times 2 = 0.098 \text{ in}^2$$

$$\text{shear strength} = \frac{(8000 \text{ lb})}{(2 \text{ rivets})(0.098 \text{ in}^2)} = 41{,}000 \text{ lb/in}^2$$

Similarly, the shear strength per rivet would be

$$\text{strength} = \frac{8000 \text{ lb}}{2 \text{ rivets}} = 4000 \text{ lb per rivet}$$

In this double-shear mode the same 10 tons could be supported by 5 rivets:

$$\frac{(10 \text{ tons})(2000 \text{ lb/ton})}{4000 \text{ lb/rivet}} = 5 \text{ rivets}$$

PROBLEM SET 2-4

1. A wooden dowel pin $\frac{1}{2}$ in. in diameter failed under a single-shear load of 920 lb. What is the shear strength of the dowel in lb/in²?
2. A brass bar $\frac{3}{8}$ in. in diameter tested in single shear failed under a load of 4200 lb. What is the shear strength in lb/in²?
3. A 1.2-in.-diameter wooden dowel tested in double shear failed at a load of 1900 lb. What is its shear strength in lb/in²?
4. A 1.2-in. square-cross-section bar tested in double shear failed at a load of 15,200 lb. What would be its shear strength?
5. A metal circular-cross-section bar with a diameter of 4.5 mm failed under a load of 2100 kg in single shear. What is the shear strength in MPa?
6. If a metal bar having a shear strength of 9000 lb/in² was required to support a load of 7200 lb in single shear, what size diameter must it have?
7. A shear pin in a lathe is designed to fail if the shear force on it exceeds 130 lb in double shear. If the shear strength of the metal is 5200 lb/in², what diameter should the pin have?
8. In an outboard boat engine, the propeller is protected by a shear pin that

will shear in two places at a load of 150 kg. The diameter of the pin is 2.5 mm. What is the shear strength of the metal in pascal?

9. A total load of 20 tons must be supported by rivets in single shear. How many $\frac{1}{4}$-in.-diameter rivets are required if the shear strength of each rivet is 2000 lb?

10. If a total load of 15 tons must be supported by rivets in single shear, how many $\frac{1}{4}$-in. rivets of a metal whose shear strength is 30,000 lb/in^2 will be needed to hold the load?

11. How many rivets would be needed for the load in Problem 10 if the rivets were in double shear?

12. A total load of 1000 kg must be supported by rivets in double shear. The rivets have a design shear strength of 90 MPa. How many 5-mm-diameter rivets are required?

13. A shear stress of 10,000 lb/in^2 applied to a specimen deformed it through an angle of 4°. What is the shear modulus of the material?

14. A shear stress of 1,000,000 Pa deformed it through an angle of 3°. What is the shear modulus of the material?

15. Poisson's ratio of a metal is 0.31 and its modulus of elasticity is 29 × 10^6 psi. What is the shear modulus of the material?

16. Poisson's ratio of a metal is 0.33 and its modulus of elasticity is 212 MPa. What is its shear modulus?

17. A load of 5000 lb on a circular-cross-section bar of metal having a diameter of 0.75 in. deformed it through an angle of 3° in shear. What is the shear modulus of that material?

18. A load of 3000 kg on a bar having a diameter of 1.5 cm deformed it through an angle of 2.5° in shear. What was the shear modulus of the material?

19. A load of 6000 lb on a bar having a diameter of 0.5 in. produced a deformation in shear of 6°. If the bar had a Poisson's ratio of 0.3 what would be the modulus of elasticity of the material?

20. A load of 40,000 kg on a bar having a diameter of 2 cm produced a deformation in shear of 5°. If the Poisson's ratio of the bar was 0.31 what would be the modulus of elasticity of the material?

TORSION TESTS

Torsion is the twisting of a material. In turning off a water faucet or tightening a screw, the material is put in torsion.

Thinker 2-3:

List as many parts as you can on your automobile that are in torsion.

In determining the torsion strength of a material, one end of a bar is held firmly while a twisting action is applied to the other. The torque applied is measured in inch-pounds, foot-pounds, or newton-meters. The angle through which the bar is twisted is measured in degrees by an instrument known as a *tropometer*. A stress–strain curve for the material can be plotted. The applied stress for a solid cylindrical bar can be calculated by the formula

$$S_t = \frac{16T}{\pi d^3}$$

where S_t = torsion strength applied (pounds per square inch or pascal)
  T = torque (inch-pounds or newton-meters)
  d = diameter of the test bar (inches or meters)

If the test sample is a pipe or other hollow tube, the formula becomes

$$S_t = \frac{16T d_{outside}}{\pi (d^4_{outside} - d^4_{inside})}$$

The torsion strain for both a hollow and a solid bar is given by the formula (Figure 2-24)

$$e_t = \frac{\theta d \pi}{360L}$$

where θ = angle applied (degrees)
  L = gauge length (inches or meters)
  d = outer diameter of the bar (inches or meters)
  e_t = torsion strain

Note: The units for the diameter and the length must be the same.

The term "inch-pounds of torque" or "foot-pounds of torque" means inches times pounds or feet times pounds and can be found by multiplying the force placed on a bar in torsion by the distance it is from the center of the shaft (Figure 2-25). For example, if a force of 75 lb is exerted on the end of a 20-in.-long wrench, a torque of 1500 in.-lb is applied.

$$T = 75 \text{ lb} \times 20 \text{ in.} = 1500 \text{ in.-lb}$$

The formulas for stress and strain for samples with other shapes become very complicated but can be found in many standard engineering handbooks or text-books.

The calculations for many engineering applications of torsion require only

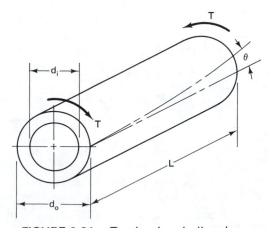

FIGURE 2-24   Torsion in a hollow bar.

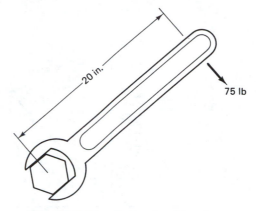

FIGURE 2-25   Example of torsion.

the maximum torsion strength of the material. This can be calculated using the formulas above at the maximum loads sustained by a test bar before failure. For example, suppose that a $\frac{1}{2}$-in.-diameter solid cylindrical bar withstood a torsion load of 900 in.-lb before failing. Its maximum shear strength would be calculated as follows:

$$S_t = \frac{16T}{\pi d^3}$$

$$= \frac{16(900 \text{ in.-lb})}{\pi(0.5 \text{ in.})^3} = 36{,}700 \text{ lb/in}^2$$

In the metric system torque is measured in newton-meters. For example, if a load of 25 kg was applied to the end of a 20-cm lever, the torque would be

$$T = (25 \text{ kg})(9.8 \text{ m/s}^2)(20 \text{ cm})(0.01 \text{ m/cm}) = 49 \text{ N-m}$$

Thinker 2-4:

Explain how a standard "torque wrench" used on an automobile could be used to find the torsion strength of a small rod.

Material: _____Solid Bar_____   Date: _____Dec. 12,--_____
Length: _____10 in._____   Inside diameter: _____--_____
Outside diameter: _____0.5 in._____   Run by: _____W.O.F._____

Reading No.	Torque (in.-lb)	Angle (deg)	Stress (lb/in²)	Strain (deg)
1	100	2	4,070	0.0009
2	200	4	8,150	0.0017
3	300	6	12,200	0.0026
4	400	8	16,300	0.0035
5	500	10	20,400	0.0044
6	600	12	24,400	0.0052
7	600	14	24,400	0.0061
8	650	17	26,500	0.0074
9	675	24	27,500	0.0105
10	680	Break	27,700	—

FIGURE 2-26   Torsion stress–strain report sheet.

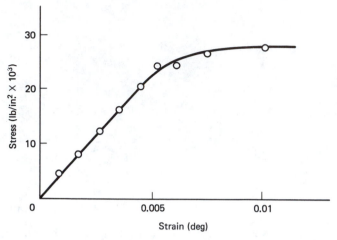

FIGURE 2-27   Torsion stress–strain curve.

The torsion stress–strain curve could be constructed if readings of torsion load and the angle at which that load occurred were obtained for several points. From the torque the stress could be calculated while the strain could be determined from the angle (Figure 2-26). The stress–strain curve would be as shown in Figure 2-27.

PROBLEM SET 2-5

1. A torsion load of 50 lb is applied to the end of a 15-in. wrench. What is the torque in foot-pounds?

2. What is the torque in Problem 1 in newton-meters?

3. A torsion load of 20 kg is applied to the end of a 50-cm wrench. What is the torque in newton-meters?

4. A solid round bar 0.75-in. in diameter has a torque of 100 ft-lb applied. What is the stress in the bar in lb/in²?

5. A hollow pipe with an inside diameter of 1.000 in. and an outside diameter of 1.250 in. has a torque of 75 ft-lb applied. What is the stress in the pipe in lb/in²?

6. A solid cylindrical bar 2.0 cm in diameter has a torque of 30 N-m applied. What is the stress in the bar in pascal?

7. A hollow pipe 2.5 cm inside diameter and 3 cm outside diameter has a torque of 20 N-m applied. What is the stress in pascal?

8. In tightening a bolt on an engine head, the specifications call for a torque of 75 ft-lb. What would be the torsion stress in a $\frac{1}{2}$-in.-diameter bolt under that load?

9. What would be the torsion stress in a 12.5-mm bolt under a torsion load of 30 N-m?

10. A solid metal bar that can withstand a torsion stress of 8000 lb/in² is needed to apply only a torque of 100 ft-lb. What is the minimum diameter that bar could be in inches?

11. A solid bar made of a metal that can withstand a stress of 50 MPa is needed to apply a torque of 50 N-m. What should the minimum diameter of the bar be?

12. A solid steel bar 2 ft 3 in. long and 0.75 in. in diameter is twisted through an angle of 3.2°. What is the strain in the bar?

Material: _____ Metal _____		Date: _____		
Length: _____ 6 in. _____		Inside diameter: _____ -- _____		
Outside diameter: _____ 0.50 in. _____		Run by: _____		
Reading No.	Torque (in.-lb)	Angle (deg)	Stress (lb/in²)	Strain (deg)
1	200	2		
2	400	4		
3	600	6		
4	800	8		
5	1000	10		
6	1200	12		
7	1400	14		
8	1600	17		
9	1800	22		
10	1900	30		
11	1900	40		
12	1900	Break		

FIGURE 2-28 Data for problem 2-5 number 20.

13. A solid steel bar 103 cm long and 3 cm in diameter is twisted through an angle of 5.1°. What is the strain in the bar?

14. Which would take more torsion at the same stress, a solid bar 0.75 in. in diameter or a hollow bar 0.5 in. inside diameter and a 0.75 in. outside diameter?

15. Which will take a greater torque under the same stress, a 0.75-in.-diameter solid bar or a hollow bar with a 0.75-in. inside diameter and a 1.000-in. outside diameter?

16. Which will take a greater torque under the same stress, a 2-cm-diameter solid bar or a hollow bar with an inside diameter of 1.75 cm and an outside diameter of 2.25 cm?

17. Which would have the greater strain, a 1-in.-diameter solid bar 20 in. long or a 12-in.-long bar with a diameter of 0.5 in., if they were both deflected 10° and had the same torque of 100 ft-lb?

18. A solid bar of metal is designed to take a stress of 30,000 lb/in². What will the bar diameter have to be to take a torsion load of 1000 in.-lb?

19. A solid bar will take a torsion stress of 50,000 lb/in². What will the diameter have to be to take a torsion of 150 ft-lb?

20. Given the data shown in Figure 2-28 for a torsion test of a solid bar, construct the complete torsion stress–strain graph.

Hardness, Impact, Fatigue, Flexure, and Creep

3

HARDNESS TESTS

The term *hardness* is rather vague and ambiguous. We speak of "hard" water, or of someone being "hardheaded." We often have to take "hard" examinations. It even seems contradictory to put a "soft" steel cap on a "hard" rubber heel of a shoe to protect it when the soft steel is much harder than the hard rubber.

This ambiguity extends even to the testing of metals. Many (at least 20) hardness test machines have been developed and standardized, yet they can measure a different property of the material. The only factor that all of them have in common is that they all measure a property of the surface of the material. Each of the commercially available hardness testers has its own strong points and limitations. Only a few of the more popular hardness testers are discussed here.

Brinell Hardness Test

One of the oldest hardness testers is the Brinell. The *Brinell hardness test* (Figure 3-1) is a static test that measures a material's resistance to surface penetration and gives a measure of the metal's surface plasticity or resistance to permanent deformation. It forces a 10-mm case-hardened steel or tungsten carbide ball into the surface of the metal with one of three standard loads: 3000, 1500, or 500 kg. Once the ball is forced into the metal, the load is released and the diameter of the indentation is measured by means of a *Brinell microscope* (see Figure 3-2). The load and the diameter of the hole or indentation left in the metal are then entered into the following equation to determine the *Brinell hardness number* (BHN):

$$BHN = \frac{2L}{\pi D(D - \sqrt{D^2 - d^2})}$$

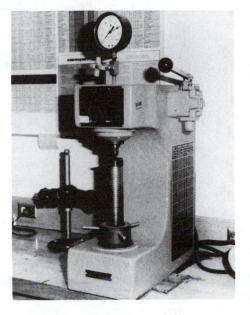

FIGURE 3-1   Brinell hardness tester.

where BHN = Brinell hardness number
      L = load (kilograms)
      D = diameter of the steel ball (10 mm)
      d = diameter of the indentation (millimeters)

In testing a soft steel, for example, a Brinell hardness test using the 3000-kg load produced an indentation of 4.5 mm. The Brinell hardness number would be

$$\text{BHN} = \frac{2(3000)}{3.1416(10)(10 - \sqrt{10^2 - 4.5^2})}$$

$$= 178$$

The Brinell hardness number will usually be a number between 90 and 630, with the higher number indicating higher hardness. The deeper the depth of pen-

FIGURE 3-2   Brinell microscope.

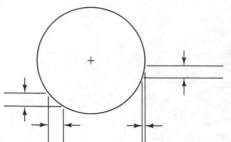

FIGURE 3-3   Depth effect of the Brinell ball.

etration, the wider the indentation left by the ball and the lower the Brinell hardness number will be. For a precise determination, the diameter of the indentation should never be over 6 mm. For the 10-mm steel ball to produce a diameter of more than 6 mm, it would have to be pushed into the metal a depth of more than 1 mm. Figure 3-3 shows that in driving the ball into the metal from $\frac{1}{4}$ to $\frac{1}{2}$ mm, the Brinell hardness changes from 382 to 191. But the change in Brinell hardness numbers in driving the ball from 1 mm to $1\frac{1}{4}$ mm only changes the Brinell hardness number from 95 to 76. Also, the deeper the ball is driven into the tested metal, the greater the tendency to leave a ridge around the hole, which tends to give a false reading of the diameter of the hole. If the ball leaves an indentation of more than 6.0 mm, a lighter load should be used.

On the higher end, the Brinell hardness numbers should not be trusted over about 650. At this end two items introduce unacceptable errors in the readings. First, the diameter of the indentation is so small that a very slight inaccuracy in reading produces a large difference in the BHN. Under a 3000-kg load the Brinell hardness number for an indentation of 2.35 mm is 681. Increasing the diameter of the indentation by only 0.05 mm changes the BHN to 653. Second, at these higher hardnesses, the ball tends to flatten out and give erroneously wide indentations. Fortunately, there are not many materials that have Brinell hardness numbers above 650.

A good rule to follow in selecting the proper loads for a Brinell test is as follows:

> For a BHN of 160 and above, use the 3000-kg load.
> For BHNs from 80 to 300, use the 1500-kg load.
> For BHNs below 100, use the 500-kg load.

The ranges of these recommended loads overlap somewhat, which gives the technician some latitude in selecting the proper loads.

Another precaution when using the Brinell test is that the minimum thickness of the test sample should be at least 10 times the depth of penetration of the ball. This implies that metals less than 8 to 10 mm thick (roughly $\frac{3}{8}$ in.) should not be used under the 3000-kg load. If thinner samples are tested, it is the hardness of the anvil that is really being measured (see Figure 3-4).

The Brinell hardness test does have several limitations. First, it is a destructive test. The fact that a 2- to 6-mm-diameter indentation is left in the part tested in most cases damages that part beyond use. The test machine is not portable. Many Brinell machines weigh in excess of 200 lb, requiring that all parts be brought to the machine. It is usually not taken into the field. Although there are different-size machines, each tester has a limit to the size of part it can test. Further, most Brinell test machines are relatively expensive to purchase. Prices for the basic

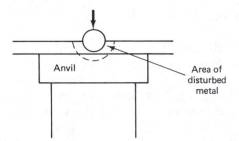

Anvil

Area of
disturbed
metal

FIGURE 3-4   Effect of the anvil.

motorized Brinell tester are now in excess of $5000 plus the accessories and the Brinell microscope. Many small companies cannot afford this purchase price. In addition, the test results are a little subjective. Since the diameter of the indentation must be read very accurately, two different operators may get slightly different readings, allowing the Brinell hardness number (BHN) to vary slightly. This error is usually not large and seldom over 10%. The fact that the BHN does not read directly but must be calculated is also considered a drawback. However, most machines have a chart giving direct conversions of the diameters of the indentation to BHNs, eliminating the need for the mathematics.

Despite these flaws in the Brinell test, it does have several advantages or strong points. The test has been around for over a half-century, so it is well accepted in industry. All engineers and technicians are familiar with it. It is accurate and not subject to minor flaws in the material tested. The 10-mm steel ball covers a large-enough area that a small defect such as a hard spot or pinhole in the material will not cause a significant change in the results. Therefore, a single test will give an average hardness number for the entire sample. Despite the large area covered by the indentation, the test is relatively precise. The test can be performed quickly. A good operator can run a Brinell test in less than 2 minutes, making this test comparatively cheap to run. It is easy to train an operator to use the Brinell tester. Usually, an operator can become proficient in making an accurate test with as little as one half-hour of training and practice.

The Brinell hardness number can sometimes be related to the tensile strength of the material. In general, the higher the BHN for the steels, the stronger the metal becomes. The relationship between the Brinell hardness numbers and tensile strength varies with different metals and alloys of those metals. There is no across-the-board relationship between strength and hardness. A soft steel with a BHN of 100 may have a tensile strength of 440,000 lb/in^2, while an aluminum alloy with the same hardness number may have a strength of 55,000 lb/in^2.

Thinker 3-1:

Would the surface roughness of a test sample affect the accuracy of the Brinell test? If so, in what way?

Thinker 3-2:

Would the Brinell test be an accurate method of testing a cylinder on the side? Explain your reasoning.

Vickers Hardness Test

The principle of the Brinell, using a ratio of load to the indented area, is used in several other types of hardness tests. The Vickers hardness testing method uses a small diamond square pyramid which has a 136° angle on the point as an indenter and loads of from 5 to 120 kg. The load is applied and removed and the diagonal of the square indentation is measured with the Vickers measuring microscope (see Figure 3-5).

The Vickers diamond pyramid hardness number (DPH), can then be calculated using the equation

$$DPH = \frac{1.8544P}{d^2}$$

where P is the load in kilograms and d is the length of the diagonal in millimeters. For example, if a load of 100 kg left an indentation of 0.78 mm on a metal, the diamond pyramid hardness number would be

$$DPH = \frac{(1.8544)(100 \text{ kg})}{(0.78 \text{ mm})^2} = 305$$

The Vickers test is used mainly in research applications. It is claimed that the diagonal of the square is easier to read through a measuring microscope than is the circular indentation of the Brinell. Further, the diamond point does not have the tendency to deform (flatten out) as does the Brinell on harder metals. The result is that the Vickers is more accurate than either the Rockwell or the Brinell, but the equipment is also much more expensive. The Vickers test is a measure of the surface plasticity of the metal and leaves a permanent indentation. It is also slower than the Rockwell and is not favored as much in production-minded industry. The Vickers test has been adapted for micro-hardness tests using loads as low as 5 g. These tests can measure the hardness of a single grain of metal. To use the micro-hardness tester, the surface of the metal must be polished, a time-consuming operation in itself.

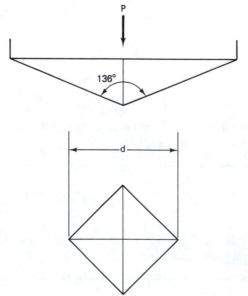

FIGURE 3-5   Vickers measurement.

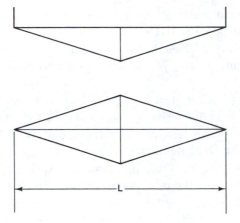

FIGURE 3-6   Knoop indenter.

Tukon Hardness Test

The Tukon hardness test uses an elongated diamond pyramid indenter (Figure 3-6) called the *Knoop indenter*. A load (P) from 25 g to 3.6 kg is lowered onto the point, driving it into the metal. The length (L) is measured through a measuring microscope. The Knoop hardness number (KHN) is calculated using the formula

$$\text{KHN} = \frac{1.43P}{L^2}$$

where P is the load in kilograms and L is the length of the longest diagonal in millimeters. The KHN will be a number between 60 and 1000, and is a measure of the material's resistance to permanent surface deformation, again a measure of surface plasticity.

PROBLEM SET 3-1

1. A Brinell test using a 3000-kg load and a 10-mm steel ball produced an indentation of 3.4 mm. Calculate the Brinell hardness number for the metal.
2. A 500-kg load using the 10-mm ball produced an indentation of 2.6 mm in the Brinell test of a piece of aluminum. What was the BHN?
3. A 1500-kg Brinell test using the standard 10-mm ball produced an indentation of 4.1 mm. Calculate the BHN.
4. A certain steel was hardened by a water quench. The diameter of the indentation from a 3000-kg test using the standard 10-mm ball was 4.0 mm before hardening and 2.9 mm after hardening. What were the "before" and "after" hardnesses?
5. A sample of brass tested under a 500-kg load and the 10-mm Brinell ball left an indentation of 3.9 mm. What was the BHN of the brass?
6. The range of BHNs is about 90 to 630. What size indentation would a 10-mm ball under a 3000-kg load leave for these two hardness numbers?
7. What are the dimensions of a Brinell hardness number? (*Hint:* Run a dimensional analysis of the BHN equation.)
8. In testing a soft metal, a 3000-kg load produced an indentation of more than 6 mm, which could not be read with the Brinell microscope. Upon reducing the load to 1500 kg, an indentation of 4.6 mm was obtained. What would have been the expected indentation from the 3000-kg load?

9. A soft metal under a 500-kg Brinell test load had an indentation of 3.10 mm. What was the BHN of this test?

10. A report was received that a 3000-kg test on a steel plate left an indentation of 2.20 mm. Would you accept this report without verification? Why or why not?

11. If a 10-kg Vickers test load produced an indentation having a diagonal length of 0.24 mm, what is the Vickers (DPH) hardness number?

12. A Vickers 10-kg test left an indentation having a diagonal of 0.36 mm. What is the DPH number?

13. A Knoop 500-g load left an indentation with a length of 0.071 mm when measured under the microscope. What is the Knoop hardness number?

14. The hardest steel available has a Knoop hardness number of about 900. What would be the length of an indentation left by a 500-g load in the Knoop test?

15. Calculate the length of the diagonal of the indentation left by a 100-kg load in a Vickers test on a material whose hardness was 50.

16. Calculate the length of the indentation left by a 500-g load by a Knoop indenter for a metal whose Tukon hardness was 700.

17. A 50-kg load on a Vickers test produced an indentation of 0.786 mm. What was the Vickers hardness number?

18. A micro-hardness test of a material using a Vickers test used a 5-g load and produced an indentation of 0.0058 mm. What was the Vickers hardness number of the material?

19. A Knoop hardness tester using a 100-kg load produced an indentation with the major length of 0.68 mm. What was the Knoop hardness number?

20. A Knoop hardness tester using a 15-kg load produced a hardness number of 494. What was the length of the longest diagonal?

Rockwell Hardness Test

The *Rockwell hardness test* works on an entirely different principle than the Brinell. The Rockwell hardness tester uses a choice of three different indenters combined with three different loads. Figure 3-7 shows a Rockwell hardness tester. In running the Rockwell test, the sample to be tested is placed on the anvil and the indenter is forced into the specimen with what is called the minor load, 10 kg. The minor load drives the point through the outer scale, surface roughness, paint, or grease on the metal to a point represented by point *A* in Figure 3-8.

The dials on the test machine are then set to zero, and the major load (60, 100, or 150 kg) is applied. The major load drives the point to a depth *B*, as shown in Figure 3-8. After the dial or digital indicator on the tester has ceased changing, the major load is removed and the surface elasticity of the metal causes the indenter to be shoved back up to point *C*. The dial reading is then taken as the Rockwell hardness number. This dial reading represents the depth of the permanent deformation caused by the major load, with each division on the dial representing 0.00008 in. of penetration (the distance from *A* to *C*). For the Brale indenter (scales A, C, and D), the readings given by the machine equal $100 \times t$, to $500 \times t$. For the $\frac{1}{16}$- and $\frac{1}{8}$-in. ball (scales B, F, G, and E), the readings equal about 130 to $500 \times t$. The Rockwell test is a measure of resistance to permanent surface deformation. As with the Brinell test, the Rockwell test measures surface plasticity.

The indenters are a $\frac{1}{16}$-in. steel ball, a $\frac{1}{8}$-in. steel ball, or a 120° conical black

FIGURE 3-7   Rockwell hardness tester.

diamond. The standard loads are 60, 100, or 150 kg. The loads are applied through a series of levers in much the same manner as a batch scale. The combination of the specific indenters and loads gives the particular Rockwell scale. If the $\frac{1}{16}$-in. ball is used with the 100-kg load, the hardness is reported as the Rockwell B scale. Similarly, the Brale point combined with the 150-kg load constitutes the Rockwell C scale. The combinations of indenters and loads for the other Rockwell scales are shown in Figure 3-9.

Some Rockwell machines are fitted for superficial Rockwell tests. The superficial test uses major loads of 15, 30, or 45 kg. The same indenters are used as with the other Rockwell tests. The superficial tests are the 15T, 30T, 45T, 15N, 30N, and 45N scales. Using the superficial test, case-hardened or nitrided steels or steels as thin as 0.006 in. can be tested. The indenter-load-scale chart for the superficial tests is shown in Figure 3-10.

Rockwell hardness should be kept in the range 0 to 100. The scales (indenters or loads) should be changed to keep the readings in that range. The hardest steel in existence will not produce a Rockwell C hardness number of more than 68. If a Rockwell C test produces a value greater than 68, the scale should be changed to the Rockwell B or other suitable scale. Rockwell numbers close to 0 or 100 should be checked on another scale to verify that they are not less than 0 or greater than 100.

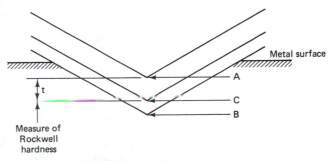

FIGURE 3-8   Rockwell test.

FIGURE 3-9   Rockwell scales.

Scale	Indenter	Load (kg)
A	Brale	60
B	$\frac{1}{16}$-in. ball	100
C	Brale	150
D	Brale	100
E	$\frac{1}{8}$-in. ball	100
F	$\frac{1}{16}$-in. ball	60
G	$\frac{1}{16}$-in. ball	150

When reporting Rockwell hardnesses it is vital that the scale on which the metal was tested be stated. To say that a metal had a hardness of 58, or even a Rockwell hardness of 58, is meaningless. A Rockwell B hardness of 58 is a relatively soft steel, while a Rockwell C hardness of 58 is a very hard steel.

Many of the same precautions must be taken with the Rockwell tester as with the Brinell. The thickness of the test specimen must be at least 10 times the depth of penetration of the indenter. Since the depth of the indentation or "hole" left by the Rockwell indenter is rarely over 0.015 in., metals as thin as 0.15 or about $\frac{1}{8}$ in. thick can be tested.

There are relationships between the various scales which can be found on charts that usually accompany the machines. For example, a Rockwell B test of 70 (reported as HRb 70) is equivalent to a Rockwell F test of 96 (HRf 96) or a Rockwell 15T test of 83.4 (HR15T 83.4). The same Rockwell B test of 70 is also about the same as a Brinell hardness number (BHN) of 126. Figure 3-11 shows a comparison of some of the scales.

Since both the Brinell and the Rockwell hardness tests are a measure of surface plasticity, it is often possible to get approximate conversions between them using the following equations. If HRc is greater than -20 and less than 40,

$$BHN = \frac{1.42 \times 10^6}{(100 - HRc)^2}$$

For HRc values greater than 40, use

$$BHN = \frac{2.5 \times 10^4}{100 - HRc}$$

If the HRb value is greater than 35 but less than 100, use

$$BHN = \frac{7.3 \times 10^3}{130 - HRb}$$

FIGURE 3-10   Superficial Rockwell scales.

Scale	Indenter	Load (kg)
15T	$\frac{1}{16}$-in. ball	15
30T	$\frac{1}{16}$-in. ball	30
45T	$\frac{1}{16}$-in. ball	45
15N	Brale	15
30N	Brale	30
45N	Brale	45

FIGURE 3-11   Hardness conversions.

HRc	HRb	HRe	HR30-N	BHN, 3000 kg	Vickers, 10 kg	Knoop, 500 g	Approximate Tensile Strength, Steel (lb/in²)
65	—	—	84	697	820	846	350,000
60	—	—	79	627	765	732	314,000
55	120	—	74	555	633	630	277,000
50	117	—	69	495	540	542	247,000
45	115	—	65	429	454	466	211,000
40	112	—	61	372	390	402	182,000
36	109	—	58	341	350	363	165,000
30	105	—	52	283	287	311	138,000
26	103	—	49	260	263	284	125,000
20	97	—	43	222	223	251	110,000
16	95	110	39	207	207	226	101,000
9	91	108	35	185	187	201	93,000
3	85	105	29	163	163	180	82,000
—	80	103	—	146	146	164	74,000
—	75	100	—	133	133	150	66,000
—	70	98	—	121	121	139	62,000
—	66	96	—	112	112	131	58,000
—	60	—	—	107	107	120	52,000
—	56	—	—	101	—	114	—
—	50	—	—	83 (500 kg)	—	107	—

As an example, an HRc value of 50 would convert to a BHN as

$$\text{BHN} = \frac{2.5 \times 10^4}{100 - 50} = 500$$

From the data in Figure 3-11 an HRc of 50 is equivalent to a BHN of 495. The equation is a good "ballpark" conversion and could be used when conversion charts are not available.

The Rockwell test, like the Brinell, has its strengths and weaknesses. The Rockwell test is very accurate and precise. In fact, since a very small indenter is used, the machine may be a little too precise and could give an erroneous reading by hitting an impurity, a hard spot, or a small hole in the metal. For this reason, Rockwell tests are usually taken at at least three different points on the sample and the average of these three readings reported. The Rockwell test is also a fast test to run. With a little practice, a Rockwell test can easily be made in 30 seconds. This fact makes the Rockwell a cheap test to run. The Rockwell test is universally accepted in industry. Like the Brinell, it has been around for many years and is used as a standard in many industrial plants. The Rockwell test is an objective test. It is also easy to train an operator to use this machine. A person can become an expert at the Rockwell test and make accurate readings with only a few minutes' experience. The Rockwell hardness tester is a very versatile machine. By using the different scales, a wide range of materials can be tested.

The disadvantages of the Rockwell test closely parallel those of the Brinell. The Rockwell machine is not portable. Many of these machines weigh in excess of 100 lb. Further, the capacity of the machines limits the size of the specimen that can be tested. The Rockwell test is considered destructive in some cases. The indenter leaves a permanent mark on the material which would render it destructive to smooth surfaces such as crankshafts, bearings, knife blades, and other applications where finish is important. The Rockwell test machines are also relatively expensive to buy.

Thinker 3-3:

a. Would surface scale or surface roughness affect the results of the Rockwell test? If so, would they raise or lower the hardness numbers obtained?

b. Would the hardness of the anvil affect the results of the Rockwell test? If so, how?

c. Outline an experiment that would determine if the anvil hardness had an effect on the results of the Rockwell test.

In an attempt to make the hardness testers more portable, the *Ames hardness* and *Teleweld hardness* testers have been developed. The Ames tester works on the same principle as the Rockwell, but is hand operated, much less versatile, slower to operate, and more difficult to use than the standard Rockwell tester. Similarly, the Teleweld tester uses the same principle as the Brinell, but is portable and hand operated. These testers are used in the field, where access to either the Brinell or Rockwell testers is lacking.

Scleroscope Hardness Test

All of the hardness tests discussed above are measures of the material's surface plasticity, or the ability of the metal to withstand permanent surface deformation, and are considered static tests. Many of them are also destructive to the surface of the metal. The Shore scleroscope is a dynamic test and measures the surface elasticity of the material. In this test, a small diamond- or sapphire-tipped plummet or hammer is dropped from a height of 10 in. onto the surface of the metal. Figure 3-12a illustrates the *scleroscope hardness tester*. The height of the rebound or bounce is read from a scale set behind the glass tube that houses the hammer (see Figure 3-12b). The scale reading is the *scleroscope hardness*. The scale readings will be between 0 and 130.

The main advantage of the scleroscope is that it is a nondestructive test. The mark left at the point on which the hammer falls is barely visible on the metal. The thickness of the material tested can be as little as $\frac{1}{64}$ in. for a hard material. The machine is very lightweight and can be taken into the field for use. The drawbacks to the scleroscope involve its subjectivity. Since the hammer bounces on the surface of the material, the height of the first bounce must be caught by the eye of the observer. This usually involves a little more training than is required for the other tests. The surface of the material must also be smooth and free of oil, scale, or other material. Surface imperfections must also be removed prior to testing. This usually requires a little more preparation of the specimen than for the other tests. The scleroscope reads only about 0.0004 in² of surface area, so several readings must be made and the average of all these readings computed to get accurate results. The scleroscope is one of the few dynamic tests accepted in industry, but it is not as popular as the Rockwell or Brinell.

One caution should be observed with the scleroscope. Since it measures surface elasticity and the other tests measure surface plasticity, there is no direct comparison between the scleroscope and the other tests. Further comparisons between two materials such as steel and rubber may not be valid.

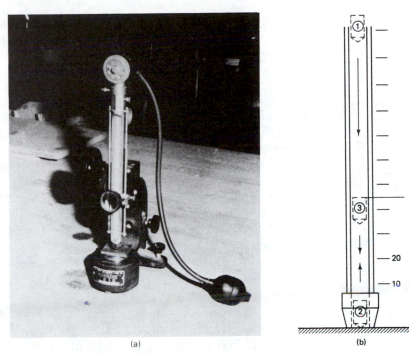

FIGURE 3-12   Shore scleroscope.

Mohs Hardness Test

A hardness test primarily used by geologists for determining the hardness of rocks and minerals is the *Mohs*. This test employs 10 different minerals, each one harder than the preceding one on the scale. These minerals and their assigned hardness numbers are given in Figure 3-13.

The test is conducted by trying to scratch an unknown with one of these minerals. If the surface is scratched, for instance by apatite but not by fluorite, the Mohs hardness number would be listed as 4.5. The Mohs hardness test is a measure of the material's resistance to surface abrasion. In all of the previous tests, the atoms were just moved around a bit. In the Mohs test the atoms are actually removed from the surface of the material.

Most precious gems are either diamond, some form of corundum, or quartz. Diamond is a form of carbon, corundum is made from aluminum oxide, and quartz is silicon dioxide. Rubies are aluminum oxide with a Mohs hardness of 9. Apache tears are quartz with a Mohs hardness number of 7.

FIGURE 3-13   Mohs hardness scale.

Mineral	Hardness Number
Talc	1
Gypsum	2
Calcite	3
Fluorite	4
Apatite	5
Feldspar	6
Quartz	7
Topaz	8
Corundum	9
Diamond	10

The problem with the Mohs hardness test is that it is not a precise test and is not recognized in industry. The change in hardness from one of these minerals to another would be several points on the Rockwell scale. Further, the difference in hardness between the minerals is not the same. It is very portable and cheap. They need not even have a complete Mohs hardness set with them to conduct a test. A fingernail usually has a Mohs hardness of about 3, a copper penny is a little over 4, and a good pocket knife will be between 5 and 6. It is included here only as an example of a hardness test that measures a property of the material other than surface plasticity or elasticity.

Closely related to the Mohs test is one that is used by machinists known as the *file test*. A standard file has a Mohs hardness of about 7 and a Rockwell C hardness of about 63. If a file will not scratch a metal, it has to be harder than these values. If a file will not cut a material, the usual method of machining it is by grinding it with an emery wheel or stone. Emery is a form of corundum and has a Mohs hardness of 9.

Sonodur Hardness Test

Of the tests listed above, only the scleroscope is a nondestructive one. The *Sonodur hardness tester* is a nondestructive test. In this test a diamond-tipped rod, vibrating at ultrasonic frequencies, is placed against the surface of the metal to be tested. Depending on the hardness of the surface of the metal, the frequency of the vibrations will be changed. This change in frequency is read directly on a meter as an indication of hardness. The readings are shown in either Rockwell or Vickers units. It is extremely portable, but its use is normally limited to steels. Aluminum and copper have too much damping in the material to make this test an accurate one for those materials.

Other Hardness Tests

Of the tests discussed above, only the Mohs will work with nonmetallic materials. Often it becomes necessary to develop tests that are applicable to specific materials. The Shore *durometer* is a hand-held instrument specifically designed to test the hardness of organic materials, such as rubber, plastics, and composites (Figure 3-14). There are two models of the durometer, the A and D. Both of these have a spring-loaded metal indenter which is pushed against the surface of the material. The resistance the material places against the spring provides a reading of its hardness. The type A durometer uses a spring loading from 56 to 822 g, while the

FIGURE 3-14   Shore durometer.

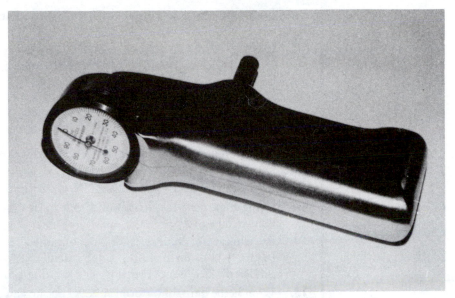

FIGURE 3-15   Barcol impresser.

type D durometer places loads from 0 to 10 lb on the indenter. Both durometers produce readings from 0 to 100. In reporting the durometer hardness of a material, the A or D scale must be noted.

Another hardness tester for plastics and composites is the *Barcol impresser* (Figure 3-15). This is also a hand-held instrument which impresses a spring-loaded needle against the material to give a reading in the range from 0 to 100. Both the durometer and the Barcol measure a material's surface elasticity as an indication of its hardness.

Thinker 3-4:

Make a list of hardness testers that define hardness in terms of the material's:

a. Surface elasticity
b. Surface plasticity
c. Resistance to surface abrasion

PROBLEM SET 3-2

1. Convert an HRc reading of 20 to a Brinell hardness number using the proper equation.
2. Convert an HRb 40 reading to a BHN using the proper equation.
3. Convert an HRc of 45 to a BHN using the proper equation.
4. If a Rockwell test used a $\frac{1}{16}$-in. ball and the 100-kg load, what scale would it be?
5. If a Rockwell test uses a brale point and a 150-kg load, what scale would it be?
6. If a Rockwell test used a 60-kg load and the brale point, what scale would it be?

7. If the Rockwell test used the $\frac{1}{8}$-in. ball and the 100-kg load, what scale would it be?

8. Use the proper equation to convert a BHN number of 250 to an approximate Rockwell hardness number. (List the correct scale.)

9. Use the proper equation to convert a BHN number of 120 to an approximate Rockwell hardness number.

10. Using the proper formula, convert a Rockwell B hardness of 78 to a Brinell hardness number.

11. Using Figure 3-11, convert an HRc reading of 30 to a Vickers hardness number.

12. Using Figure 3-11, convert a BHN of 283 to a HR30-N number.

13. Using Figure 3-11, a Rockwell C hardness of 45 would equal what on the Knoop 500-g scale?

14. By what percent does the value obtained from the conversion chart in Figure 3-11 differ from the value calculated from the proper equation for a HRc of 45 when converting it to a Brinell hardness number?

15. What is the percent difference between the calculated value and the Figure 3-11 value when converting a Rockwell B hardness of 70 to a BHN?

16. If you were testing plastics for hardness, which hardness tester or testers would be appropriate?

17. If you needed a nondestructive hardness test of a crankshaft, which hardness tester or testers would be appropriate?

18. If you were going to test only very hard steels, which tester or testers would you choose?

19. List three advantages and three disadvantages of the Rockwell hardness tester.

20. List three advantages and three disadvantages of the Brinell hardness tester.

IMPACT TESTS

Most materials are somewhat strain-rate dependent. A material that could withstand a load of 100 lb if the load were gently lowered onto it might break if it was struck sharply with same 100-lb load. A classic example of this is Silly Putty. This gum-like substance will stretch into a very thin string when gently pulled, yet will fracture when yanked or jerked sharply. A test to find the energy required to break a material under a sharp blow is called an *impact test*.

Energy is defined as the ability to do work. Work is a force taken through a distance and is given by the equation

$$W = F \times D$$

where W = work (foot-pounds or newton-meters)
F = force (pounds or newtons)
D = distance (feet or meters)

Since energy is the ability to do work, it also has the units of foot-pounds or newton-meters.

The information found in the impact test is the energy required to fracture the part under a sudden blow. One such test would simply be to drop a known weight on a test specimen from a known height. The energy generated by the falling object would be

$$E = W \times h$$

where E = energy (foot-pounds or newton-meters)
  W = weight of the object (pounds or newtons)
  h = height of the object before falling (feet or meters)

If a steel shotput weighing 16 lb fell from a distance of 13 ft, it would hit the ground with an energy of

$$E = 16 \text{ lb} \times 13 \text{ ft} = 208 \text{ ft-lb}$$

A metric system problem might be for a mass of 10 kg falling from a height of 10 m. The energy dispelled upon impact would be

$$E = 10 \text{ kg} \times 9.8 \text{ m/s}^2 \times 10 \text{ m} = 980 \text{ N-m}$$

Two standard impact tests presently used in industry are the *Izod* and the *Charpy*. Both of these tests use the concept of energy shown above by using a swinging pendulum to strike a test sample. The differences between these two tests are in the design of the test specimen and the velocity with which the pendulum hits the test bar. The machine used in the Izod and Charpy tests is shown in Figure 3-16(b).

In the standard Izod test, the pendulum is set to strike the test specimen at a velocity of 11.5 ft/sec, while in the Charpy, the standard velocity is 17.5 ft/sec. In the Charpy test a notched bar is held horizontally at both ends as it is hit by the pendulum. The Izod test anchors the notched bar vertically and holds it in a vise only at the bottom end. Figure 3-17 shows these standard test specimens for both the Izod and Charpy tests. The purpose of the notch is to act as a stress riser or stress concentrator to allow all the energy to be applied to a single point on the bar. Without this notch, the energy would be spread evenly along the bar and could cause it to plastically deform by bending rather than breaking.

The pendulum is set at an angle A (Figure 3-16a), then released to hit the test sample. In falling through the angle A, it picks up momentum. If no test bar was in the machine, the pendulum would theoretically swing to an angle A on the other side. Actually, there is some loss due to the friction in the machine, which must be accounted for in the final calculations. With the test sample in place, some of the energy generated by the falling pendulum is absorbed in breaking the bar. The remaining energy will cause the pendulum to swing to an angle B on the opposite side of the original set. The energy absorbed by breaking the bar is given by the equation

$$E = W \times R \times (\cos B - \cos A)$$

where E = energy required to break the bar (ft-lb or N-m)
  W = weight of the pendulum (pounds or newtons)
  R = radius of the swing (length of the pendulum; feet or meters)
  A = initial angle of the set (angle of fall)
  B = angle to which the pendulum swings after breaking the bar (angle of rise)

To illustrate this problem, suppose that a standard Izod test machine with a 60-lb pendulum, 31.5 in. long, is set at an angle of 76° and after striking the test

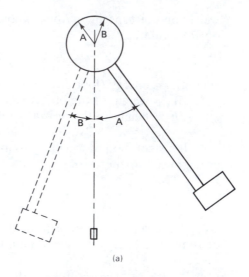

(a)

(b)

FIGURE 3-16 (a) Impact test concepts; (b) impact test machine.

specimen, proceeds to rise to an angle of 29° past the center. What would be the energy absorbed by the test specimen?

$$E = W \times R \times (\cos B - \cos A)$$

$$= 60 \text{ lb} \times (31.5 \text{ in./12 in./ft})(\cos 29° - \cos 76°)$$

$$= 60 \text{ lb} \times 2.63 \text{ ft} \times (0.8746 - 0.2419)$$

$$= 99.8 \text{ ft-lb}$$

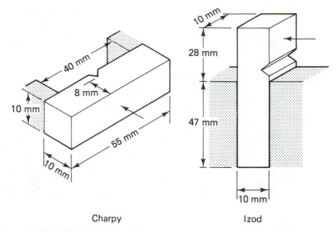

Charpy   Izod

FIGURE 3-17   (a) Izod and (b) Charpy test bars.

In a metric system impact test the problem would be as follows. A pendulum of 30 kg weight, 80 cm long was set at an angle of 75° and came to rest at an angle of 22°. The energy absorbed in the impact would be

$$E = (30 \text{ kg})(9.8 \text{ m/s}^2) \left(\frac{80 \text{ cm}}{100 \text{ cm/m}} \right) (\cos 22° - \cos 75°)$$

$$= 157 \text{ N-m}$$

Most materials are somewhat temperature dependent in their impact resistance. Usually, the colder the material, the less impact resistance it shows. For some materials there is a critical temperature below which they should not be used in an impact situation. It is fortunate, however, that these critical temperatures are very low and not normally encountered. Besides the Izod and Charpy tests, which test the flexure impact of a material, straight tension and compaction impact tests are conducted. Many machine parts and parts in aircraft, automobiles, and weapons must be designed to withstand impact loads.

There are many factors that affect a material's impact strength. The composition of the material is of prime importance. The addition of alloying elements in steels can increase its impact strength enormously. The addition of about 2% nickel to a steel will greatly increase its impact strength. The impact strength of a steel can also be increased by heat treatments. A steel that has been heated into the red-hot range, then rapidly cooled by plunging it into cold water, will probably become brittle and lose much of its resistance to impact. An annealed or slowly cooled steel will experience an increase in impact strength. Heat treatment of steels is discussed in Chapter 6. The impact strength of materials can also be lowered by manufacturing techniques. A notch or scratch on a brittle material will concentrate external stresses and significantly lower the impact resistance. Flaws in welds, bubbles or imperfections in the metal, and corrosion can cause some materials to fail in impact.

Thinker 3-5:

List at least three examples where impact resistance should be considered in the design of a part.

PROBLEM SET 3-3

1. If a 94-lb sack of portland cement fell from the second floor of a building (20 ft), what would be its impact energy?

2. A pile driver using a 500-lb hammer is raised 2 ft above the end of a piling before being released. How much energy will be imparted to the piling per blow?

3. A 5-kg sledgehammer fell from the top of a 3-m ladder onto a block of concrete that was designed to withstand an impact of 200 N-m. Would the concrete block be expected to break in this event?

4. If a Navy carrier-based aircraft weighing 20 tons is tested by dropping it on its landing gears from a height of 20 ft, what impact energy must the landing gears absorb?

5. A standard 31.5-in. arm impact test machine with a 60-lb pendulum is set with an initial angle of 76°. After impact the angle of rise is 24°. What is the energy absorbed by the test sample?

6. A standard 31.5-in. arm on an impact test machine using a 60-lb pendulum is set to test a specimen in tension impact. A $\frac{1}{2}$-in.-diameter bar, 3 in. long, stops the pendulum at a 0° rise. The pendulum was initially set at 76°.
    (a) What is the energy absorbed by the test sample?
    (b) What would be the energy absorbed per cubic inch of the material?

7. A 3-in.-long $\frac{3}{8}$-in.-diameter bar of metal is set up in tension impact with a 60-lb 31.5-in. arm set at 76°. The rise angle after the impact is 12°.
    (a) What is the energy absorbed by the test sample?
    (b) What is the energy per cubic inch absorbed by the sample?

8. Using the material tested in Problem 6, what diameter rod 3 in. long would be needed to absorb 7000 in.-lb of energy?

9. A 30-kg 80-cm-long impact test hammer is set initially at 76°. After impact in a Charpy test the rise angle is 17°. What is the energy absorbed in newton-meters?

10. A 30-kg 80-cm-long impact test hammer set initially at 70° is stopped completely by a test sample. How much energy did it absorb in newton-meters?

11. What would be the energy absorbed if a 31.5-in. 60-lb pendulum was released from an angle of 90° and stopped by a superalloy without breaking it?

12. What would be the energy absorbed if a 60-lb, 31.5-in. pendulum was released from an angle of 50° hit a test bar, then rose to an angle of 20° after impact?

13. Some plastics are tested with low impact by setting the pendulum at small angles. If a 60-lb pendulum on a 31.5-in. arm was released from an angle of 30° and rose to an angle of 10° after impact, what was the energy absorbed?

14. If the 60-lb pendulum on a 31.5-in. arm was released from an angle of 30° and was stopped by the specimen, how much energy was absorbed?

15. A pendulum of reduced load can also be used to reduce the impact for some plastic specimens. If a 30-lb pendulum on a 31.5-in. arm was released from an angle of 24° and rose to an angle of 12° after impact, how much energy was absorbed?

16. A golf club "putter" was allowed to swing freely from a point 30 in. from the head of the club and hit a golf ball. If the head of the club weighed 0.6 lb, the swing started from an angle of 30°, and the follow-through was 15°, how much energy was imparted to the golf ball?

17. In Europe a putter similar to the one in Problem 16 weighed 300 g and was swung from a point 80 cm from the head of the club. The swing started from 30° and ended at a point 10° past the point of impact. How much energy was absorbed by the ball?

18. A test sample is listed as being able to absorb 50 ft-lb of energy. If a 31.5-in. 60-lb hammer was started at an angle of 76°, how far would it swing after hitting the specimen?

19. A 30-lb hammer on a 31.5-in. arm swung from an angle of 36° hit a plastic with an impact strength of 120 in.-lb of energy. How far would the pendulum swing after hitting the plastic?

20. A material was listed as having an impact strength of 650 in.-lb. What would be the minimum angle at which a 60-lb 31.5-in. hammer could be started to just break the material?

FATIGUE TESTS

Most people are familiar with the technique of breaking a wire by repeatedly bending it back and forth. The reason the wire breaks after being bent many times is known as *fatigue* in the metal. Fatigue in materials is defined as the failure of a material due to repeated or cyclic stresses being applied to it. The number of cycles or times a piece of metal can be bent and straightened depends on the stress or amount of pressure on the piece. If one were to bend a coat hanger wire only 5° each time, it would last a long time before finally breaking. However, if the coat hanger wire were bent double and straightened, it might last only a few bendings before breaking.

There are several types of fatigue testing machines. In all of them the number of cycles until failure is counted and the applied stress used to cause the failure is set. Many materials will not break below a given stress no matter how many times they are bent. Above that *critical stress* the number of cycles to failure decreases rapidly. For example, a metal sample might be set in the test machine and be bent only 0.5° each cycle. The machine is started and the specimen is bent at 60 cycles per second and the time taken to fail the material noted. The number of seconds times the 60 cycles per second gives the number of cycles to fail the material. A million cycles at 60 cycles per second would take about 4 hours 38 minutes. A sample set at a stress that will cause a strain of 1° might also take thousands of cycles to fail. If many samples of the same material are tested to failure, each at a different stress, and the number of cycles to failure are plotted against the stress, a graph similar to the one shown in Figure 3-18 would be obtained.

The graph shown in Figure 3-18 can be divided into two almost straight lines.

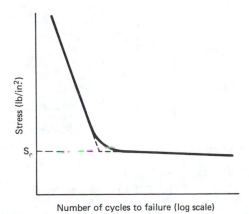

Number of cycles to failure (log scale)

FIGURE 3-18 Typical fatigue curve.

If these two lines are extrapolated until they meet, the critical stress of the material can be noted (point S_c in Figure 3-18). Below this stress, the material will flex or bend almost forever before it breaks. At stresses greater than this critical stress, the fatigue life of the material is limited. If possible, the parts should be designed not to exceed the critical stress. If this is not possible, a strict maintenance plan must be enforced in which these parts are replaced at a time interval not over half the fatigue life of the part.

Fatigue stresses can be put on a material by vibration, torsion, shear, or bending. Any parts that undergo flexure or cyclic stresses must be carefully designed to withstand fatigue. Parts should not have sharp corners or stress risers designed into them. Sharp corners concentrate the energy of the stresses and eventually cause cracks to form there. Once a crack is started, no matter how shallow, at least 80% of the life of the part is gone. Fillets and rounds provide stress relief at these corners and should always be used (see Figure 3-19). Parts subject to fatigue should be inspected on a periodic basis and replaced when necessary for safe operation of the machine. Aircraft are an example of parts that are subject to fatigue. The Federal Aviation Administration and the military air arms have strict rules governing the maintenance and replacement of critical parts.

There are several causes of fatigue in metals. The mechanisms of stress-corrosion cracking, Frank–Read sources, and thermal cracking are discussed in Chapter 4. Parts broken by fatigue are easy to identify. Parts broken through stress-corrosion will have part of the fractured surface oxidized or rusted with only the part of the material that was the last to fail being clean. Since cyclic strains cold work the material, the grains will be broken up into very small grains. This fine-grained structure will show up on the fractured surface. Figure 3-20 illustrates a fatigued surface.

Thinker 3-6:

List at least three examples or instances where fatigue in metal parts is a critical factor and must be considered in design.

Thinker 3-7:

Outline, or design, a method whereby a common metal lathe could be converted into a fatigue testing machine.

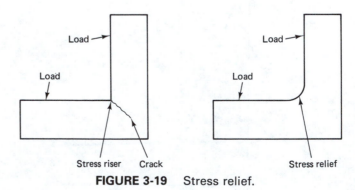

FIGURE 3-19   Stress relief.

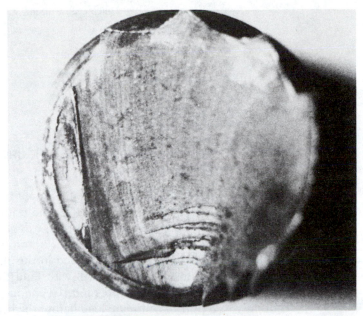

FIGURE 3-20 Fatigued metal.

FLEXURE TESTS

Flexure is the bending of a member. The strength that a beam shows in flexure is not only a function of the inherent strength of the material, but also of the geometry of the part. A rectangular beam on edge having the same cross-section area of metal as a square beam will have more flexure strength but will be equal in weight. Therefore, the mathematics for calculating the flexure stress are quite involved and will not be presented here. The parameters needed in these calculations—the modulus of elasticity, ultimate tensile strength, yield point, and so on—are the same as the ones discussed in the tensile test. It is not necessary to test a beam in flexure to determine these values. Beams are often tested in flexure to determine which geometry is best for given situations. A simple apparatus for such a test is shown in Figure 3-21.

FIGURE 3-21 Flexure apparatus.

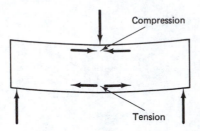

FIGURE 3-22 Typical beam in flexure.

If a beam is simply loaded, that is, supported at both ends and a load applied in the middle as shown in Figure 3-22, the bottom edge of the beam will go into tension. The top edge, being pushed on, will go into compression. There is a line somewhere in between which is neither in tension nor compression and is known as the neutral axis. For a beam loaded in this manner, it is necessary to design it in such a way that the bottom will withstand the required tension while the top can take the necessary compression. In conducting a flexure test, it is necessary not only to report the ultimate load before the beam fails, but to report the manner and type of failure. Did the beam fail in tension or compression, did it shear, or did it buckle? Photographs are often taken of the broken members and included in the report to corroborate the rupture mode.

Wood, tested in flexure, will fail in shear parallel to the grain (illustrated in Figure 3-23a). Flange beams, channels, and box beams often show some buckling of the web or flanges when failed in flexure (Figure 3-23b). Concrete always breaks where there is tension along the bottom edge of a beam in flexure (Figure 3-23c).

Weld Bend Tests

One modification of the flexure test is the weld bend test. Any good weld will be stronger than the parent metal around it. The weld bend tester is shown in Figure 3-24. This machine can be used on metal samples up to $\frac{3}{8}$ in. thick.

After a weld is completed, the excess weld metal is machined off, leaving the surfaces flat. The weld is then placed in the tester over a 1.5-in.-diameter mandrel and welded bar bent through an angle of 180°. Welders must be able to produce welds that will pass this test to be certified as a licensed welder.

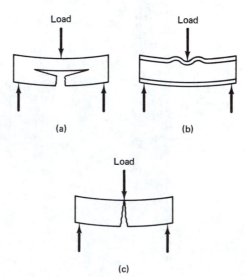

FIGURE 3-23 Failure in flexure: (a) wood; (b) metal; (c) concrete.

FIGURE 3-24 Weld bend test.

CREEP TESTS

If a structural member is placed under a constant load for a long period, it may deform even though its design load is well below the elastic limit of the material. This phenomenon is known as *creep*. Creep is a concern of engineers in the design of bridges, buildings, and other permanent structures in which the structural members have relatively large dead or static loads. Creep is also a function of temperature. The higher the temperature, the faster the materials will creep under the same loads. These facts lead to several methods of determining the amount of creep a beam or other part will undergo over a period of time.

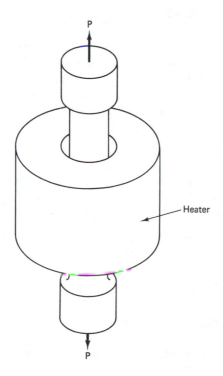

FIGURE 3-25 Creep test setup.

A simple creep test would be to place scale markings on a member actually installed in a bridge or other structure, then to compare these readings annually over a period of years. This has been done, but the obvious drawback is the length of time required to obtain useful data. Often 50 years or so are required to accumulate significant data by this method.

In the *long-time creep test* a test specimen is placed under a constant load in tension by applying a dead weight through a series of levers. The specimen is enclosed in a furnace in which the temperature is held at a preset constant temperature (see Figure 3-25). The amount of elongation is read periodically. The test is repeated at different temperatures and different loads. The amount of elongation per unit of time (hour, day, week, month, etc.) can be found for each temperature and load. A graph is then plotted of the stress in pounds per square inch per unit of time versus the temperature. By extrapolating the line back to average daily temperatures the expected creep per unit of time can be determined (see Figure 3-26).

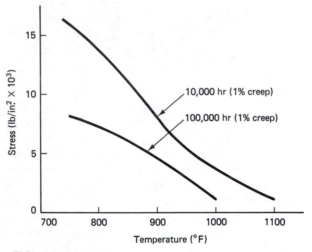

FIGURE 3-26   Stress versus temperature graph.

Figure 3-27 shows the typical creep versus time graph for a single temperature. It can be seen that the initial creep, or primary creep, can be rather rapid. There then follows a region of relatively constant elongation. It is this region of secondary creep that is of interest in the design of structures and is often used to estimate the expected creep over the lifetime of the part. The members are designed so that

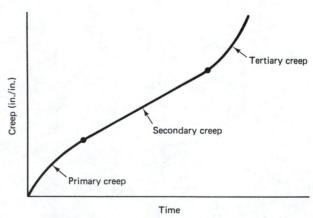

FIGURE 3-27   Creep versus time graph.

the working stress will produce a deformation which will be less than 1% over 100,000 hours (11.4 years) at average daily temperatures.

NONDESTRUCTIVE TESTS

Any method of testing that does not render the part unusable for its intended purpose is considered a nondestructive test. Nondestructive testing (NDT) is often used to find defective under- or oversized parts, or to determine if the product does not meet specifications. Flaws in metals can be found in several ways. If the flaws are on the surface, visual inspection may be sufficient. A simple "chip hammer" can often detect poor welds. Many times, however, these surface flaws are too small to be seen or may be obscured by surface roughness or scale. In such instances a process of *magnafluxing* is used. Magnafluxing involves applying a strong magnetic field through the part, and then sprinkling finely ground or powdered iron filings on the surface. A flaw will channel the lines of magnetic flux along the crack and cause the iron filings to accumulate there (see Figure 3-28).

Instances in which magnafluxing are used include shops where large locomotive engines are rebuilt and places where human safety may depend on a given part. Many aircraft parts are magnafluxed. It is a requirement that all parts on the race cars at the Indianapolis 500 Memorial Day Race be magnafluxed or X-rayed prior to their use in the car. Drivers have been injured and killed because critical parts in the steering, braking, or suspension have failed.

Another method of detecting surface flaws in parts is the *dye penetrant*. In this technique a red dye dissolved in a penetrating oil is sprayed onto the surface of the piece to be tested. This dye will penetrate or sink into any crack or flaw that outcrops the surface. The excess dye is removed and a chalk developer is sprayed onto the surface. The dye is then absorbed from the crack by the chalk and will be plainly visible in the white background.

For flaws that are beneath the surface, the parts must be tested in other ways. Here sonic tests, eddy currents, or X-rays are often used. In the sonic test, a transducer is placed on the surface of the part and high-frequency sound waves are projected through the metal. Any flaw will act as a reflector to these waves

FIGURE 3-28 Magnafluxing.

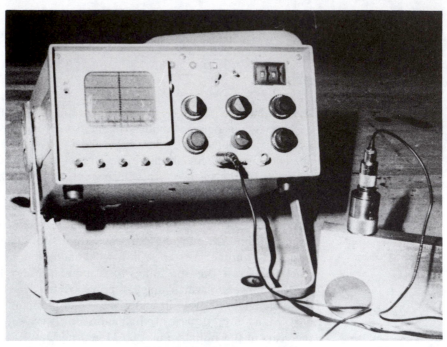

FIGURE 3-29   Sonic tester.

and will be picked up through a microphone. The reflected signal is then shown on an oscilloscope and the depth of the flaw can be determined. Figure 3-29 shows a typical small sonic tester.

Metals can also be *X-rayed* or gamma-rayed to find flaws. A film is placed on one side of the metal and an X-ray is beamed through the part. Flaws will show up on the film. In the field, a high-intensity gamma-ray source (from a radioactive element) is often used instead of the bulky X-ray machines. Critical welds in pipelines, bridges, and water tanks are usually subjected to X-ray or gamma-ray tests.

Eddy current tests can be used on electrically conductive metals. In this test an electric current is either induced in a part or sent through it. Flaws in the part interrupt the flow of the electricity which can be detected by sensitive instrumentation. Eddy current testers are used not only to detect flaws in the metals but can also be used in nondestructive evaluation (NDE) of the material's electrical conductivity, its magnetic permeability, heat treatment condition, and other properties. NDE use of eddy current instruments is also used in measuring the thickness of materials and the thickness of coatings on parts. Eddy current apparatus is also connected to computers and used to sort materials by differences in their compositions and microstructures.

The relatively recent science of *holography* is now finding its way into nondestructive testing and evaluation. A laser is used to project an interference pattern onto the surface of a specimen. If a light load is then placed on that object, it will deform and the interference pattern will be changed. If there are flaws in the part, the deformation of the part will not be uniform and it will be slightly more deformed along the flaw. The interference pattern will be concentrated along the flaw. This is an extremely sensitive technique and is usable for nonmetals as well as metals. To illustrate the sensitivity of this method, if a holographic interference pattern is projected onto a concrete block wall, it will be distorted by a person leaning or pushing against the wall even though the wall was flexed only an infinitesimal distance. Further, any cracks or other flaws in the concrete will show up in the interference pattern. An interference pattern projected on the side of a coffee cup will detect how much coffee is in it. The hot coffee causes the cup to expand slightly as it enters the cup, changing the pattern projected on the cup. From the change in the interference pattern it is possible to tell from a distance the depth of the coffee. Holographic interferometry is now used to detect flaws in composite aircraft parts. It is used to "see" if honeycombed laminations in helicopter blades are becoming separated. From vibrational analysis to solder joint failure to leak testing in hermetically sealed batteries, the applications of holographic interferometry are endless. Besides being an inspection tool, holography is a very useful instrument in many aspects of materials and engineering research.

Many other nondestructive testing techniques are also being used. Radioactive tracers, thermal paints, and remote sensors, often using telemetry, are now on the market. NDT and NDE is a growing field and many new techniques are being developed. Whenever possible, nondestructive tests are preferred over their destructive counterparts in industry.

Thinker 3-8:

What nondestructive tests do you routinely perform on your own automobile?

Structure of Matter

THE ATOM

To fabricate and test materials properly, it is necessary to understand the structure of them. All matter is composed of *atoms*. An atom is the smallest unit into which an element can be divided and still retain the characteristics of that element.

The idea that all matter is composed of very small, indivisible particles is not new. Thales of Miletus first suggested that all matter is composed of the "seeds of life" about 590 years before the birth of Christ. The Greek Scholar, Democritus, coined the word "atom" about the year 460 B.C. The word "atom" is derived from the Greek words *a-tomos*, meaning "not-cuttable." A setback for atomic theory occurred around 360 B.C. when Aristotle proposed that all matter was composed of only four elements: Earth, Air, Fire, and Water. This erroneous concept prevailed for over 2000 years, but the atomic theory of matter was kept by a few scholars. Periodically throughout the era of the Aristotelian doctrine, treatises were published resurrecting the idea of the atom. Finally, by the late nineteenth century, experimental evidence became available to confirm the existence of the atom.

There is still much to be learned about the atom and considerable research is being conducted on it. We do know that the atom is extremely small. The atom has been measured and weighed. It takes roughly 4.92 heptillion or

$$4,900,000,000,000,000,000,000,000 = 4.90 \times 10^{24}$$

atoms to make a pound of iron. To give an idea of the size of this number, if each atom in that pound of iron was the size of a Ping-Pong ball, 1 inch in diameter, this number of atoms would occupy a volume of 14.8 trillion cubic miles (14,800,000,000). This would be a sphere with a diameter of about 3050 miles, which is almost one and one-half times the diameter of the moon.

During the last century, physicists have determined that the atom is composed of several types of particles. In the center of the atom is the *nucleus*. The nucleus contains almost all of the mass of the atom and is composed of protons and *neutrons*. The neutron is an electrically neutral particle which has been assigned an atomic mass unit of 1.* *Protons* carry a single positive electric charge and have an atomic mass unit of 1. In 1903, Albert Einstein proposed the theory that mass and energy are related through his now famous equation

$$E = mc^2$$

where E is the energy equivalent to the mass (m) multiplied by the speed of light (c) squared. Since almost all the mass of the atom is in the nucleus, most of the energy is also there. Physicists are interested primarily in the energy of the atom and have succeeded in using very high energy machines to break the nucleus into further subatomic particles, such as mesons, quarks, gluons, and others, in an attempt to determine the answers to such questions as "What holds the neutrons and protons together" and "Is there a single particle common to all types of atoms?"

Outside the nucleus is the third atomic particle, the *electron*. The electron has a mass of about 1/1836 that of the neutron but carries a negative electric charge. In a neutral atom, one with no net electric charge, the number of electrons is equal to the number of protons. It is possible to add or subtract electrons from an atom. If an electron is added to the atom, there would be more electrons than protons, so there would be a net negative charge to the atom. Similarly, the removal of an electron from the atom would create a net positive charge on the atom. These charged particles are called *ions*. Positively charged ions are known as *cations*, while negatively charged ones are called *anions*.

There is considerable space between the nucleus of the atom and the electrons. A hydrogen atom contains only one proton in the nucleus, no neutrons, and one electron outside the nucleus. The diameter of the nucleus of a hydrogen atom is about 1×10^{-13} cm, and the diameter of the entire atom is about 1×10^{-8} cm. The diameter of the atom is 100,000 times the diameter of the nucleus. If the nucleus of the hydrogen atom was enlarged to the size of a basketball, for instance, the electrons would be about the size of a pea approximately 1 mile away. For a weight comparison, if the proton weighed 1 ton, the electron would weigh a little more than a pound.

The electrons are arranged about the nucleus in definite patterns or energy levels. It is this arrangement of electrons that allows the electrons to react or combine with other atoms to form *molecules*. A molecule is the smallest unit to which any material or *compound* can be divided and still remain that compound. Any further dividing of the molecule would cut it into atoms. Water, plastics, table salt, sugar, and every compound in the universe, including the ones in animal tissues and plants, are composed of these molecules. Compounds no longer have the characteristics of the individual atoms from which they are made but take on entirely new characteristics of their own. For instance, sodium is a metal that is very soft and very reactive with other compounds, especially water. If eaten, it is fatal in small concentrations. Chlorine is a deadly yellow gas. Yet these two types of atoms combine to form sodium chloride, common table salt, which is a compound necessary for one's good health.

Physicists study energy and therefore focus on the nucleus of the atom. Chemists, materials scientists, and engineers are interested in the ways atoms combined to form matter and are interested in the electronic structure of the atom. The nucleus of the atom determines the type of atom or element it is and carries the

---

*Presently, the mass of the neutron is defined in relation to the mass of the carbon-12 (^{12}C) atom and is equal to 1.008665 atomic mass units. It is rounded off to 1.0 for purposes of discussion here.

mass of the atom. The mass of the atom is one factor in establishing the density or weight per unit volume of the material.

Elements and compounds can be mixed and do not necessarily react chemically. Compounds become new materials with their own new characteristics, such as melting points, color, electrical conductivity, and so on. *Mixtures* retain the characteristics of the individual components. Mixtures can be separated physically. Sand and water, for instance, will form a mixture but not a compound. The sand can be separated from the water by filtering or the water can be removed by boiling it off. Metals can form mixtures or compounds.

ELEMENTS

Everything in the universe is formed from the combination of elements. At present, over 105 elements have been identified but only 92 of them occur in nature. The type of element depends only on the number of protons in the nucleus. If an element has only one proton, it is hydrogen. The element that has two protons is helium; three protons makes the element lithium. The element with the most protons that occurs in nature is uranium, with 92 protons. Elements that have more than 92 protons are artificial and are made in high-energy laboratories throughout the world. These transuranic elements include neptunium, plutonium, americium, curium, berkelium, californium, einsteinium, fermium, mendelevium, nobelium, lawrencium, unilquadium, and others that may be produced in the future.

The number of protons in the nucleus is referred to as that element's *atomic number*. The atomic number of hydrogen is 1, while helium's is 2. Carbon has an atomic number of 6 since it has six protons. The nucleus of an atom may also have neutrons. Most hydrogen atoms do not have any neutrons. About one atom out of every 1000 hydrogen atoms will have one neutron. The hydrogen atom with one neutron is known as *deuterium*. Hydrogen atoms with two neutrons in the nucleus are known as *tritium*. Hydrogen, deuterium, and tritium are *isotopes* of each other. Hydrogen is the only element that has its isotopes named individually. Isotopes are atoms of the same element; that is, they have the same number of protons, hence the same atomic number, but they differ in the number of neutrons in the nucleus. Isotopes have the same atomic number but a different atomic weight. Most elements have several isotopes. Uranium, one atom from which the fuel for nuclear power plants is made, has 92 protons but has isotopes that can have either 135, 136, 137, 138, 139, 140, 141, 142, 143, 145, 146, 147, or 148 neutrons. In all, the 92 naturally occurring elements have more than 1360 isotopes.

The sum of the number of protons and neutrons in an atom is the *atomic weight* of that element. The atomic weight for hydrogen is 1 (1 proton + 0 neutrons). Helium has 2 protons and usually 2 neutrons and would have an atomic weight of 4. The number of neutrons in an element can be found by subtracting the atomic number (the number of protons) from the atomic weight (the number of protons plus neutrons).

In 1811, Amedeo Avogadro advanced the theory that one atomic weight in grams of any element would have the same number of atoms. One gram-atomic weight or *gram-atom* has 6.023×10^{23} atoms. Thus 55.8 g of iron or 24.3 g of magnesium have the same 6.023×10^{23} atoms. This number of atoms per gram-atom is known as *Avogadro's number*. A sample problem might be: How many atoms are there in 10 g of iron?

$$(10 \text{ g}/55.8 \text{ g/g-atom})(6.023 \times 10^{23} \text{ atoms/g-atom}) = 1.08 \times 10^{23} \text{ atoms}$$

An often confusing point is that the tables of elements show the atomic weights of the elements as fractions. For example, aluminum is listed as having an atomic weight of 26.9815. This occurs for two reasons: (1) the atomic weights of the neutrons and protons are not exactly 1, and (2) the listed atomic weights of the elements are the weighted average of all the isotopes of the element.

To simplify the writing of these elements, each has been assigned a chemical symbol. Many of the elements are very rare and have no practical use. Therefore, not all of them will be discussed here. A list of the more common and useful elements is provided in Figure 4-1. The student of materials should become thoroughly familiar with this list and memorize the chemical symbols for these elements. Most of the symbols are of two letters, with the first letter being a capital and the second letter lowercase. The majority of the elements use the first one or two letters of the name of the element as the symbol, but this is not always the case. For some of the elements, the symbols are taken from the Latin or other language name for that element. For example, the symbol for the element sodium was taken from the Latin name for that element, *natrium*. The source of these names is listed in Figure 4-1.

To identify a specific isotope of an element, it is customary to write the atomic weight of the particular isotope as a superscript preceding the chemical symbol. The isotope of uranium that has an atomic weight of 235 could be differentiated from the uranium isotope with an atomic weight of 238 by the symbols

$$^{235}U \quad \text{and} \quad ^{238}U$$

By the year 1870, about 68 of the chemical elements had been discovered. Dmitri Ivanovich Mendeleev observed that many of these elements displayed similar chemical characteristics. Sodium and potassium are very active metals that oxidize (corrode) extremely rapidly. Oxygen and sulfur both react chemically with hydrogen and other elements in the same manner. Chlorine, bromine, and iodine all attack metals violently. By arranging these elements in order of increasing atomic weights, he observed that for the first two dozen or so elements, every eighth element had similar characteristics. He simply put these elements with similar properties beneath each other and developed what is now known as the *periodic table of the elements*.

Using this periodicity of the elements, Mendeleev was able to show that there were "holes" in the table which must be for elements that were not then known. From his table he was also able to describe what the characteristics of these unknown elements would have to be. Carbon, silicon, and tin were all known by his day, but a gap existed between the silicon and tin. In 1871, Mendeleev published a paper in the *Journal of the Russian Chemical Society* in which he predicted that the element which would fill this gap, which he called ekasilicon, would have an atomic weight of 72; would have properties "intermediate between silicon and tin"; would be a gray, brittle metal that would decompose water vapor with difficulty; and would have almost no action on acids. Later, in 1886, the German chemist Clemens Alexander Winkler discovered the element in nature and named it after his homeland, germanium. Germanium is one of the fundamental elements used in the semiconductor industry.

Chemists set out to find all the elements missing from the periodic table. All of those elements have now been found and identified, but a few of them are so rare in nature that the entire world's supply is a few grams. Any new elements will have to be made artificially.

Some of the elements with similar properties have been grouped into families of elements. The elements lithium, sodium, potassium, rubidium, cesium, and francium are known as the *alkali metals*. Beryllium, magnesium, zinc, cadmium, and mercury are the *alkaline earth metals*. On the right side of the chart, the

FIGURE 4-1   Chemical elements.

Element	Symbol	Atomic Number	Atomic Weight	Source of chemical symbol
Hydrogen	H	1	1	
Helium	He	2	4	
Lithium	Li	3	7	
Beryllium	Be	4	9	
Boron	B	5	11	
Carbon	C	6	12	
Nitrogen	N	7	14	
Oxygen	O	8	16	
Fluorine	F	9	19	
Neon	Ne	10	20	
Sodium	Na	11	23	Latin, *natrium*
Magnesium	Mg	12	24	
Aluminum	Al	13	27	
Silicon	Si	14	28	
Phosphorus	P	15	30	
Sulfur	S	16	32	
Chlorine	Cl	17	35	
Potassium	K	19	39	German, *kalium*
Calcium	Ca	20	40	
Titanium	Ti	22	48	
Vanadium	V	23	51	
Chromium	Cr	24	52	
Manganese	Mn	25	55	
Iron	Fe	26	56	Latin, *ferrum*
Cobalt	Co	27	59	
Nickel	Ni	28	59	
Copper	Cu	29	63	Latin, *cuprum*
Zinc	Zn	30	65	
Germanium	Ge	32	72	
Arsenic	As	33	75	
Selenium	Se	34	79	
Bromine	Br	35	80	
Strontium	Sr	38	88	
Molybdenum	Mo	42	96	
Silver	Ag	47	108	Latin, *argentium*
Cadmium	Cd	48	112	
Tin	Sn	50	119	Latin, *stannum*
Antimony	Sb	51	121	Latin, *stibium*
Tellurium	Te	52	128	
Iodine	I	53	127	
Barium	Ba	56	137	
Tungsten	W	74	184	German, *wolfram*
Platinum	Pt	78	195	
Gold	Au	79	197	Latin, *aurum*
Mercury	Hg	80	200	Latin, *hydrargyrum*
Lead	Pb	82	207	Latin, *plumbum*
Bismuth	Bi	83	209	
Radium	Ra	88	226	
Uranium	U	92	238	
Plutonium	Pu	94	242	

elements fluorine, chlorine, bromine, iodine, and astatine are called the *halogens*. Helium, neon, argon, krypton, xenon, and radon are the *inert gases*. These gases do not react with anything. Gold, silver, platinum, and palladium are known as the *noble metals* and tend not to react quickly with other elements. Since they do not react easily, gold, silver, platinum, and even copper can be found as pure metals in such forms as nuggets and fine metallic particles. The elements cerium through lutetium, atomic numbers 58 through 71, are referred to as the *lanthanide series* since they follow lanthanum in the periodic table. This series is also called the *rare earths* because of their scarcity. Thorium through lawrencium comprise the *actinide series* and follow actinium on the chart. All elements that follow uranium (have atomic numbers greater than 92) are called *transuranic* elements.

PROBLEM SET 4-1

Refer to Figure 4-1 for information to work these problems.

1. Suppose that an element had 23 protons and 28 neutrons in its nucleus. What are
    (a) the atomic number,
    (b) the atomic weight,
    (c) the name, and
    (d) the chemical symbol of the element?

2. Suppose that an element had an atomic weight of 56 and an atomic number of 26.
    (a) What is the name of the element?
    (b) How many protons does it have in the nucleus?
    (c) How many neutrons does it have in the nucleus?

3. An element used in the nuclear power industry is ^{235}U (235 is the atomic weight). How many
    (a) protons and
    (b) neutrons does this element have in the nucleus?

4. Which element has 19 protons in the nucleus?

5. Which element has more neutrons, mercury or gold?

6. The ^{235}U used in nuclear power plants can split into ^{141}Ba and ^{92}Kr. Are there the same number of protons and neutrons in the barium and krypton that there were in the uranium?

7. How many electrons are there in a neutral atom of aluminum?

8. A cation of potassium has one net positive electric charge. (It is missing one electron.) How many electrons are there in this potassium ion?

9. How many electrons are there in a neutral iron atom?

10. Consider the element silver.
    (a) What is the chemical symbol for this element?
    (b) How many protons are in the nucleus?
    (c) How many neutrons are in the nucleus?
    (d) How many electrons would be in the neutral atom?
    (e) How many electrons would be in a singly charged cation?

11. How many atoms are there in 20 g of silver?

12. How many atoms are there in 40 g of magnesium?

13. How many atoms are there in 500 g of copper?

14. Which has more atoms 30 g of iron or 20 g of aluminum?

15. What is the mass of 1 atom of iron?

16. What is the mass of 1 atom of gold?

17. A small nugget of gold had a mass of 1 g. How many atoms were there in that nugget?

18. Each silver atom loses one electron to become an ion. How many electrons are given up in ionizing 108 g of silver?

19. Each aluminum atom can give up three atoms to become an ion. How many electrons would be given up in ionizing 54 g of aluminum?

20. How many gram-atoms or gram-atomic weights are there in 1 lb of iron?

BONDING

The atoms go together to form molecules. There are four different mechanisms or *chemical bonds* by which they can join together. Around each nucleus of the atom there are several electronic shells. The closest shell to the nucleus can hold only two electrons. The next shell is completed with eight more electrons. The lowest-energy states occur in the atoms with completed shells. Completed shells have 2, 10, 18, 36, 54, or 86 electrons. The electrons in the elements will always go to as low an energy state as possible. Atoms with one, two, or three electrons above a completed shell tend to give up these electrons fairly easily to become ions. Since there are always the same number of electrons in the neutral atom as there are protons in the nucleus, the removal of electrons will leave a net positive charge on the atom. These positively charged atoms are called cations. Cations always come from metallic elements. The atoms that need one, two, or three more electrons to fill out a complete shell tend to attract more electrons. In doing so they take on a net negative charge and are known as anions. Anions are always formed by nonmetallic elements.

If a metal and nonmetal are placed together, the metal tends to give up its electrons which are in excess of a completed shell to a nonmetal which needs electrons to complete a shell. In nature "opposites attract," so the positive cation will be attached to the negative anion by a mechanism known as an *ionic bond* (Figure 4-2). A common example of this type of bond occurs between the elements sodium and chlorine. Sodium has 11 electrons about its nucleus, while chlorine has only 17. The sodium releases one electron to the chlorine and both elements go to completed shells and lower-energy states. The product formed by the attachment of the sodium to the chlorine is sodium chloride. The chemical reaction for this is

$$2Na + Cl_2 \rightarrow 2NaCl$$

You should notice that the chlorine is shown as Cl_2. Gases that end in the letters "-gen" or "-ine" exist by themselves as diatomic molecules. If no other element is around with which they can react, they will join together in molecules having two atoms. Let another atom come by and the diatomic molecule will split and combine with it. The diatomic molecules include hydrogen, oxygen, nitrogen, fluorine, chlorine, bromine, and iodine. The equation above is a balanced equation in that there are the same number of each type of atoms on the left and right sides of the reaction. The sodium ion is nothing like the sodium atom, nor does the chloride ion have the same properties of the chlorine atom. Sodium metal is a very

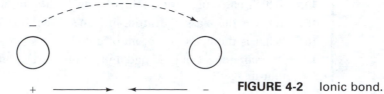

FIGURE 4-2   Ionic bond.

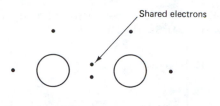

FIGURE 4-3   Covalent bond.

active metal that will react with anything with which it comes in contact. The sodium ion exists only in compounds.

Elements that are near the middle of a completed electronic shell can neither gain enough electrons nor lose them to go to a complete shell. In this case they will share their electrons with another element so that each element can show completed electronic shells. Elements such as carbon, silicon, and others will always form this *covalent bond*. Figure 4-3 illustrates this concept.

Metals always have an excess of electrons over a completed shell. They hold onto these electrons rather loosely. When in the neighborhood of like atoms the theory says that these electrons are pooled or form an "electron cloud" to which the positive nuclei are attracted. This is the formation of the *metallic bond* (Figure 4-4). According to this theory, it is easy to see why metals are good conductors of electricity. In this electron cloud it would take little energy to cause the electrons to move.

If one looks up the boiling points of hydrogen telluride (H_2Te), hydrogen selenide (H_2Se), hydrogen sulfide (H_2S), and hydrogen oxide (H_2O) (water) (all elements of the same chemical families) and plots their boiling points against their molecular weights, the first three will fall on a line which, if extrapolated to the molecular weight of water, would have water boiling at about $-65°C$. Yet the actual boiling point of water is $100°C$. Figure 4-5 shows this relationship.

The reason for this phenomenon is that water forms what is known as *hydrogen bridge* or *Van der Waals bond* between the molecules (Figure 4-6). Hydrogen shares only two electrons about it, while the oxygen has eight. This makes the hydrogen ends of the molecule slightly more positive (or less negative) than the oxygen and there is a slight attractive force or bond between the hydrogen of one molecule and the oxygen of the neighboring molecule.

Seldom is there a material with just one type of bond. Usually, materials are made up of several types of bonds. Water could be thought of as having covalent bonds with van der Waals bonds between the molecules. Each type of bond has its own characteristics. Ionic bonds produce optically transparent materials, fairly high melting points, and electrical insulators at low temperatures. At high tem-

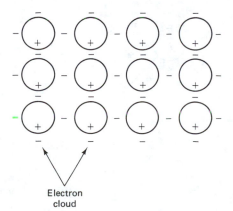

FIGURE 4-4   Metallic bond.

FIGURE 4-5   Relationship between the boiling points and molecular weights of H₂Te, H₂Se, H₂S, and H₂O.

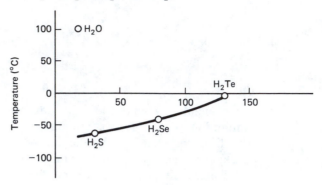

Molecule	Boiling Point (°C)	Molecular Weight
Hydrogen telluride	−2	130
Hydrogen selenide	−41.5	80
Hydrogen sulfide	−60.7	32
Hydrogen oxide	100	18

peratures ionically bonded materials will conduct electricity electrolytically but not electronically. That is, the ions move to carry the current not the electrons. In general, ionic materials are brittle, but the strength, ductility, and hardness vary tremendously.

Thinker 4-1:

Common table salt is ionically bonded. What optical, mechanical, electrical, and thermal properties would you expect it to have? Does table salt actually have these properties?

Covalent materials are optically transparent, opaque to X-rays, and are electrical insulators. The electrical conductivity goes up as the temperature rises. They produce very hard, brittle materials.

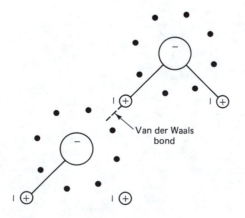

Van der Waals bond

FIGURE 4-6   Van der Waal's bond.

Thinker 4-2:

   Table sugar is covalently bonded. What optical, mechanical, and electrical properties would you expect table sugar to have? Does it actually have these properties?

Metallic bonded materials are optically opaque and have a metallic luster. They are very good electrical conductors, but the resistance generally goes up as the temperature rises. They have moderate to high melting points. The strength, hardness, and ductility vary considerably and depend on factors other than the bond type. Van der Waals bonds always occur with other types of bonds (with minor exceptions such as liquid helium) and are much weaker than the other types. The properties depend on those other bond types, not the Van der Waals.

MOLECULES

When atoms combine to form molecules, they do so in definite ratios. For instance, two atoms of hydrogen will react with one atom of oxygen to form water. To represent this molecule, the chemist uses the symbols for the individual atoms in the molecule and subscripts to represent the number of each type of atom in the molecule. Water would have the formula.

$$H_2O$$

meaning that there are two atoms of hydrogen for every atom of oxygen in the molecule. Emery is aluminum oxide, which has two atoms of aluminum joined to three atoms of oxygen. Its formula would be

$$Al_2O_3$$

Similarly, sulfuric acid, with two atoms of hydrogen, one atom of sulfur, and four atoms of oxygen would have the formula

$$H_2SO_4$$

Sometimes, entire groups of atoms are repeated in a molecule. These repeated groups are called *radicals*. These radicals are shown in parentheses and subscripts are used in the same manner as with individual atoms. An example of this is alum (potassium aluminum sulfate), which has the formula

$$KAl\,(SO_4)_2$$

Here the entire sulfate radical (SO_4) is repeated. Some of the more common radicals are shown in Figure 4–7.

All the radicals in Figure 4–7 have negative charges with the exception of ammonium. These negative radicals will combine with positive elements in such a manner that there will be no net negative charge. All metals take on positive charges when they become ions. The alkali metals all assume one positive charge.

FIGURE 4-7   Common chemical radicals.

Radical	Name
$(NH_4)^+$	Ammonium
$(HCO_3)^-$	Bicarbonate
$(HSO_4)^-$	Bisulfate
$(CO_3)^{2-}$	Carbonate
$(CN)^-$	Cyanide
$(OH)^-$	Hydroxide
$(PO_4)^{3-}$	Phosphate
$(NO_3)^-$	Nitrate
$(NO_2)^-$	Nitrite
$(SO_4)^{2-}$	Sulfate
$(SO_3)^{2-}$	Sulfite
$(S_2O_3)^-$	Thiosulfate

The sodium ion having one positive charge will react with the single negative charge of the hydroxide to form sodium hydroxide:

$$NaOH$$

It will take two sodium ions to cancel the two negative charges of the sulfate radical. The formula for sodium sulfate is

$$Na_2SO_4$$

It would take three ammonium radicals, each having a single positive charge, to combine with one phosphate to form the compound ammonium phosphate. The formula for ammonium phosphate is

$$(NH_4)_3(PO_4)$$

If a positive ion or radical having a charge of $+3$ combines with a negative radical with only a -2 charge, two of the positive ions would combine with three of the negative radicals so that there would be a total of six positive and six negative charges. Aluminum sulfate is such an example. The aluminum ion has a $+3$ charge, while the sulfate radical has a -2 charge. The formula for aluminum sulfate is

$$Al_2(SO_4)_3$$

PROBLEM SET 4-2

1.   Rust is iron oxide. Its molecule has two atoms of iron combined with three atoms of oxygen. What is its chemical formula?
2.   Sodium chloride, common table salt, is composed of one atom of sodium combined with one atom of chlorine. Write the formula for table salt.
3.   Table sugar has 12 carbon atoms, 22 hydrogen atoms, and 11 oxygen atoms per molecule. Write a chemical formula for sugar.
4.   How many atoms of nitrogen, hydrogen, sulfur, and oxygen are in ammonium sulfate, $(NH_4)_2(SO_4)$?
5.   Sodium bicarbonate has one sodium atom and one bicarbonate radical in its molecule. Write its formula.

6. Sodium hydroxide (lye) has one sodium atom and one hydroxide radical per molecule. What is its formula?

7. Nitric acid has one hydrogen atom and a single nitrate radical. State its formula.

8. What is the formula for silver nitrate if it has an atom of silver and one nitrate radical per molecule?

9. Ammonium citrate has the formula $(NH_4)_2H_6C_5O_7$. How many atoms of nitrogen, hydrogen, carbon, and oxygen are there in the molecule?

10. Match the names of the compounds below with the correct formula.

   _____ 1. Trisodium phosphate   a. $CaSO_4$
   _____ 2. Barium chloride   b. Cr_2O_3
   _____ 3. Ferric iodide   c. PbI_2
   _____ 4. Calcium sulfate   d. Na_3PO_4
   _____ 5. Strontium nitrate   e. $Mg(OH)_2$
   _____ 6. Sodium sulfate   f. $BaCl_2$
   _____ 7. Magnesium hydroxide   g. H_2S
   _____ 8. Lead iodide   h. $Sr(NO_3)_2$
   _____ 9. Chromic oxide   i. FeI_3
   _____ 10. Hydrogen sulfide   j. $Na_2(SO_4)$

11. Would you expect carbon dioxide (CO_2) to have ionic, metallic, or covalent bonds?

12. Would potassium bromide (KBr) have ionic, metallic, or covalent bonds?

13. Aluminum bromide has ionic bonds. Each aluminum atom gives up three electrons, while each bromine atom accepts one electron. Write the formula for aluminum bromide.

14. Calcium chloride has ionic bonds. Calcium gives up two electrons per atom, while chlorine accepts only one electron per atom in becoming ions. Write the formula for calcium chloride.

15. Calcium carbonate has ionic bonds between the calcium and the carbonate radical. From the information in Figure 4-7 and Problem 14, write the formula for calcium carbonate.

16. Barium gives up two electrons in becoming an ion. The phosphate radical (PO_4) accepts three electrons in becoming an ion. Write the formula for barium phosphate.

17. The ammonium radical has one positive charge. Write the formula for ammonium sulfate.

18. (a) If gold is $400 per ounce, what is the cost of 1 atom of gold?
    (b) How many atoms of gold would you get for $1.00?
    Note: Gold is weighed in troy ounces which have 32.15 g/oz.

19. Antimony nitrate has the formula $Sb(NO_3)_3$. How many oxygen atoms are there in the molecule?

20. Aluminum sulfate has the formula $Al_2(SO_4)_3$. How many
    (a) sulfur atoms and
    (b) oxygen atoms are there in each molecule?

Every atom has its own atomic weight assigned to it. The *molecular weight* of a compound (sometimes called the mol-weight or just mol) is the sum of all of the atomic weights in the molecule. Water has two hydrogen atoms and one oxygen. The atomic weight of hydrogen is 1 and the atomic weight of oxygen is 16. The molecular weight of water is 18 and is figured as follows:

Formula: H_2O

$$2\,H = 2 \times 1 = 2$$
$$1\,O = 1 \times 16 = \underline{16}$$
$$18$$

The molecular weight of sulfuric acid can be figured as follows:

Formula: H_2SO_4

$$2\,H = 2 \times 1 = 2$$
$$1\,S = 1 \times 32 = 32$$
$$4\,O = 4 \times 16 = \underline{64}$$
$$98$$

The molecular weight of more complicated molecules involves a little more arithmetic but is figured in the same way. The molecular weight of aluminum sulfate would be calculated as follows:

Formula: $Al_2(SO_4)_3$

$$2\,Al = 2 \times 27 = 54$$
$$3\,S = 3 \times 32 = 96$$
$$12\,O = 12 \times 16 = \underline{192}$$
$$342$$

There are the same number of molecules in a *gram-molecular weight* of a compound as there are atoms in a gram atom of an element. There are 6.023×10^{23} molecules of aluminum sulfate in 342 g of the compound. One gram-molecular weight of a compound is often called a *gram-mol*.

PROBLEM SET 4-3

Use Figure 4-1 for information in working these problems.

1. What is the molecular weight of sodium hydroxide (NaOH)?
2. What is the molecular weight of aluminum chloride ($AlCl_3$)?
3. What is the molecular weight of ammonium sulfate [$(NH_4)_2(SO_4)$]?
4. What is the molecular weight of nitric acid (HNO_3)?
5. What is the molecular weight of table salt (NaCl)?
6. What is the molecular weight of common table sugar ($C_{12}H_{22}O_{11}$)?
7. What is the molecular weight of benzene (C_6H_6)?
8. What is the molecular weight of potassium dichromate ($K_2Cr_2O_7$)?
9. Which has a higher molecular weight, hydrogen bromide (HBr) or nitric acid (HNO_3)?

10. A compound used in nuclear energy is uranium hexafluoride (UF_6). What is its molecular weight?

11. What is the molecular weight of aluminum hydroxide [$Al(OH)_3$]?

12. How many molecules of sodium hydroxide are there in 40 g?

13. How many molecules of table salt ($NaCl$) are there in 100 g?

14. How many molecules of sugar ($C_{12}H_{22}O_{11}$) are there per pound?

15. Ethyl alcohol has the formula C_2H_6O. What is its molecular weight?

16. How many molecules are there in 100 g of ethyl alcohol?

17. Aluminum sulfate has the formula $Al_2(SO_4)_3$. What is its molecular weight?

18. How many molecules of aluminum sulfate are there in 200 g?

19. If 2.41×10^{24} molecules of a compound weigh 128 g, what is its molecular weight?

20. If 5×10^{24} molecules of a compound weigh 200 g, what would be the molecular weight of the compound?

STATES OF MATTER

Atoms and molecules make up the matter of the universe. This matter can exist in four states: *gas, liquid, solid,* and *plasma.* If the molecules are not held together and are free to move about, the matter is in the gaseous state. If the molecules are randomly bound to each other without much order in their arrangement the matter is in the liquid state. The solid state is defined as that state of matter which has a definite long-range crystal structure. A *crystal* is defined as a definite array of atoms to occupy space. Figure 4-8 illustrates these three states of matter. *Plasma* is an ionized gas. Although the majority of the universe is in the plasma state, including our sun and the stars, there is not much of it on the earth, nor is it a primary concern of engineers, so it is not discussed here.

Atoms are never standing still except at absolute zero (0 K or $-273°C$). The amount of motion of the atoms is a measure of their energy and can be related to the internal heat of the solid. The more motion of the atoms, the more collisions per unit of time and the higher the temperature. At low temperatures the bonds between the atoms will be strong enough to hold the material in a definite crystal structure. The atoms will vibrate about a fixed point. At higher temperatures, the atoms vibrate with greater amplitude. Finally, a temperature is reached where the

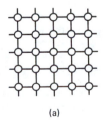

(a)

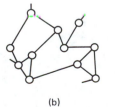

(b)

(c)

FIGURE 4-8 States of matter: (a) solid; (b) liquid; (c) gas.

bonds between the atoms are too weak to hold them in their assigned place and the bonds break apart. This is the temperature of the melting point. All solids will have a definite melting point. Above the melting point, there may be a random structure with some short-range crystals left.

Throughout the entire solid and liquid temperature range, some of the vibrating atoms will have enough energy to escape from the surface of the material. These escaping atoms or molecules hit the atoms of the atmosphere or the walls of their container and cause what is known as the *vapor pressure* of the material. The higher the temperature, the more atoms will escape from the surface and the higher the vapor pressure will be. The atoms in the atmosphere above a liquid can collide with these escaping atoms and drive them back onto the liquid. The atmospheric pressure above a liquid is a result of the number of atoms per unit volume in that atmosphere. The more atoms per unit volume the greater the atmospheric pressure and the more atoms there will be to "push" escaping atoms back onto the surface of the liquid. The temperature at which the vapor pressure of the liquid equals the atmospheric pressure of the gas above it is the *boiling point* of that liquid. The boiling point of any liquid is a function of the atmospheric pressure above it.

Thinker 4-3:

Glass is a liquid. What is evidence that will support this conclusion?

CRYSTAL STRUCTURES

In the solid state, molecules (or other particles such as atoms, ions, or radicals) are bonded together to form a *crystal structure*. There are only seven different crystal systems or ways in which atoms or particles can be arranged in a definite array to occupy space. There are a total of 14 modifications of these crystal systems.

The simplest of these crystal structures is the *cubic system*. The cubic crystal system has the particles equally spaced, with their planes of atoms being at right angles to each other. Figure 4-9 is a diagram of the *unit cell* or *primitive* of the simple cubic structure. No pure element is found in the simple cubic structure. However, many compounds, such as common table salt (NaCl), have this structure.

One modification of the cubic structure has an atom in the middle of the cube. This is the *body-centered cubic* (BCC) structure. The unit cell of the body-centered cubic structure is shown in Figure 4-10a. Many elements, such as pure

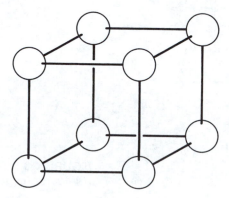

FIGURE 4-9   Simple cubic structure.

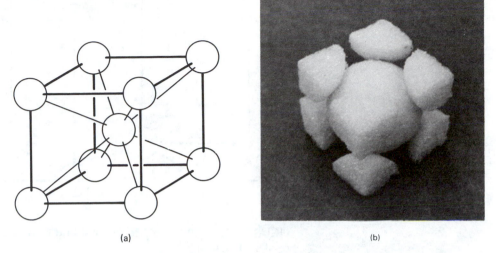

(a)   (b)

FIGURE 4-10   Body-centered cubic structure.

iron and sodium, are found in this arrangement. A model of the BCC is shown in Figure 4-10b.

Still another modification of the cubic system is the *face-centered cubic* (FCC) structure. In the face-centered cubic structure, an extra particle is in the middle of each side or face of the cube (see Figure 4-11a). Figure 4-11 is a photo of the model of a FCC structure. Copper, lead, silver, gold, and other elements are found in the face-centered cubic structure.

The second crystal structure is the *tetragonal system*. In the tetragonal system, all the atomic planes are still at right angles to each other, but in one of the dimensions, the planes of the atoms are farther apart than in the other two directions. There are two modifications of the tetragonal structure, the simple tetragonal (Figure 4-12a) and the body-centered tetragonal (see Figure 4-12b). Pure tin is probably the best example of the tetragonal structure found in nature.

The third crystal structure, the *orthorhombic system*, still has the planes of atoms at right angles to each other, but the unit lengths on the three axes are all different. There are four modifications of the orthorhombic system. Besides the simple orthorhombic crystal structure (see Figure 4-13), there is the body-centered

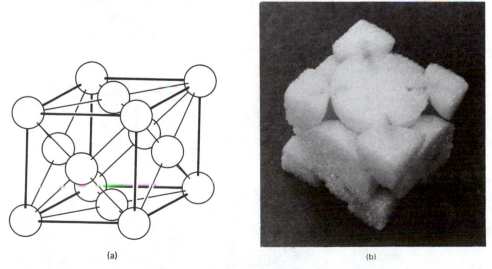

(a)   (b)

FIGURE 4-11   Face-centered cubic structure.

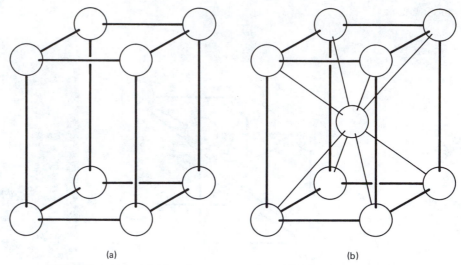

(a)   (b)

FIGURE 4-12   (a) Simple tetragonal and (b) body-centered tetragonal systems.

orthorhombic system, which has an atom in the center of the tetrahedron (see
Figure 4-14). The base-centered orthorhombic system has an extra atom in both
the top and bottom faces of the unit cell (Figure 4-15). The face-centered orthorhombic system has an extra atom in each side or face of the crystal (Figure 4-16).

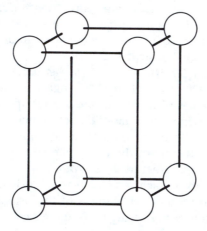

FIGURE 4-13   Simple orthorhombic system.

In the cubic, tetragonal, and orthorhombic crystal systems, the planes of the
atoms are all at right angles to each other. In the *rhombohedral system*, all the
distances between the crystal planes are equal but none are at right angles to each
other (see Figure 4-17).

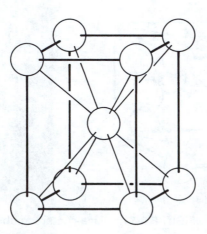

FIGURE 4-14   Body-centered orthorhombic structure.

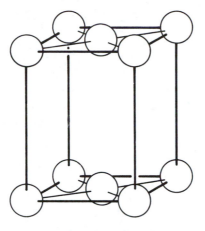

FIGURE 4-15   Base-centered orthorhombic structure.

A closely related system to the rhombohedral is the *hexagonal*. In the hexagonal system, the distances between the atoms in the base are the same and the base is perpendicular to the sides of the crystal, but the angle between the side faces of the crystal is exactly 120° (Figure 4-18a). The distance between the bases may or may not be equal to the distances between the atoms in the base. It may not be obvious why this arrangement is called the hexagonal system until three of the unit cells are put together (see Figure 4-18b).

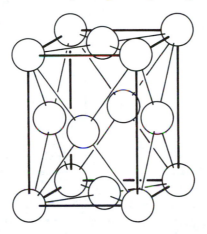

FIGURE 4-16   Face-centered orthorhombic structure.

In many of these crystal systems, the individual primitive cells can often overlap. This gives rise to the close-packed crystals. The *hexagonal close-packed* system is shown in Figure 4-19. Graphite is carbon in the hexagonal close-packed system, while diamond is carbon in the face-centered cubic close-packed or *complex cubic system*. The complex cubic system is often referred to as the "diamond structure."

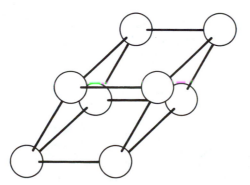

FIGURE 4-17   Rhombohedral system.

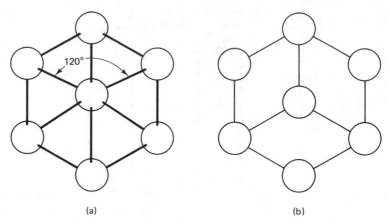

FIGURE 4-18   Hexagonal system: (a) unit cell; (b) assembled unit cells.

In the sixth system, the *monoclinic*, none of the bond lengths, or distances between the atoms in the crystal planes, are equal, but two of the planes are perpendicular to each other. The third angle is not a right angle (see Figure 4-20). The monoclinic system also has a base-centered modification.

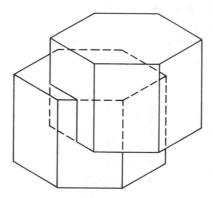

FIGURE 4-19   Hexagonal close packed system.

The last crystal system is the *triclinic*. In this system, no two bond distances are equal and no planes are at right angles to each other, nor are the angles between the planes equal (see Figure 4-21). This structure can make the crystals look like long needles. Many compounds are in this system.

A general crystal system would have the axis and lengths as shown in Figure 4-22. Let the lengths of the axes, or lattice constants, be a, b, and c and the angles between the crystal planes be designated by the lowercase Greek letters alpha, beta, and gamma, (α, β, and γ). The crystal systems are summarized in Figure 4-23.

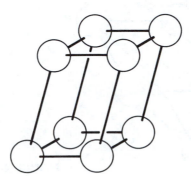

FIGURE 4-20   Monoclinic system.

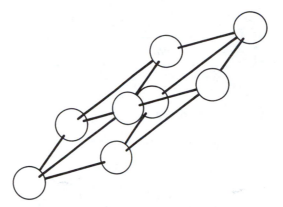

FIGURE 4-21   Triclinic system.

Density, ductility, and other properties of materials often depend on its crystal structure. In general, for pure metals, the most ductile and malleable materials will have a face-centered cubic crystal structure. Body-centered cubic structures are a little less ductile, while simple cubic materials are quite brittle. Most metals are either face-centered cubic, body-centered cubic, or hexagonal close-packed. A few, such as antimony and bismuth, are rhombohedral, and tin is tetragonal. Figure 4-24 shows a partial list of the crystal structures of some metals.

In the diagrams shown to illustrate the crystal structures, only the unit cell was shown. The atoms continue to follow this orderly array indefinitely until the crystal runs into another crystal or it comes to the end of the material. Crystals always break in a manner that breaks the fewest number of bonds. In the simple cubic example shown in Figure 4-25, for the crystal to break along path A, 8 bonds would be broken in the area shown. Path B would break 8 bonds, while path C would require only 6 bonds to be broken. The crystal will therefore break along path C since less energy will be required to break the fewer number of bonds. This phenomenon ensures that large crystals will always shatter or break into smaller, but same-shaped crystals. Examples of this occur in table salt and common table sugar. If large crystals of these materials are grown, they will form cubes. Breaking these cubes into smaller pieces will produce smaller cubes (see Figure 4-26).

In Figure 4-25 the direction C is called a *slip plane*. The plane of atoms above line C will slip over the plane below the line. All planes parallel to this plane are also slip planes. In a simple cubic structure, the same number of bonds per centimeter would be broken in the planes perpendicular to line C and would slip just

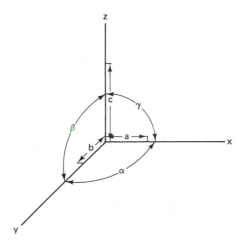

FIGURE 4-22   Crystal axis.

FIGURE 4-23   Summary of crystal systems.

System	Axis Lengths	Angles	Modifications
Cubic	$a = b = c$	$\alpha = \beta = \gamma = 90°$	Simple Body centered Face centered
Tetragonal	$a = b \neq c$	$\alpha = \beta = \gamma = 90°$	Simple Body centered
Orthorhombic	$a \neq b \neq c$	$\alpha = \beta = \gamma = 90°$	Simple Base centered Body centered Face centered
Rhombohedral	$a = b = c$	$\alpha = \beta = \gamma \neq 90°$	Simple
Hexagonal	$a = b \neq c$	$\alpha = \beta = 90°, \gamma = 120°$	Simple
Monoclinic	$a \neq b \neq c$	$\alpha = \beta = 90°, \gamma \neq 90°$	Simple Base centered
Triclinic	$a \neq b \neq c$	$\alpha \neq \beta \neq \gamma \neq 90°$	Simple

as easily. Since the longer bonds are the weakest, slip will always occur between planes which are the farthest apart. In simple cubic crystals these planes are parallel to the edges of the cube. In face-centered and body-centered cubic crystals other planes will slip first. As shown in Figure 4-27, these slip planes are the faces of the cube. Each face can slip in two directions. Since there are three different slip planes and two directions of slip in each plane, there are six ways or *mechanisms of slip* in the simple cubic crystal.

Large crystals have larger planes which can slip over each other. Larger crystals will be more ductile than their smaller crystals of the same material. In steels and other metals the size of the crystals can be controlled by the heat treatments of the metal. The slower the metal is cooled, the larger the crystals or grains will be and the more ductile the metal becomes.

In the face-centered cubic crystal there are 12 mechanisms of slip while the

FIGURE 4-24   Crystal structures of some metals.

Metal	Symbol	Crystal Structure
Aluminum	Al	FCC
Antimony	Sb	Rhombohedral
Bismuth	Bi	Rhombohedral
Cadmium	Cd	HCP
Calcium	Ca	FCC and BCC[a]
Chromium	Cr	BCC and HCP[a]
Copper	Cu	FCC
Gold	Au	FCC
Iron	Fe	BCC and FCC[a]
Magnesium	Mg	HCP
Molybdenum	Mo	BCC
Nickel	Ni	FCC
Platinum	Pt	FCC
Silicon	Si	Complex cubic
Silver	Ag	BCC
Tin	Sn	Tetragonal, cubic, and rhombohedral[a]
Tungsten	W	BCC
Vanadium	V	BCC
Zinc	Zn	HCP

[a]Structure depends on temperature.

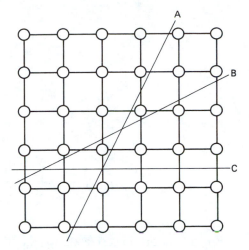

FIGURE 4-25 Possible fracture directions.

body-centered cube has a total of 48. The hexagonal close-packed system usually has only three mechanisms of slip. It stands to reason, therefore, that if other factors do not interfere, the more slip planes a crystal has the more ductile and more malleable it becomes. The "other factors" could be imperfections, which are discussed later. Body-centered cubic structures of metals in the pure state will, in fact, produce more ductile metals than their face-centered counterparts.

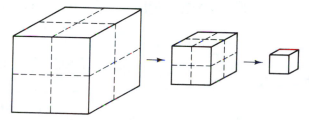

FIGURE 4-26 Crystal breakage.

Miller Index

In discussing crystals and crystal structures it is often necessary to discuss a particular plane of atoms or molecules. This is done by what is called the Miller index of the plane. The Miller index of a plane can be found in three steps.

1. Find the point at which the plane intercepts the x, y, and z axes.

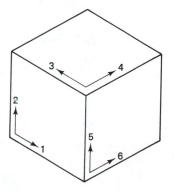

FIGURE 4-27 Mechanisms of slip in a simple cube.

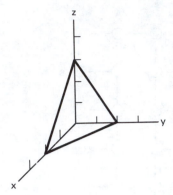

FIGURE 4-28   Crystal plane.

2. Take the reciprocal of those intercepts.
3. Multiply the reciprocal by the lowest common multiple so that the Miller index is in small whole numbers.

Consider the plane shown in Figure 4-28.

Intercepts: $x = 2$, $y = 2$, $z = 3$

Reciprocals: $\frac{1}{2}$, $\frac{1}{2}$, $\frac{1}{3}$

The lowest common multiple is 6, so the Miller index of the plane would be

(3, 3, 2)

Miller indices are always shown in parentheses.

PROBLEM SET 4-4

1. A crystal structure has two axes of equal unit length, while the third is longer. All three axes are at 90° to each other. What is the crystal system?
2. A crystal structure has no axes of equal unit length, and no angles between the axes are equal. What is the crystal structure?
3. Which crystal systems have all axes at 90° to each other?
4. Which crystal system has the angle between the axes exactly 120°?
5. Which crystal systems have all three axes having equal unit lengths?
6. Which crystal system has all axes perpendicular to each other, but none of the unit lengths on the axes are equal?
7. Which crystal system has all the unit lengths equal, but none of the axes are perpendicular?
8. Which crystal systems have body-centered modifications?
9. Which crystal systems have face-centered modifications?
10. A crystal plane has x, y, and z intercepts at 1, 2, and 4, respectively. What is the Miller index of that plane?
11. A crystal plane has x, y, and z intercepts at 1, 3, and 2, respectively. What is the Miller index of the plane?
12. A crystal plane has x, y, and z intercepts at 2, 1, and 1, respectively. What is the Miller index of the plane?
13. A crystal plane has x, y, and z intercepts at 2, ∞ (infinity), and ∞, respectively. What is the Miller index of the plane?

14. A crystal plane is parallel to the z axis and intercepts the x and y axes at 2 and 4, respectively. What is the Miller index of the plane?

15. A crystal plane is parallel to both the x and z axes and intercepts the y axis at 3. What is the Miller index of the plane?

16. What is the Miller index of a plane through three face diagonals of a cube?

17. What is the Miller index of a plane passing through one face diagonal of a cube but parallel to the z axis?

18. A crystal plane passes through the atoms on one edge of a body-centered cube and the atom in the middle of the cube. What Miller indices could it have?

19. A crystal plane passes through the points $\frac{1}{2}$, $\frac{1}{4}$, and 1. What are its Miller indices?

20. A crystal plane passes through the points $\frac{1}{4}$, $\frac{1}{2}$, and ∞. What are the Miller indices of the plane?

IMPERFECTIONS IN MATERIALS

No material in nature is perfect. In fact, the *imperfections* in many materials are an asset. The sparkle in rubies, sapphires, and diamonds is a result of natural imperfections. Synthetic rubies lack these random imperfections and do not have nearly the sparkle of the natural ones. Were it not for imperfections in the crystal structures of materials used in semiconductors, they would not work and transistors and integrated circuits would not exist. Some imperfections in materials are detrimental. Steel, for instance, has a theoretical tensile strength of nearly 3 million pounds per square inch, yet the best that we can produce commercially is about one-tenth of that, or 300,000 lb/in². The reason for this loss of strength is the imperfections in the crystal structure of steel.

An imperfection is anything that upsets or changes the orderly array of atoms or molecules in the crystal structure. Some imperfections affect the mechanical properties, such as tensile strength, hardness, ductility, and malleability of the material. Other imperfections can affect such electrical properties as resistance and electronic conductivity in semiconductors. Optical properties such as color and optical transparency are also affected by imperfections. Photochromic glasses, which darken in the sunlight, are a result of deliberately introducing imperfections into the glass. The melting point of a material is an example of thermal properties that are affected by imperfections.

There are several types of defects of crystals or ways by which the orderly array of particles can be upset. *Point defects* are those which involve only a single atom or radical per defect. A common point defect is the *vacancy*. A vacancy is an atom or radical missing from its normal position (see Figure 4-29). Since an atom is missing from the normal position or *lattice point*, the bonds normally provided by that atom are also missing. This would result in weakening the material. In aluminum, for example, about 1 out of every 10 billion atoms are missing. At room temperature, this amounts to about 1×10^{13} vacancies per cubic inch. The number of vacancies depends on the temperature of the metal and rises rapidly as the temperature increases. At the melting point of aluminum, there is about 1 vacancy per 1000 atoms. It is therefore easy to understand why aluminum is not a good material to use at elevated temperatures. A second point defect affecting the mechanical properties of a material are *interstitials*. An interstitial is an atom or molecule at a point other than a lattice point in the crystal structure (see Figure 4-30).

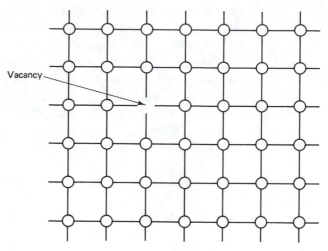

Vacancy

FIGURE 4-29   Vacancy.

Ductility in a crystal is a result of one plane of atoms sliding over another plane. An interstitial is shown between planes A and B in Figure 4-31. The effect of this interstitial is to hinder or even prevent plane A from sliding over plane B. This is called *pinning* the slip planes. Plane A might slip over plane B up to a point near the interstitial, then have to "detour" around it (path C), which means that more bonds must be broken and more energy required. This would tend to make the material stronger, with a resultant loss of ductility. The more interstitials there are in the crystal structure, the fewer planes there are left to slip and the more brittle the material becomes.

One example of the pinning of slip planes occurs in steel. Steel is iron with carbon contained interstitially in the crystal lattice. Pure iron is very ductile. A low-carbon steel with less than 0.2% carbon is much stronger than pure iron, but is still relatively ductile. However, in high-carbon steels, of 0.8% carbon, most of the slip planes are pinned and the resulting steel is very brittle, with somewhat higher strength than the lower-carbon steel.

Substitutionals are imperfections resulting from the replacement of an atom of the parent crystal by a different atom (see Figure 4-32). Since the size of the substituted atom may be different from that of the other atoms of the crystal, and

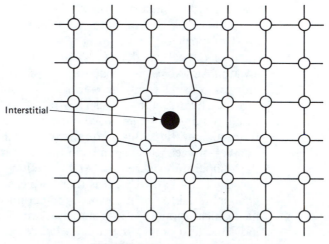

Interstitial

FIGURE 4-30   Interstitial.

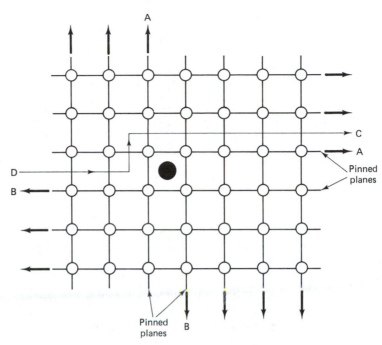

FIGURE 4-31   Pinned slip planes.

since it may have a different bond strength than the surrounding atoms, the melting point is changed. The effect of substitutional atoms in alloys is often the lowering of the melting point. This will be discussed later in the section on phase diagrams.

By judiciously and carefully introducing substitutional atoms into a crystal, a semiconductor material for transistors and integrated circuits can be formed. This will be discussed later. Further, if sufficient substitutionals are added to the material, the characteristics of the material can be changed. Bronze is an example of copper with a large number of substitutionals of other metals that react chemically with the parent matrix. In brasses, many of the copper atoms are replaced by zinc atoms. The result is a material that resembles neither copper or zinc but has characteristics all its own.

Under certain conditions, interstitials can fall into vacancies creating a substitutional. Since substitutionals do not embrittle the material as much as intersti-

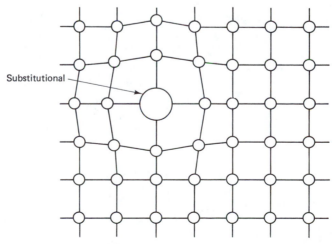

FIGURE 4-32   Substitutional.

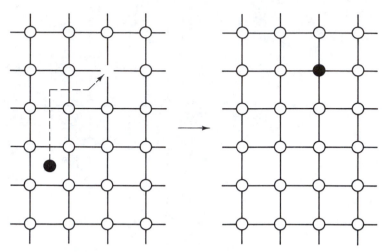

FIGURE 4-33   Migration of an interstitial to fill a vacancy.

tials, the result of this migration would be to soften the material (see Figure 4-33). Although this migration would occur normally over a period of time, higher temperatures which cause the atoms to move faster will increase the speed of this change. This is one of the principles underlying the heat treatment of steels.

A mechanism that involves substitutionals, vacancies, and interstitials is *diffusion*. Diffusion allows one material to penetrate another material atom by atom or molecule by molecule. A drop of perfume or incense placed in one corner of a room will diffuse through the air and can eventually be detected throughout the room. If a crystal having many vacancies is immersed in an atmosphere of another material, the outside atoms can "seep" into the vacancies and the interstices of the surface of the crystal. The atomic motion of the atoms in the crystal can then allow the vacancies to move about in the crystal. These new substitutionals can migrate from vacancy to vacancy into the interior of the crystal (see Figure 4-34). This is the principle used in "case hardening" of steel, which will be discussed in Chapter 6. Interstitials can also move about in the crystal by the diffusion process.

In the crystal structure, not only is there an orderly array of atoms, but the number of electrons about those atoms is also fixed. In a silicon crystal, for instance, each atom normally has 14 electrons about it. If an atom of phosphorus is substituted for a silicon atom, the phosphorus atom carries with it 15 electrons. To the crystal lattice, the extra electron from the silicon atom is an imperfection. The imperfection of an additional electron in the lattice is called, oddly enough, an *electron* (see Figure 4-35). It is important not to confuse the imperfection "electron" with the atomic particle "electron." The imperfection electron is an extra atomic electron in the crystal lattice. (You should be thoroughly confused by now. Work on it.)

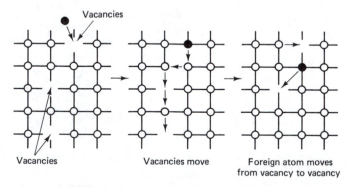

FIGURE 4-34   Process of diffusion.

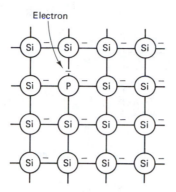

FIGURE 4-35   Imperfection "electron."

If the silicon atoms were replaced with an aluminum atom, having only 13 electrons about the nucleus, the crystal would view this missing electron as an imperfection known as a *hole* (see Figure 4-36). Do not confuse holes with vacancies. Holes refer to the electronic structure of the atom, while vacancies refer to the missing nucleus. Make sure that you have the definitions of these imperfections clearly understood.

Since electrons carry a negative charge, the imperfection electron would impart an additional negative charge to that portion of the crystal. Similarly, a hole, lacking a negative charge, would be a relatively positively charged spot in the matrix. This is the principle upon which the semiconductors used in transistors and integrated circuits work. Semiconductors that have many holes in them are known as P type, while those with electrons are N-type semiconductors. The holes and electrons can move from atom to atom in much the same manner as interstitials and vacancies. *P- and N-type semiconductors* are fused together to make diodes, transistors, and integrated circuits. P- and N-type semiconductors are made by *doping* parent materials, such as silicon and germanium, with elements that have more or fewer electrons than do the atoms of the parent crystals. The mechanism whereby these materials can be doped is the diffusion already discussed.

Not all imperfections involve single atoms. Some of these defects in the crystals involve entire planes of atoms. This type of imperfection is known as a *line defect*. As crystals grow, atom by atom and layer by layer, the atoms are kept in the orderly array. Sometimes, the atoms get confused as to where their correct place in the array should be. This gives rise to two imperfections: *stacking faults* and *twinning*. In face-centered cubic crystals, there are three layers of atoms which are repeated (Figure 4-37a). In hexagonal close-packed systems, there are only two layers which are repeated (Figure 4-37b). Both the FCC and the HCP put the

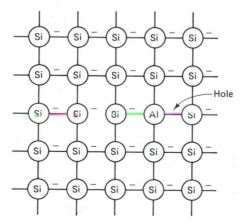

FIGURE 4-36   Imperfection "hole."

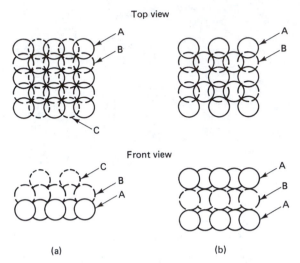

FIGURE 4-37 Stacking of atoms: (a) face-centered cube; (b) hexagonal close-packed.

atoms in their densest packing or put them as close together as they can possibly be.

In the face-centered cubic crystal the stacking would be an A layer on top of a B layer followed by a C layer. Then another A, B, and C; A, B, C; and so on. In the hexagonal close-packed crystal the stacking would be A, B, A, B, A, B, ... A *stacking fault* occurs by having a few layers or planes of the hexagonal closed-packed structure (HCP) slipped into a face-centered cubic crystal (see Figure 4-38). The stacking of the layers might be A, B, C, A, B, C, A, B, A, B, A, B, C, A, B, C, ... The presence of the A, B, A, B stacking of a HCP mixed in with the A, B, C, A, B, C of the FCC indicates a stacking fault.

Twinning is the formation of a mirror-image crystal within a grain or crystal. Atoms that may be arranged in a right-handed structure may suddenly be found in a left-handed arrangement (see Figure 4-39). Twinned crystals may be formed in two ways. Twinned crystals, which are formed as the solid is being cooled from the molten metal, are called *annealing twins*. The twinning in the crystals due to mechanical deformation, such as cold working, are known as *mechanical twins*.

Twinning occurs often in nature. It is difficult to find copper or brass that does not have twinned structures in the grains. Figure 4-40 shows examples of twinning in copper. Very large twinned crystals of quartz and other minerals are very prevalent (Figure 4-41). Regardless of how they are formed, twinned crystal structures interrupt the crystal slip planes and result in lowering the ductility of the material.

One of the most common line defects are *dislocations*. There are two types

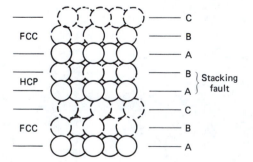

FIGURE 4-38 Stacking fault.

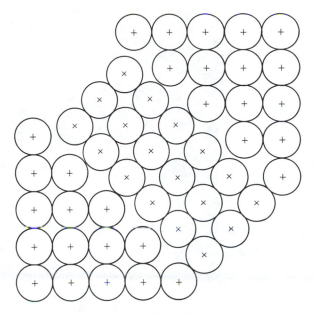

FIGURE 4-39   Twinning.

of dislocations, *line dislocations* and *screw dislocations*. Line dislocations are an extra partial plane of atoms in the crystal structure (see Figure 4-42). As with twinning, dislocations result from the way that atoms arrange themselves upon solidifying, or they can be introduced by mechanical deformation such as bending or cold working.

Line dislocations weaken the material. It can be seen from Figure 4-42 that there is some empty space at the end of the partial plane of atoms or dislocation. This space is a weak spot on the material.

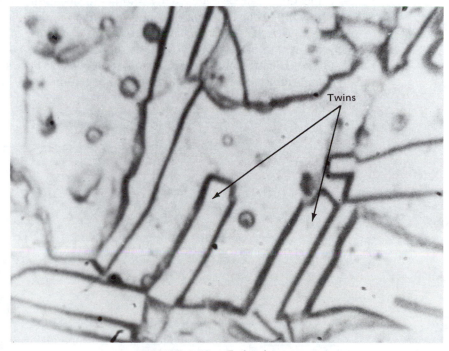

FIGURE 4-40   Twins in copper.

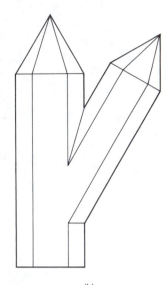

(a)   (b)

FIGURE 4-41   Twinned quartz crystals.

Line dislocations can move through the crystal. A force on the bottom part of the plane, shown as line B in Figure 4-43, can cause that partial plane to move to the left under the dislocation. The dislocation would then be moved to the right by one atom plane. The next plane could shift under the dislocation moving the partial plane of atoms farther right (Figure 4-43). This shifting can continue as long as nothing blocks the movement of the planes.

The movement of dislocations can be blocked in several ways. An interstitial atom from some place else in the crystal can fall into the space below the dislocation and become attached to the dislocation. When this happens, the end of the dislocation becomes pinned and cannot move (Figure 4-44). The movement of dislocations can be blocked by another dislocation, coming up from the bottom, moving in the opposite direction, and overlapping the top one.

Screw dislocations are the result of one plane of atoms being twisted slightly above the plane just below it, as shown in Figure 4-45.

Electron microscopic examination of steels have shown that there are as many as 1 million line dislocations per square centimeter of surface area. Imagine a cube

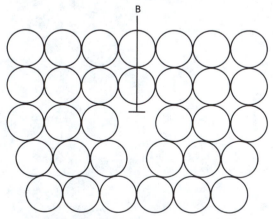

FIGURE 4-42   Line dislocation.

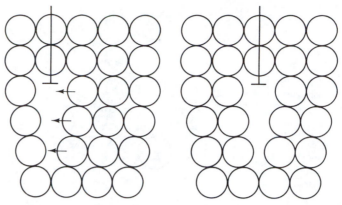

FIGURE 4-43   Movement of line dislocations.

1 centimeter on an edge. If each one of those million line dislocations went completely through the cube, there would be 3 million centimeters of dislocations per cubic centimeter (1 million in each direction). This is illustrated in Figure 4-46. Three million centimeters is equal to

$$\frac{3 \times 10^6 \text{ cm}}{(2.54 \text{ cm/in.})(12 \text{ in./ft})(5280 \text{ ft/mi})} = 18.6 \text{ mi}$$

that is, 18.6 miles of dislocations per cubic centimeter. With all of these dislocations tending to weaken the material, it is little wonder that steel does not approach its theoretical tensile strength of 3 million pounds per square inch.

Dislocations play an important part in the behavior of materials. Fatigue in metals is the failure of the material due to repeated or cyclic stresses being placed on the metal. One cause of fatigue in metals is due to dislocations. The cause of fatigue by dislocations was first proposed by two researchers, Frank and Read, and is known as the Frank–Read source.

One method of visualizing a Frank–Read source is to imagine a dislocation that is pinned at two places (Figure 4-47a; follow through the explanation of the Frank–Read source in this figure). If a force is applied against that dislocation, it will move in the center but cannot move at the ends, due to the pinning (Figure 4-47b). Further pressure on the dislocation will bow it out even further, taking in more and more atoms (Figure 4-47c and d). Finally, the dislocation loops back on itself (Figure 4-47e) to the point where it touches or reconnects. Since nature always prefers the path of lowest energy, the dislocation will divide itself into the old one between the original two pins and a new circular dislocation around the original one (Figure 4-47f).

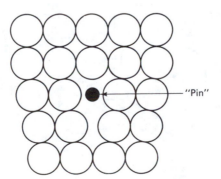

FIGURE 4-44   Blocked movement of dislocations.

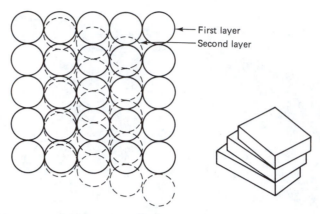

FIGURE 4-45 Screw dislocations.

The result of this routine is to generate a new dislocation while retaining the old one. Another cycle of stress on the material could cause the process to repeat generating another circular dislocation inside the first one. If enough dislocations are generated, they will be forced very close together, causing a dislocation pile-up. If several of these dislocations are forced together, a crack is initiated (see Figure 4-48). At this point most of the life of the metal is gone. All that is necessary to fail the material is for the crack to propagate by breaking one bond at a time through the material. The effect would be much the same as a zipper opening and the material would soon break.

This method of fatigue can be restated in six steps.

1. Cyclic stresses acting on
2. Frank–Read sources (pinned dislocations) generate
3. new dislocations which cause
4. dislocation pile-up which
5. initiate a crack.
6. Further stresses cause the crack to propagate to failure.

Stress corrosion cracking as a cause of fatigue in metals was mentioned in Chapter 3. In this mechanism, a crack on the surface, no matter how slight, can open up slightly as the part is stressed or bent. This allows oxygen and water into

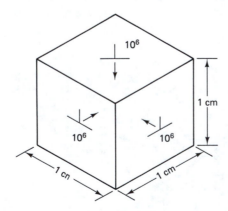

FIGURE 4-46 Dislocations per cubic centimeter.

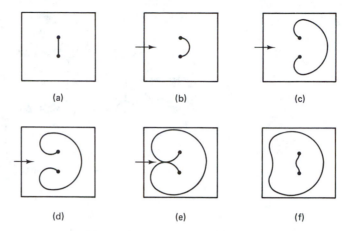

(a)   (b)   (c)

(d)   (e)   (f)

FIGURE 4-47   Frank–Read source.

the crack. The oxygen will attack one bond between the metal atoms, react with the metal to form an oxide, and effectively deepen the crack by one bond. Each time the crack is strained, more bonds are oxidized. Eventually, the crack will be corroded all the way through the part. Figure 4-49 illustrates this mechanism.

Fatigue can also occur due to cyclic thermal stresses. Repeated heating and cooling in some materials will cause it to break. In glass and ceramics there are many surface cracks. As the material is heated, it expands, causing the crack to widen and break a few bonds around it. Each time the material is heated, the crack will extend. The crack will eventually work its way through the piece and the material breaks.

Besides point defects and line defects, there are plane defects. Large-volume defects that replace many atoms are called *inclusions*. Inclusions may be large enough to be visible. Air bubbles and pieces of foreign matter are considered inclusions. There are other crystal imperfections that affect electrical, thermal, and optical properties of materials, but these will not be discussed here.

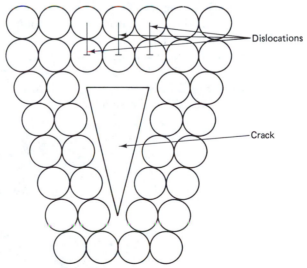

FIGURE 4-48   Dislocation pileup and crack.

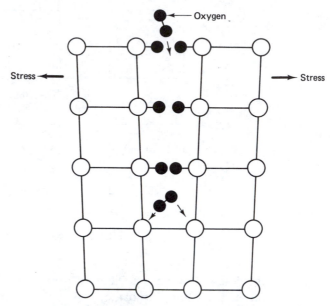

FIGURE 4-49 Stress corrosion cracking.

GRAIN STRUCTURE

As heat is removed from the molecules in the liquid state, their motion is slowed to the point where a few of the particles will bond together for the beginning of a crystal structure. This process is known as *nucleation*. A certain amount of energy is required to form the nucleus of a crystal. Without this energy, the liquid can be supercooled below its normal melting point without the material solidifying. Once a crystal is nucleated, the removal of more energy from the melt will cause the atoms or molecules in the liquid to join the crystal. Each particle will "fall in" to a particular crystal lattice point and the crystal will grow. In the solidification of metals from the liquid, many points of nucleation are started at the same time. As the individual minute crystals continue to grow, they will eventually grow together to form one solid mass (refer to Figure 4-50). The individual crystals have become a *grain* in the metal.

As the material continues to cool, smaller grains will be swallowed up by the larger ones. That is, the atoms from the smaller grains will leave the structure of the little grains and attach themselves to the larger ones. The reason for this lies in the fact that the smaller the crystal or grain face, the faster it will grow outward. Figure 4-51 illustrates this growth of the smaller surfaces, which allows the larger grains to eat up the smaller ones.

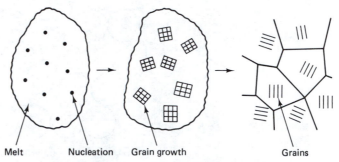

Melt   Nucleation   Grain growth   Grains

FIGURE 4-50 Process of solidification.

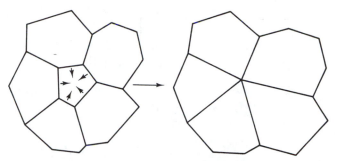

FIGURE 4-51   Grain growth.

The speed at which these crystals grow is also a function of the temperature. The higher the temperature, the larger the grains will grow. If the cooling rate is very slow, the atoms can move about easily and grain growth will continue, leaving only large crystals or grains. Conversely, if the material is cooled very rapidly, the millions of small grains initially nucleated will not have time to grow and the grains will remain small. As a rule of thumb, the slower the cooling rate, the larger the grains. Examples of grain growth occur naturally in rocks and ores. On the surface of a lava flow, the cooling is so rapid that even nucleation is prohibited. The result is obsidian, nature's glass. A little deeper into the magma or flow the cooling rate was slow enough to allow nucleation, but grain growth was limited. A very fine grained material, rhyolite, was formed. Still deeper into the volcanic eruption, some grain growth occurred and fine-grained granite will be found. The grains in the granite are easily seen by the unaided eye. In areas where cooling is extremely slow, sometimes requiring years to cool, very large grained porphyries and pegmatites can be found.

A similar phenomenon is found in metals. If a metal is allowed to cool very slowly, large grain sizes will result. Since the larger grains have longer slip planes, the slowly cooled material will be more ductile and malleable. Small grains, resulting from very rapid cooling, will produce a less ductile metal. This principle forms the basis of one type of heat treatment for some metals.

PHASES

Some materials will not mix, or dissolve completely, with other materials. The classic example is oil and water, which separate into *phases*. A phase is a macroscopically, physically distinct portion of matter. The word "macroscopically" implies that the individual phases can be seen optically using at most a low-powered microscope. Phases do not refer to mixtures on the atomic or molecular scale. There will always be an interface or sharp boundary between the separate phases. One can tell by eye exactly where the oil phase stops and the water phase starts.

Phases may be made from one or more *components*—a component being any material needed to make up the system being studied. In the oil–water example, oil and water are the individual components making up the oil–water system. (In science, the word "system" means "everything needed to conduct an experiment or to have an effect on it.")

In a single-component system, water for example, there can be three phases: solid, liquid, and gaseous. If the water is at its melting temperature, two phases can exist in equilibrium. If an ice cube is placed in a glass of water, and the ice is not melting or the water freezing, a two-phase system, solid and liquid, exists at equilibrium. This solid-liquid equilibrium can only exist at the melting temperature

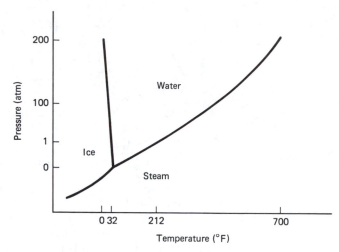

FIGURE 4-52   Phase diagram for water.

of the single component. Equilibrium implies that there is no overall or net change in the quantities of either phase.

Other two-phase systems can also exist. Steam is an example of water drops suspended in a gas (air), while the bubbles in champagne, soft drinks, and beer illustrate a gas suspended in a liquid. Snow and soot are examples of solids suspended in a gas. Two liquids, such as the oil and water, can exist in equilibrium. One has only to examine steel under a microscope to find that it is an example of two solid phases existing together.

A *phase diagram* is a map of the parameters, such as pressure, temperature, or composition, over which the various phases exist. A single-component phase diagram can be drawn for water, using pressure and temperature as the axes (refer to Figure 4-52). One can see that at 1 atmosphere pressure and 212°F, water is in the gaseous phase. (An atmosphere, the pressure exerted by dry air at sea level, is equal to 14.7 lb/in.² *) There is one point, at 0.006 atm pressure and 32°F, where all three phases can be present at once in equilibrium. This is called the triple point for water. If the temperature and pressure give points that fall on the lines, two phases can exist at the same time. At all other points, only one phase can be present. Since the boiling point of a liquid is the temperature at which the vapor pressure of the liquid equals the atmospheric pressure above the liquid, if the pressure on the water is lowered to almost a vacuum, 0.008 atm, the water will boil at 41°F. At this temperature it is possible to hold boiling water in your bare hand.†

Phase diagrams for mixtures of two or more materials are also possible. If two metals, for instance, are mixed together, temperature can be plotted against their percent composition for a phase diagram. In a two-component phase diagram, temperature is plotted on the ordinate (vertical axis) while the percent composition is plotted on the abscissa (horizontal axis). A typical example of this is shown in Figure 4-53. The left axis represents pure component A (100% A–0% B), while

---

*A column of air with a base 1 in. by 1 in. square, which rises from sea level to outer space, will weigh 14.7 lb.

†An experiment to probe this is as follows. Fill a round-bottomed Pyrex flask half full of water, then bring it to a rolling boil. Plug the flask with a rubber stopper and cool the flask of water under cold running water. Soon the flask can be held in the bare hands and the water will still be at a rolling boil inside the flask.

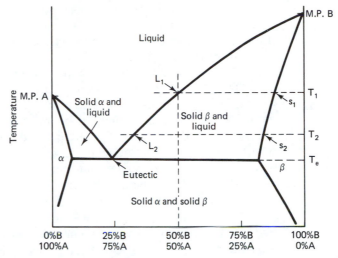

FIGURE 4-53   Generalized liquid–solid, two-component phase diagram.

the right axis is pure component B (0% A–100% B). The middle of the horizontal axis would be 50% A–50% B.

It is possible to mix two materials together without them separating into two phases. If sugar is added to a glass of tea, a certain amount of the sugar (the solute) will dissolve into the liquid (the solvent) and a single-phase system (all liquid) will remain. If more sugar is added, eventually some of it will separate from the solution and precipitate to the bottom of the glass. A two-phase system (solid sugar and the tea–sugar liquid) now exists. The same phenomenon exists with other materials. If oil and water are mixed, they will separate into two distinct phases. However, some of the oil will have dissolved into the water and some of the water will have gone into the oil. One only needs to draw off some of the water and taste it to ascertain that there is oil in the water. The water in the oil can be detected by heating the oil phase and the water will be evaporated before the oil begins to boil.

Metals too can dissolve into each other. The atoms of one metal can dissolve into another metal either substitutionally or interstitially without separating out as a different phase. This is called a solid solution. In Figure 4-53 the area in which B will dissolve completely into A is called the alpha (α) phase. The area shown as the beta (β) phase is that region in which A will completely dissolve into B. There is a limit to the number of atoms that can dissolve into another. There are only a fixed number of sites per unit mass in a solid material where substitution or interstitials can occur. After all these sites are occupied, the second material (solute) must form its own separate crystal or grain. The result will be a two-phase composition.

In the generalized phase diagram shown in Figure 4-53, the melting point of pure A is shown as M.P. A and the melting point of 100% B is indicated by M.P. B. One would expect to mix components A and B together and get a mixture with a melting point that fell on a straight line between M.P. A and M.P. B. This is never the case. It is possible to mix two components in the proportions and get a melting point lower than either of the two components. The lowest melting point of the mixture is defined as the *eutectic temperature* which occurs only at the *eutectic composition*. On Figure 4-53 the eutectic temperature is labeled as T_e and the eutectic composition is 75% A and 25% B.

There are several other terms used in reference to phase diagrams. The curved line from M.P. A to T_e to M.P. B is the *liquidus line*. The horizontal line running through the eutectic point is the *eutectic temperature line*. The lines running downward from the melting points to the eutectic temperature line are referred to as the *solidus lines*. The eutectic temperature line is also a solidus line.

In the areas between the liquidus lines and the solidus lines a two-phase system of liquid and solid will be formed. In the area to the right of the eutectic point, above the solidus line but below the liquidus line, a two-phase system of liquid and solid beta ($L + \beta$) will be found. To the left of the eutectic point there will be liquid and solid alpha ($L + \alpha$). Below the eutectic temperature line and between the alpha and beta solid solutions there is a two-phase solid consisting of solid alpha and solid beta ($\alpha + \beta$).

Consider the mixture of 50% A and 50% B in Figure 4-53. When cooling a mixture of this composition from the liquid, the first solid will appear when the temperature is reduced to the liquidus line (labeled T_1 on the diagram). The composition of that solid will be that of the solidus line at the same temperature (marked point s_1 on the diagram). Since the newly formed solid is rich in B, the composition of the liquid will be lowered in B. At a lower temperature (marked T_2) the composition of the liquid will be L_2 and the composition of the solid will be s_2. The composition of the liquid will continue to be reduced in B as the temperature is lowered. Finally, at the eutectic temperature, the eutectic composition will have been reached in any of the liquid that is left. The liquid that is left will solidify at the eutectic composition forming what is called the *eutectic matrix* in the solid. By this time a considerable amount of solid beta (β) will have already formed. If observed through a microscope, the eutectic matrix and the excess beta can easily be seen.

One interesting mixture is that of lead and antimony. The eutectic composition of this system is at 11.2% antimony. If a few grams of a mixture of 20% antimony and 80% lead is melted and poured onto a heated microscope slide and allowed to solidify, it can be pried from the slide, dipped in a solution of acetic acid and glycerine, and observed through a metallurgical microscope. The microstructure of this mixture is shown in Figure 4-54. The triangles are the excess antimony (the β phase). The rest of the picture shows the lead–antimony eutectic matrix.

A common two-component system is tin and lead. Figure 4-55 is the phase diagram of this system. A considerable amount of information can be obtained from this phase diagram. The melting point of pure lead is shown to be 621°F, while the melting of 100% tin is 450°F. The eutectic composition of the tin–lead system is 61.9% tin and 38.1% lead, with the eutectic temperature being a low 361°F. The alpha (α) phase shows that up to a maximum of 19.2% tin will dissolve into the lead if the temperature is set at 361°F. The beta phase indicates that a maximum of 2.5% of lead will dissolve into the tin at the eutectic temperature.

The mixture of tin and lead is commonly called solder. Plumbers use a solder that has a composition of 50% lead and 50% tin. From the phase diagram, it is easy to see that this mixture must be heated above 400°F to melt. Upon cooling the liquid will start to solidify at about 400°F but will not be completely solid until the eutectic temperature of 361°F is reached. Between 400 and 361°F the material will be in a two-phase system and look like slushy metal. This two-phase region allows the plumbers to "wipe" or smooth on the solder with a rag before it completely solidifies. Electricians and electronic technicians use a 60–40 solder (60% tin and 40% lead), which is close to the eutectic composition. This solder will melt at slightly over 360°F and completely solidifies at one temperature.

Phase diagrams are always *equilibrium diagrams*. The relative amounts of any phase at any given temperature are always constant. The equilibrium temperature

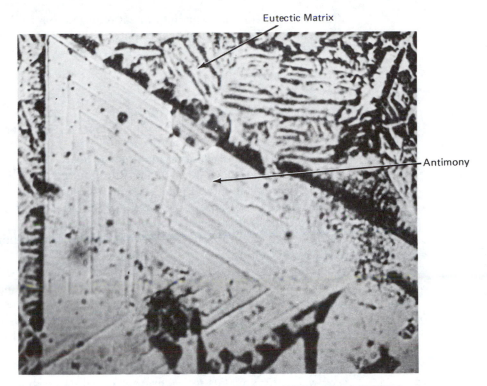

FIGURE 4-54   15% antimony–85% lead mixture.

for a two-phase ice–water system is 0°C (32°F). At that temperature the amount of solid and liquid will not change. This is also a *dynamic equilibrium* as far as the molecules are concerned. If one molecule of the solid phase leaves the ice to become liquid water, a molecule of liquid will replace it on the solid. In dynamic equilibrium, the individual particles may change, but the overall picture stays the same.

Cooling Curves

One method of developing the phase diagrams is through the use of *cooling curves*. It is obvious that for a material to undergo a change in temperature, heat must be added or removed. Heat is measured in *calories* or *British thermal units (Btu)*. A calorie is defined as the amount of heat required to raise 1 gram of water 1 degree

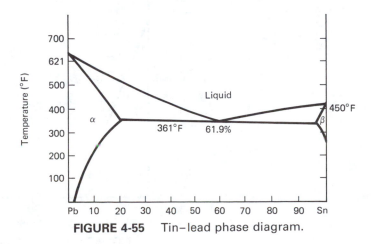

FIGURE 4-55   Tin–lead phase diagram.

Celsius. The Btu is the amount of heat required to raise 1 pound of water 1 degree Fahrenheit. The amount of heat required to raise 1 liter of water 10 degrees Celsius is 10,000 cal.

$$(1 \text{ L})(1000 \text{ mL/L})(1 \text{ g/mL})(10°C)(1 \text{ cal/°C/g}) = 10,000 \text{ cal}$$

If 1 gal of water is lowered 20°F, 167 Btu must be removed.

$$(1 \text{ gal})(8.33 \text{ lb/gal})(20°F)(1 \text{ Btu/lb/°F})$$

The number of calories required to heat 1 gram of the material 1 degree Celsius in the metric system, or the Btu required to heat 1 pound of a material 1 degree Fahrenheit in the English system is called the *heat capacity* of the material. The heat capacity of water has already been stated as 1 calorie per gram per degree Celsius. The heat capacity of iron is only about 0.1 calorie per gram per degree Celsius. The heat capacity of steam and ice are both about 0.5 calorie per gram per degree Celsius.

Another value used in the phase diagram study is the *heat of fusion*. The heat of fusion is the heat (in calories or Btu) required to change 1 gram of a material from a solid at its melting temperature to a liquid at the same temperature. For water, the heat of fusion is 76.4 calories per gram. This means that to change 1 gram of ice at 0°C to water at 0°C, 76.4 calories would be required or removed from its surroundings. The heat required to change 1 gram of a material from the liquid to gaseous state, at the boiling point of that material is its *heat of vaporization*. For water this amounts to 540 calories per gram or about 970 Btu per pound. The heat required to change 1 gram of ice at 0°C to steam at 100°C is 716.4 calories.

$$(1 \text{ g})(76.4 \text{ cal/g}) + 100°C(1 \text{ g})(1 \text{ cal/g/°C}) + 1 \text{ g}(540 \text{ cal/g}) = 716.4 \text{ cal}$$

Every pound of water evaporated must take 970 Btu away from its surroundings. This is why perspiring in animals and humans keeps them cool. There is little wonder why drying off after a dip in the swimming pool cools the skin. It gets the heat to evaporate the water from your body.

To remove the heat necessary to change a vapor into a liquid, cool the liquid to its freezing temperature, then remove the heat of fusion to cause it to solidify requires time. If the number of calories removed per second is held constant, the temperature will change with respect to time. For example, if a sample of water weighing 100 grams is cooled and frozen from 20°C by removing 10 calories per

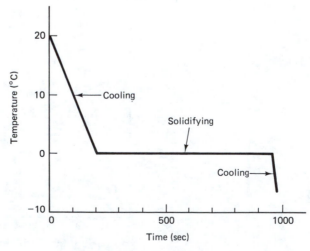

FIGURE 4-56   Cooling curve.

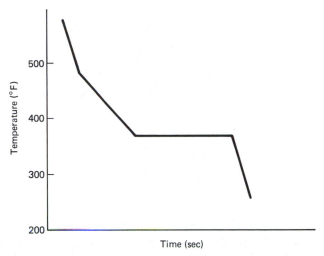

FIGURE 4-57   Cooling curve for 30% tin–70% lead.

second, it would require 200 seconds (3 minutes 20 seconds) to lower the temperature from 20°C to 0°C, and another 764 seconds (12 minutes 42 seconds) to freeze the water. The graph plotting the temperatures versus the time elapsed is called the *cooling curve* of the material. Figure 4-56 shows the cooling curve for the problem just described.

Returning to the tin–lead phase diagram (Figure 4-55). If a mixture of 30% tin and 70% lead is melted and cooled at a constant rate with the temperature noted every few seconds, the resulting cooling curve will be as shown in Figure 4-57. The melt will have a constant rate of cooling until it reaches the liquidus line (approximately 480°F). At that temperature some of the lead-rich alpha (α) phase will begin to solidify. Since the heat of fusion of the alpha phase must also be removed as the remaining liquid cools, there will be a change in the slope of the cooling curve. This new cooling rate will be maintained until the eutectic temperature (361°F) is reached. The temperature will then remain constant until all the material is solidified. After the entire mixture is solidified, the temperature will again begin to fall.

The cooling curve for the eutectic composition is also shown in Figure 4-58.

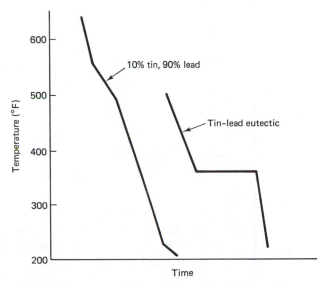

FIGURE 4-58   Other tin–lead cooling curves.

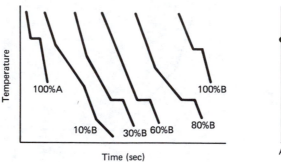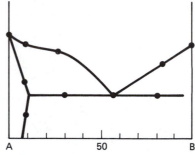

FIGURE 4-59   Phase diagram constructed from cooling curves.

Since cooling the melt of the eutectic composition does not go through a two-phase (liquid–solid) region as did the cooling curve previously shown, there will be no change in the slope in the temperatures above the eutectic, and the cooling curve resembles that of a pure metal. The cooling curve through the alpha region (10% tin) is also shown in Figure 4-58. It can be seen that anytime there is a change of state, there is a point of inflection or a change of slope on the cooling curve. The phase diagram can be constructed simply by running cooling curves of several different compositions and plotting all the points of inflection on a graph. Once these points are plotted on the temperature versus composition graph, lines connecting the points can be drawn to complete the phase diagram. An example of the cooling curves converted into a phase diagram is shown in Figure 4-59.

PROBLEM SET 4-5

Problems 1 through 8 are based on the following phase diagram.

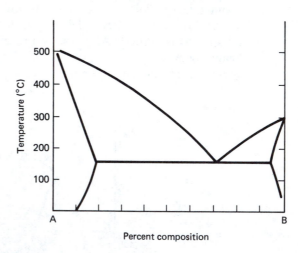

1. What is the eutectic temperature?
2. What is the eutectic composition?
3. What is the melting point of element B?
4. At what temperature would a composition of a 50% A–50% B have to be heated to make it completely liquid?
5. What is the maximum amount of A that could be dissolved in B without it separating into a two-phase system?

6. Draw the cooling curve that would be obtained by cooling a mixture of 70% A–30% B from the liquid phase.

7. Draw the cooling curve that would be obtained by cooling a mixture of 50% A–50% B from the melt.

8. At a temperature of 300°C and a composition of 70% A, what phases would exist?

9. Given the set of cooling curves below, plot the phase diagram.

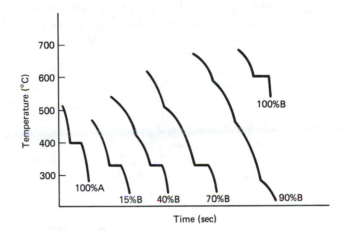

10. In the data given in Problem 9, what are the composition and temperature of the eutectic point?

11. Refer to the tin–lead phase diagram (Figure 4-55). What is the eutectic composition of the system?

12. In the tin–lead phase diagram, what is the maximum solubility of lead in the tin?

13. In the tin–lead phase diagram, what is the maximum solubility of tin in the lead?

14. In the tin–lead phase diagram, would a composition of 55% lead show excess tin or lead over the eutectic matrix?

15. In the tin–lead phase diagram, construct the expected cooling curve for a composition of 70% tin. (Show the temperatures at which changes occur.)

16. In the tin–lead phase diagram, construct the expected cooling curve for a composition of 60% lead. (Show the temperatures at which changes occur.)

17. At a temperature of 150°F and a composition of 40% lead in the tin–lead system, what phases would be present?

18. If the heat of vaporization of water is 540 cal/g, the heat of fusion is 76 cal/g and the heat capacity of water is 1 cal/g/°C, how long would it take to cool 100 g of steam from 100°C to ice at 0°C if the heat was removed at a rate of 300 cal/s?

19. If the heat of vaporization of water is 973 Btu/lb, the heat of fusion of water is 137 Btu/lb, and the heat capacity of water is 1 Btu/lb/°F, how long would it take to cool 2 lb of steam at 212°F to ice at 32°F if the heat was removed at 10 Btu/sec?

20. The heat capacity of steam is $\frac{1}{2}$ cal/g/°C, the heat of vaporization is 540 cal/g, the heat capacity of water is 1 cal/g/°C, the heat of fusion of water is 76 cal/g, and the heat capacity of ice is $\frac{1}{2}$ cal/g. How many calories would it take to heat 50 g of ice at −10°C to steam at 120°C?

ACTIVITY SERIES

All the metals in the periodic table have different chemical activities. Some of them will react very rapidly with other elements, while others combine very slowly. Sodium is more active than aluminum, which is more active than iron, which is more active than copper, and so on. Figure 4-60 shows the relative activity of several metals, listed from most active to least active.

If a metal of a higher activity is placed in contact with a molecule containing a metal ion of lower activity, the more active element will replace a metal of lower activity in the compound. The less active element will be given up as a free metal. As an example, if metallic copper is placed in a solution of silver nitrate, the copper will go into solution, producing copper nitrate, and the silver will precipitate out as metallic silver. This can be expressed by the chemical equation

$$Cu + AgNO_3 \longrightarrow Cu(NO_3)_2 + Ag$$

This method could be used for refining and purifying silver from its ore, but it is rather expensive. This principle is used in the refining of iron ore and other metals. This is discussed in Chapter 5.

FIGURE 4-60   Activity series.

Element	Symbol
Lithium	Li
Potassium	K
Rubidium	Rb
Sodium	Na
Calcium	Ca
Magnesium	Mg
Thorium	Th
Aluminum	Al
Manganese	Mn
Zinc	Zn
Chromium	Cr
Iron	Fe
Thallium	Tl
Cadmium	Cd
Cobalt	Co
Nickel	Ni
Tin	Sn
Lead	Pb
Hydrogen	H
Antimony	Sb
Bismuth	Bi
Arsenic	As
Copper	Cu
Mercury	Hg
Silver	Ag
Platinum	Pt
Gold	Au

ELECTROPLATING

Metals form cations. Since they are positively charged they will be attracted to a negative charge or to electrons. If a negative charge is placed on a piece of metal (called the *cathode*), which is submerged in a solution having the metal ions, the ions will absorb the electrons from the metal and become atoms of metal that can stick to the cathode. This depletes the cations in solution. The anions in solution will migrate to the positively charged pole (called the *anode*) and cause the metal atoms there to become ions that go into solution. The solution that has the cations to be plated out is called the *electrolyte*. An electrolyte is any solution that will conduct electricity. Figure 4-61 shows an electroplating cell.

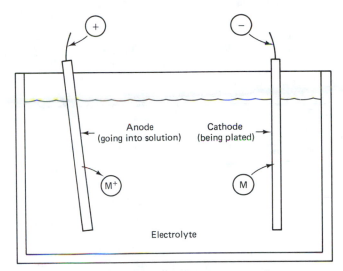

FIGURE 4-61   Electroplating cell.

If two dissimilar metals connected by a wire are placed in an electrolyte, the more active metal will go into solution, giving up electrons. These electrons will flow to the metal having a lesser chemical activity. The flow of electrons is an electric current that can be used to power light bulbs, radios, and other instruments requiring low voltage. This is an *electric cell*. The two metals are called *electrodes*. The voltage generated by this electric cell depends on the metals used as the anode and cathode.

The electroplating cell and the electric cell are opposites. To plate out a metal on the anode, an electric current is applied to the anode. If this electric current is removed, the anode will go back into solution, giving up electrons, and it becomes an electric cell. This is the principle of the rechargeable battery. A battery consists of two or more electric cells operating in series or parallel.

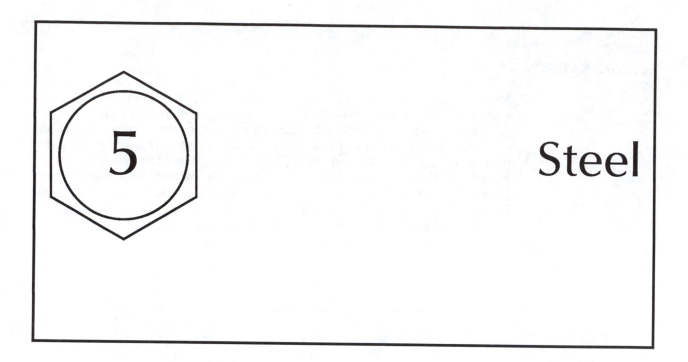

Steel

5

INTRODUCTION

Contrary to popular belief, steel is not an element. There is no atom known as "steel." *Steel* is an iron–carbon alloy that has less than 2% carbon. If the iron contains more than 2% carbon, it becomes *cast iron*. Steel and cast iron have been a part of our materials inventory for many centuries, but its fabrication was more of a craft than a science. People knew that certain iron mongers made a hard "steel" while others made soft iron, but no one knew why. It was learned over five centuries ago in Toledo, Spain, that a soft steel could be pounded together with a hard steel to form Damascus steel. This steel is tough enough to withstand impact and of sufficient hardness to keep an edge on a cutting blade.

Iron is the primary element in steel. In its pure form iron is a silvery-looking metal that weighs 7.86 g/cm^3 or 490 lb/ft^3. Its melting point is 1535°C (2795°F) and it boils at 3000°C (5432°F). It is the fourth most abundant element by weight in the crust of the earth (behind oxygen, silicon, and aluminum) and makes up most of the core of the earth and the other planets. It reacts easily with oxygen and other nonmetallic elements, so it is usually found in the form of a compound. Iron is never found naturally in its pure form (except as meteoritic iron), so it must be refined from its ores. Pure iron is soft, malleable, and ductile. It also has the very useful property of being magnetic. Although not as good a conductor of electricity and heat as silver or copper, iron does conduct heat and electricity very well.

REFINING OF IRON ORE

There are several ores of iron. The major ones are *magnetite*, *hematite*, and *taconite*. Magnetite is a combination of ferric oxide and ferrous oxide (Fe_2O_3 and FeO). It is a black ore containing about 65% iron. Magnetite is strongly magnetic. One form of magnetite found in nature is a permanent magnet and is called *lodestone*.

Hematite is almost blood red in color (hence its name) and is basically ferric oxide or rust (Fe_2O_3). It is a little lower grade ore than magnetite having only about 50% iron. Taconite is a green, low-grade ore that only recently became commercially usable. It contains less than 30% iron. Other ores of iron that have been used are *limonite, siderite, marcasite* and *iron pyrites*. Limonite is hydrated ferrous oxide ($FeO \cdot H_2O$), siderite is ferrous carbonate ($FeCO_3$), and iron pyrites and marcasite are iron sulfides (FeS). These ores are of such poor quality in iron content, often less than 10%, that they are seldom used as yet. Future needs may dictate that these low-grade ores be used commercially.

The main iron ore deposits in the United States are in the Mesabi mountain range near Duluth, Minnesota. The ore is removed from the earth by strip mining. One of the largest mines in that area is now a pit more than 3 square miles in area and over 500 ft deep. The United States now imports more iron ore than it produces. At present, much of the iron ore refined in this country comes from the Hudson Bay region of Canada.

Since iron is always found as ores or in the combined state, it must be refined. The principle on which iron ore is refined is based on the fact that some chemical elements are more active than others in a chemical reaction. This concept was discussed in Chapter 4. Iron could be removed from its ore by trading aluminum or magnesium for the iron in the oxides. The equations for these reactions are

$$Al + Fe_2O_3 \rightarrow Al_2O_3 + Fe$$

$$3Mg + Fe_2O_3 \rightarrow 3MgO + 2Fe$$

The reaction of the powdered aluminum and iron oxide is used. This reaction liberates a tremendous amount of heat when it occurs, being hot enough to melt through steel. It is called the *thermite* reaction and is used in incendiary bombs and for certain types of welding. The problem with using it as a refining technique for iron is that the aluminum or magnesium must first be refined, and the cost would be prohibitive.

One element that is more active than iron and is cheap and easy to get is carbon. The chemical equation for this reaction is

$$(FeO \text{ or } Fe_2O_3) + C \rightarrow Fe + CO$$

Carbon can be obtained from wood or coal. The earliest recorded smelting of iron ore used charcoal. The charcoal was mixed in a furnace with chunks of the iron ore. Bellows were used to blow hot air from a fire underneath or outside the furnace through the charcoal–iron ore chamber. The temperatures were not hot enough in these early furnaces to melt the iron (the melting point of iron is 1535°C or 2795°F) but could reach the 1200°C (about 2200°F) required to react with the ore. The product of this technique was a spongy mass or "bloom" which had to be hammered to squeeze out the slag and other impurities.

The first iron mill in the United States was built at Saugus, Massachusetts, in 1647, just 25 years after the Pilgrims landed at Plymouth Rock. This mill produced only a few pounds of iron a day, but it supplied the local farmers with tools, pots and pans, and other necessary implements. Today over 150 billion tons of steel is used annually in the United States.

To get the carbon used in today's steel mills, coal is heated in a furnace from which most of the oxygen (air) is removed. Hydrogen and the other elements in the coal are driven off and the carbon is left in the form of *coke*. Once the iron ore is mined and concentrated to remove the soil and other undesirable materials, it is shipped to the refinery. There it is mixed in layers in a huge *blast furnace* with

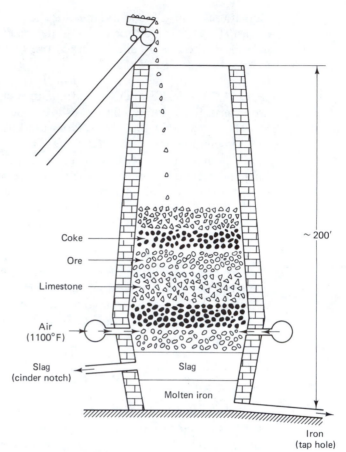

FIGURE 5-1 Typical blast furnace.

coke and limestone (Figure 5-1). The limestone is required as a *flux*, which helps remove the impurities formed in the refining process.

Air, heated to 1100°F, is blown from the bottom of the blast furnace, through the iron ore, coke, and limestone layers of the charge. At this temperature, the carbon in the coke will react with the oxygen in the iron ore and start to burn off the oxygen from the iron oxides. This, in turn, causes the temperature to rise above the melting point of the iron, to about 3000°F. After about 5 to 6 hours, the molten iron is drawn off through a tap hole at the bottom and poured into ingots called *pigs*. Each pig weighs about a ton. Later, the slag consisting of the silicates and other scrap materials from the ore will be taken off through the cinder notch. The pig iron is not yet steel and has many impurities, such as silicon and sulfur, left in it. The carbon content of this metal is about 4% at this stage. If cast iron is the desired end product, the pigs can be remelted and used directly. The addition of alloying elements to the pig iron is now a common practice to give the metal better properties. Some of these cast irons are discussed later.

The blast furnaces are truly awesome in size. A typical furnace may be 125 to 200 ft tall and have a diameter of over 30 ft. The furnaces are lined with firebrick. The charge (ore, limestone, and coke) is fed into them at the top by means of skip buckets which carry the material to the top of the furnace. The molten iron and slag are removed from the bottom. It is a continuous-flow process and the furnaces, once started, are operated 24 hours a day and shut down only once every 2 to 5 years to be repaired and rebricked.

The furnace may weigh in excess of 10,000 tons and cost millions of dollars each. However, each furnace is capable of producing from 1000 to 4000 tons of pig iron daily, depending on its size. A typical day's run on a blast furnace might be as follows:

Pig iron produced (tons per day)	1,400
Coke used (tons per day)	1,200
Limestone used (tons per day)	550
Air blown (cubic feet per minute)	87,000
Cooling water (gallons per day)	13,000,000

The tremendous cost of these furnaces and their operating costs require that as little as possible be wasted. The air that is pumped through the furnace reacts with some of the coke to become carbon monoxide. The carbon monoxide can be burned to provide heat for the furnace. The slag that is drawn off can be crushed and used as aggregate for concrete and roadbeds. Even with this, it is surprising that the cost of steel is still very low.

CONVERSION OF IRON TO STEEL

The pig iron contains too much carbon and other undesirable materials in it for use as a strong metal. It must now be converted into steel. This is done in one of several types of *converters*. Although they differ in design, all the converters do the same thing: they burn off the carbon in the iron.

The oldest converter was invented about 1860 and is called the *open hearth converter* (Figure 5-2). The open hearth converter is primarily a large firebrick bowl which may measure some 15 by 46 ft across and stand up to 40 ft high. The pigs from the blast furnace process are mixed with scrap steel and melted. Hot air is blown across the surface and the excess carbon and other impurities are burned out of the melt. A typical melt would be from 150 to 250 tons of metal and would take about 11 hours to refine. At the end of this time, the desired alloying elements and the exact carbon content are added and the steel is drawn off into large firebrick-lined crucibles and cast into ingots.

About the same time as the development of the open hearth furnace, Henry Bessemer was looking for a method of producing better metal for gun barrels. He developed a pear-shaped steel vessel, lined with firebrick and mounted on trunnions to convert the pig iron into steel. These crucibles are of various sizes but may contain from 5 to 30 tons of charge per blow. Scrap steel along with the new pigs are placed into the *Bessemer converter* and hot air is blown through the melted metal as shown in Figure 5-3. Again, the carbon is burned out and the desired amount of carbon and alloying elements are added. About 12 to 15 minutes is required to refine 25 tons of steel. The converter can then be tipped on its trunnions and the metal poured into a large firebricked-lined ladle, which in turn pours the steel into ingots for further fabrication.

One problem with both the open hearth and the Bessemer process is that the

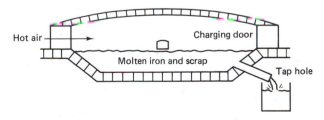

FIGURE 5-2 Open hearth converter.

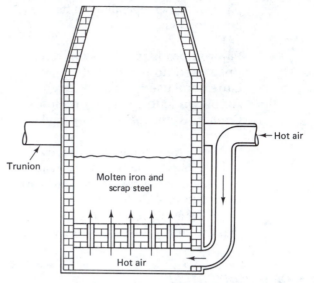

FIGURE 5-3   Bessemer converter.

steel produced contains many impurities. Sulfur, phosphorus, oxygen, nitrogen, and other elements in the steels produced in this manner weaken the product. The need for a clean steel was apparent. Around the turn of the twentieth century, the *electric arc* furnace was invented (Figure 5-4). The electric arc furnace is basically a small open-hearth crucible mounted on trunnions. A typical furnace may have a capacity of from 1 to 100 tons. From the top, carbon electrodes are brought close to the scrap steel and pig iron charge and an arc is struck. The arc melts the metal and burns out the carbon. Since no air is brought into the system, the result is a steel with fewer air bubbles, and interstitial oxygen or nitrogen atoms.

The main drawback to the electric arc production of steel is the tremendous power requirements. Three phase 200-V electricity is used. The transformers and power transmission lines which supply the electricity must be capable of handling up to 30,000 kW. The electrodes in the furnaces may range from 12 to 24 in. in diameter, and draw up to 25,000 A each. The carbon in the electrodes is burned off and they must be fed into the melt automatically. Each ton of steel requires about 15 lb of electrode. Each ton of steel requires about 400 kW or about 200 W

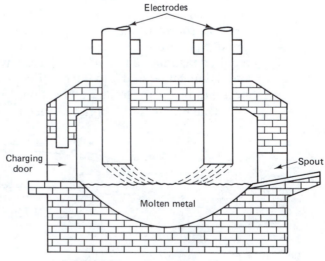

FIGURE 5-4   Electric arc furnace.

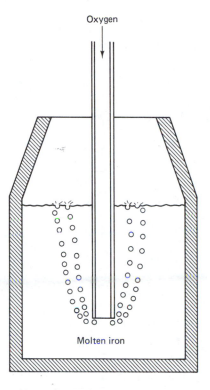

Oxygen

Molten iron

FIGURE 5-5   Oxygen lance converter.

per pound of steel. Sometimes the electric arc furnace is used in conjunction with other converters to produce high-quality steels. Iron from the blast furnace may be run through an open hearth or Bessemer converter, then on to the electric arc furnace for the final product.

One of the newest converters was invented in Europe and first used in the United States in 1954. The *oxygen lance* method uses a furnace similar to the Bessemer (Figure 5-5). It is charged with premelted iron and scrap and a water-cooled pipe or lance is lowered into the melt. Pure oxygen is blown through the liquid metal under a pressure of about 150 lb/in². The oxygen burns out the carbon very rapidly (about 45 minutes per charge). After the carbon is burned out, the lance is removed and the proper amounts of alloying elements and carbon is added to get the desired steel. The steel is then cast into ingots for further use. Since air is not used in the oxygen lance converter, there is no nitrogen contamination in the steel as with the other processes.

Casting the steel into fixed-size ingots puts a size limitation on the product made from that ingot. There is just so much steel in an ingot. A *continuous-pour* process has been developed which allows much larger beams or other structural steel to be formed. Figure 5-6 illustrates this process. The molten steel is poured through a form which is water cooled from the outside. As the steel proceeds through the form, it solidifies and is then carried downward by gravity and turned into a horizontal flow by rollers. The entire melt from a converter or ladle could be poured into one billet if desired. The billet could be cut into desired lengths while hot and the metal rolled or formed into predetermined shapes. This continuous-pour process is not limited to steels. Cooper and bronze alloys are now being manufactured by a continuous pour so that bars and wires of very long lengths can be fabricated.

Upon completion of the conversion process, the steel is ready to be formed into any of several thousand products of the steel industry. The steel can be sent to the rolling mills while still red hot or can be formed later after cooling. The ingots (either fixed size or continuous pour) are formed into billets, blooms, or

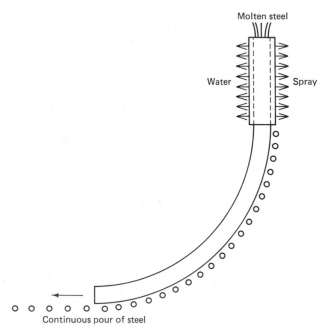

FIGURE 5-6   Continuous pour process.

slabs, depending on the shape of the final product needed. These shapes are, in turn, rolled or forged. Some of the forms produced are described below.

Bars: Solid shapes that can be hot or cold rolled in rounds, squares, or flats in sizes ranging from $\frac{3}{4}$ to 12 in. thick.

Billets: A section of an ingot that is suitable for rolling.

Blooms: A slab of steel or other metal of width generally equaling the thickness.

Plates: Large flat slabs thicker than $\frac{1}{4}$ in.

Shapes: Can be in the form of beams of several types, channels, angles, and others. The types of beams include:

> Standard (I) beams in sizes ranging from 3 by 2.3 in. to 24 by 8 in.
>
> Wide-flange beams, which can run from an 8- by 5.25-in. cross section to as large as 36 by 16.5 in. The 36- by 16.5 in.-wide flange beam weighs 300 lb per linear foot.
>
> Channels can be purchased in sizes as small as 1.3 in. deep by 3 in. wide to as large as a 4- by 18-in. one. There are smaller metal channels available, but these are extruded or drawn not rolled. Many of the smaller channels are not made of steel.
>
> Angles can have either equal-length sides or unequal lengths. They range in size from the little 1 by 1 in. to an 8 by 8 in. in equal-length legs to a 9 by 4 in. in the unequal-length legs.

The cross sections of these shapes are shown in Figure 5-7.

Sheets: Can be either hot or cold rolled and vary in thickness from 0.01 to 0.25 in. Sheets are usually larger than 24 in. wide and can be cut into uniform lengths or rolled into coils.

Wires: Usually formed from bars that have been rolled down to a relatively small diameter, then drawn into wires of $\frac{1}{4}$ in. and smaller. Steel wire can be sent to mills for further fabrication into nails, wire rope, cables, fencing, screen wire, and much more.

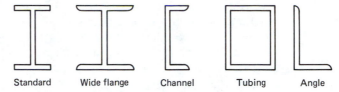

Standard   Wide flange   Channel   Tubing   Angle

FIGURE 5-7   Steel shapes.

Special shapes can be special ordered from steel mills if the buyer needs a quantity sufficiently large to warrant the steel producer to tool up for the job. Railroad rails, structural members, and machine parts stock can be made on request by the steel companies. To order nonstandard shapes is quite expensive and can generally be avoided by careful engineering.

The mill products can be further manufactured into countless final products. Square, rectangular, or round tubing and pipe can be made from the sheet. Structural rectangular tubing in sizes from 2 by 3 in. to 12 by 20 in. is finding many applications in construction and machinery design. Square tubing in sizes from 2 by 2 in. to 16 by 16 in. is made by several of the steel companies. Round tubing and pipe in sizes as small as $\frac{1}{4}$ in. to several feet in diameter in several thicknesses are also a part of the steel industry's inventory.

Thinker 5-1:

Look around for a full day and identify examples of steel shapes that you find.

TYPES OF STEEL

The type of steel depends on two factors:

1. The carbon content
2. The alloying elements

Carbon is in steel as interstitial atom in the iron crystal. The effect of the carbon is to pin or anchor the slip planes of the crystal so that they cannot slide easily over each other. As little as 0.1% of carbon in the steel can change the strength, ductility, and other properties of the steel drastically. An increase in the amount of carbon in steel will:

1. Decrease the ductility of steel
2. Increase the tensile strength of steel
3. Increase the hardness of steel
4. Decrease the ease with which steel can be machined
5. Lower the melting point of steel
6. Make steel easier to harden with heat treatments
7. Lower the temperature required to heat treat steel
8. Increase the difficulty of welding steel.

Pure iron is a very ductile and relatively low strength material. The tensile strength of a single crystal of pure iron is about 4000 lb/in.². If as little as 0.2%

carbon is added, the tensile strength approaches 60,000 lb/in². The tensile strength of a plain carbon steel with 0.8% carbon is over 110,000 lb/in².

The maximum hardness that a steel with only 0.2% carbon can attain is about a Rockwell C hardness of 40, while an 0.8% carbon steel can be hardened to a Rockwell C of 65. As the steels add carbon and get harder, they are necessarily more difficult to machine. Some of the cutting tools are of high-carbon steel themselves and would not be much harder than the high-carbon steel it is supposed to cut. (There are cutting tools made of tungsten carbide and ceramics which will cut these hard steels, but these tools are quite brittle and not suitable for large or roughing cuts. The carbide and ceramic tools are used extensively for finish work.)

The melting point of a 0.2% carbon steel is about 2800°F, while the 0.8% carbon steel will melt near 2700°F. The effects of carbon on the hardenability and heat treatment of steel are discussed in Chapter 6. Steels are often classified by their carbon content. Carbon will dissolve into iron without producing a steel up to about 0.026% at 1330°F. Low-carbon or mild steels contain less than 0.3% carbon. Medium-carbon steels have from 0.3 to 0.55% carbon. Steels containing more than 0.55% carbon are referred to as high-carbon steels. Remember that iron with more than 2% carbon is no longer steel but cast iron.

Besides the carbon content, steels are often typed by their alloying elements. Silicon, aluminum, manganese, niobium, molybdenum, tungsten, and nickel are commonly added to steels. Each alloying element is added for a specific purpose.

In steel production, oxygen, sulfur, and other elements are left in the steel. These elements can be detrimental to the quality of steel. Rather than go to the expense of removing all of these undesirable elements, they are often tied up and rendered harmless by the addition of other elements. Oxygen, for instance, which embrittles steel, can be tied up by the addition of a small amount of aluminum or silicon to steel in the converter. The blowing of air through or over the melt lets the oxygen react with some of the hot iron, producing sparks or flames that may shoot many feet above the converter. Since aluminum is more chemically active than iron, once it is added to the melt, the excess oxygen reacts with the aluminum rather than with the iron and no sparks are produced. This steel, which has aluminum added, is often called *killed steel*.

Sulfur, introduced into iron in the refining process by the limestone and ore, can accumulate at the grain boundaries in the form of iron sulfide. The low-melting iron sulfide causes a condition in steel known as *hot short*, which means that steel loses its strength at high temperatures. Sulfur can effectively be tied up by the introduction of small amounts of manganese. Manganese in excess of the amount needed to combine with the sulfur can improve its properties further by dissolving in the steel, making it more malleable.

Vanadium is added to give steel a fine-grained structure and to increase its toughness. It is often used in tool steels because of its increased resistance to impact.

Chromium is added to steels to provide corrosion resistance and to increase its hardenability or the depth to which the steel can be hardened. Niobium (formerly called columbium) greatly increases the tensile strength of steel. Only 40 lb of niobium per ton of steel will increase the tensile strength by 10,000 to 15,000 lb/in².

Tungsten in the form of tungsten carbide gives steels high hardness even at red heats. The use of carbide tool steels is widespread.

Titanium is a very strong, very lightweight metal that can be used alone or alloyed with steels. Titanium is added to steels to give them high strength at high temperatures. Without titanium steels, the modern jet engine, which operates at white-hot temperatures, would be impossible.

Phosphorus and lead are sometimes added to steels to increase their machineability.

Steel for springs must have at least 0.45% carbon to attain the required hardness. Plain carbon steels such as 1045, 1060, 1074, 1080, or even 1095 can be used for springs but become rather brittle at higher carbon contents. Usually, plain carbon steels are used for flat springs. The alloyed steels, such as 6150, 9260, and 8650, are recommended because of their added toughness. A 302 stainless steel is often used for springs that have higher-temperature requirements. When electrical conductivity is needed, nonferrous metals such as beryllium bronzes and brass can be used as springs, but they do not have the strength of steels.

Nomenclature of Steels

The Society of Automotive Engineers (SAE) and the American Iron and Steel Institute (AISI) have developed a method of naming steels based on their alloying elements and carbon content. Steels are referred to by a four-number designation, examples being 1060, 8620, 4140, and 4340 steels. In each of these designations, the first two numbers refer to the alloying elements present in that steel. A 10-- steel is a plain carbon steel. A 41-- steel contains from 0.40 to 1.10% molybdenum and 0.08 to 0.35% chromium. Figure 5-8 is a partial table of the numbering system for steel alloys.

The second two numbers of the four-character numbering system refers to the percent carbon in hundredths of a percent. Therefore, a --18 steel has 0.18% carbon. A high-carbon steel having 0.80% carbon would be designated as --80. A steel designated "1080" would be a plain carbon steel with 0.8% carbon. A 4118 steel would have 0.8 to 1.10% chromium and 0.15 to 0.25% molybdenum with 0.18% carbon.

FIGURE 5-8 Steel nomenclature.

AISI Number	Type of Steel	Alloying Elements (%)
10--	Plain carbon	None (0.4 Mn)[a]
11--	Free machining	0.7 Mn, 0.12 S
13--	High manganese	1.60–1.90 Mn
2---	Nickel steels	3.5–5.0 Ni
3---	Nickel–chromium	1.0–3.5 Ni, 0.5–1.75 Cr
40--	Molybdenum	0.15–0.3 Mo
41--	Chrome–molybdenum	0.80–1.10 Cr, 0.15–0.25 Mo
43--	Nickel–chrome–molybdenum	1.65–2.0 Ni, 0.4–0.9 Cr, 0.2–0.3 Mo
46--	Nickel–molybdenum	1.65 Ni, 1.65 Mo
5---	Chromium	0.4 Cr
61--	Chromium–vanadium	0.5–1.1 Cr, 0.10–0.15 Va
81--	Nickel–chrome–molybdenum	0.2–0.4 Ni, 0.3–0.55 Cr, 0.08–0.15 Mo
86--	Nickel–chrome–molybdenum	0.4–0.7 Ni, 0.4–0.6 Cr, 0.15–0.25 Mo
92--	Silicon	1.8–2.2 Si

[a]All steels have a fraction of a percent manganese, phosphorus, sulfur, and silicon.

PROBLEM SET 5-1

Give the carbon content and the alloying elements for the steels listed in Problems 1 through 8.
1. 1040
2. 4340
3. 8620
4. 4140
5. 4120
6. 6160

7. 1330
8. 1120
9. Which of the following is not a steel?
    (a) An iron–carbon alloy having 0.8% carbon;
    (b) an iron–carbon alloy having 8% carbon;
    (c) an iron–carbon alloy having 0.1% carbon;
    (d) an iron–carbon alloy having 0.4% carbon and 1% chromium;
    (e) an iron–carbon alloy having 0.08% carbon.
10. If a low-carbon steel suitable to be turned on a lathe was needed, what would be a good steel to specify?
11. One of the better steels for use in aircraft landing gears has 1.8% nickel, 0.80% chromium, 0.25% molybdenum, and 0.4% carbon. What is the designation for this steel?
12. Instead of using aluminum powder to replace the iron in iron oxide in the thermite reaction, list three other metals that would work.
13. What is the difference between iron and steel?
14. What is the difference between cast iron and cast steel?

IRON–CARBON PHASE DIAGRAM

Iron and carbon form a two-component system for which a phase diagram may be developed. The common phase diagram for iron and carbon (shown in Figure 5-9) has pure iron plotted on the left side of the diagram and the percent carbon, up to 7%, is plotted on the abscissa (x axis). The temperature is plotted on the ordinate (y axis). Carbon will dissolve into iron without separating into a separate phase

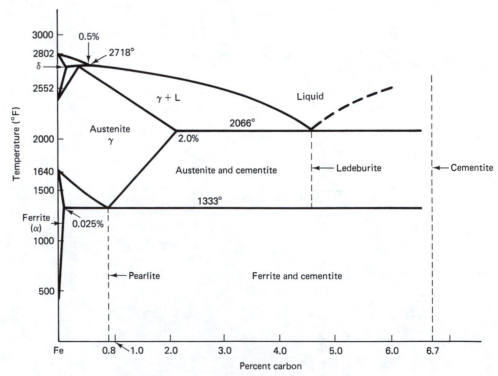

FIGURE 5-9   Iron–iron carbide phase diagram.

up to a concentration of 0.025% at a temperature of 1330°F. This single-phase region is known as *ferrite* or *alpha iron* (α iron). The single-phase region at elevated temperatures between 2552° and 2802°F is called *delta iron* (δ iron). Ferrite and delta iron are considered pure iron for all practical purposes. The small amount of carbon that is in the metal dissolves into the crystal lattice interstitially.

On the right side of the phase diagram is an intermetallic compound, iron carbide. Iron carbide has the chemical formula Fe_3C and is called *cementite*. The melting point of pure iron is 2802°F. The eutectic between pure iron and cementite is found at 4.3% carbon at a temperature of 2066°F. At concentrations of less than 2% carbon and above a temperature of 1333°F, a single-phase solid is formed. This high-temperature single-phase solid solution is named *austenite or gamma* iron (γ iron). Austenite is the face-centered cubic (FCC) structure of steel. The lowest temperature at which austenite will occur is 1333°F (727°C) at a concentration of about 0.8% carbon. Below the 1333°F the steel will be in a body-centered cubic crystal structure and is a two-phase system. The lowest temperature at which a single-phase solid will exist before it turns into a two-phase solid is called its *eutectoid*. (The eutectoid looks similar to a eutectic on a phase diagram, but the eutectic is the lowest temperature at which a liquid of a multiple-component system will exist before solidifying.) Just as eutectic composition will solidify into a two-phase solid system with no excess alpha or beta, the eutectoid composition is a two-phase system with no excess of the solid solutions on either side of it. If austenite is heated above 2718°F, it will decompose into a liquid and solid delta iron. The temperature at which this decomposition occurs is known as the *peritectic* temperature.

The iron—carbon phase diagram is so well used that almost every part on it is given a name. The eutectoid composition of steel is called *pearlite*. In a photograph of the microstructure of steel, pearlite shows up as alternating layers of ferrite and cementite. Figure 5-10 is an example of a steel containing 0.8% carbon and is entirely pearlite. The eutectic composition of the iron—carbon phase diagram is known as *ledeburite*. The boundary line between austenite and the two-phase region of ferrite and austenite is known as the Ac_3 line, while the boundary between austenite and austenite—cementite two-phase region is labeled the Acm line. The eutectoid temperature line is referred to as the Ac_1 line.

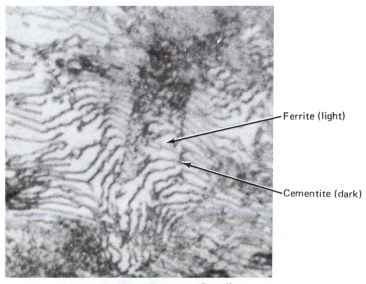

FIGURE 5-10   Pearlite.

Steel was defined earlier as "an iron–carbon alloy having less than 2% carbon." With an understanding of the iron–carbon phase diagram, steel can also be defined as "an iron–carbon alloy that can be heated into austenite." Although steel technically can have up to 2% carbon and still be heated into austenite, steels having more than 1% carbon are seldom used. They would be as brittle as cast iron but would have melting points above 300°F, above the melting point of cast iron. The higher melting temperature would make them more expensive. Steels having more than 0.8% carbon would be called *hypereutectoid steels*, while whose with less than 0.8% carbon are known as *hypoeutectoid steels*.

Metals can be examined under a microscope to determine their grain structure, grain size, composition, and to determine the imperfections and flaws in the specimen. To view a specimen under high magnification, it must be perfectly flat. The preparation of a metallurgical specimen for microscopic examination involves sanding it to a fine, flat surface by using a series of finer and finer grit sandpapers, then polishing the sample to a mirror surface. Since the grain boundaries are only one atom wide and not visible under optical magnification, the polished metal samples must be etched before viewing. The etching erodes the grain boundaries to a point where they can be seen with the aid of the optical microscope.

Hypoeutectoid steels will consist of a two-phase system having pearlite with an excess of ferrite. As seen under the microscope, a low-carbon steel will have much more ferrite than pearlite. Ferrite appears as a white grain while the pearlite will show the alternate layers of ferrite and pearlite. Pearlite will often show up as a dark gray grain in the picture. Figure 5-11 is an example of a low-carbon steel. A medium-carbon steel will have about equal amounts of ferrite and pearlite, as shown in Figure 5-12. High-carbon steels are mostly pearlite. Figure 5-10, the 100% pearlite, would be considered a high-carbon steel. Hypereutectoid iron–carbon alloys include the cast irons, which are discussed later in the chapter.

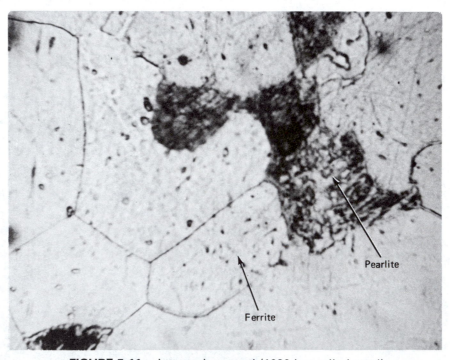

FIGURE 5-11   Low-carbon steel (1020 hot-rolled steel).

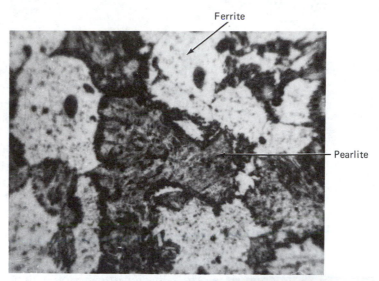

FIGURE 5-12   Medium-carbon steel (1040 hot-rolled steel).

PROBLEM SET 5-2

1. What is ferrite?
2. What is pearlite?
3. What is austenite?
4. What is cementite?
5. What is ledeburite?
6. What is the eutectoid composition of steel?
7. What is the eutectic composition of the iron–iron carbide system?
8. What is steel?
9. Construct a theoretical cooling curve, starting from the liquid phase, which would be obtained by cooling a composition of 0.4% carbon in iron.
10. Construct a theoretical cooling curve that would be obtained by cooling a composition of 3% carbon in iron from the melt.
11. Why is cast iron cheaper to produce than a plain carbon steel?
12. About what temperature would a 1040 steel have to be heated to get it entirely into the austenite range?
13. If a steel having 100% pearlite was to be heated 50°F into the austenite range, what would that temperature be?
14. A steel having 0.2% carbon is to be heated 100°F into the austenite region. What would that temperature be?
15. What is the peritectic temperature of steel?
16. What two phases are present in an iron–carbon alloy that has 3% carbon at a temperature of 1500°F?
17. What phases are present in an iron–carbon alloy having 2.5% carbon at a temperature of 2200°F?
18. What is the eutectic temperature of the iron–iron carbide alloy?
19. What is the maximum solubility of carbon in iron?
20. What phases are present at a composition of 0.4% carbon and a temperature of 1400°F?

CAST IRON

Other products of the iron and steel industry include gray and white cast iron, ductile cast iron, malleable iron, and wrought iron. Long before the processes of making steel was discovered, wrought iron was used. *Wrought iron* is made from the pig iron produced from the blast furnace. The pig iron is heated in an oxidizing flame to burn out the carbon, but it still retains much of the slag, as silicates, from the refining process. After casting the molten wrought iron into ingots, it is rolled into billets, bars, or other shapes. In this very low carbon (0.02 to 0.08%) iron the slag is forced into long fibers by the rolling. Wrought iron has relatively good corrosion resistance and ductility, yet is relatively inexpensive. It has a tensile strength of about 40,000 lb/in^2 and a yield strength of nearly 24,000 lb/in^2, much lower than steels.

Carbon will go into iron interstitially up to about 0.8% carbon. Above that concentration, the carbon in the form of iron carbide or graphite appears as a separate phase in the steel. The excess carbon in the metal takes the form of platelets of graphite which form minute cracks in the metal. These cracks are of different sizes. The behavior of cast iron is unpredictable. The cast iron, already brittle from the high carbon content, breaks at the largest crack.

Two basic types of cast iron are available, *gray cast iron* and *white cast iron*. Gray cast iron is the most common. It has about 4% carbon, which appears as flakes of graphite. Since the quality of this iron is not as controlled as steel, the ASTM simply divides the several types of gray cast iron into "classes" based on their minimum tensile strengths. A class 20 gray cast iron would have a minimum tensile strength of 20,000 lb/in^2. The highest class, class 60, has a minimum tensile strength of 60,000 lb/in^2. Despite its brittleness and low tensile strength, gray cast iron does find considerable use. It is good in compression, has good machinability, and has good vibration damping characteristics. This fact alone makes it ideal for machine tool bases. The graphite in the structure can even act as a lubricant, so its sliding wear resistance is good. It is also one of the cheapest irons available. Figure 5-13 is an example of gray cast iron. Note the flakes of graphite.

White cast iron has only about 2.5 to 3.5% carbon and 0.5 to 1.5% silicon. With the addition of a little chromium (1 to 3%), nickel, or molybdenum and very slowly cooling the melt in the molds, the white cast iron can be produced. In this material the excess carbon appears not as graphite as in gray cast iron, but as iron carbide or cementite. The tensile strengths of white cast iron are around 20,000 lb/in^2, but compressive strengths run up to 200,000 lb/in^2. White cast iron parts are hard (BHN > 400) and brittle. The uses of white cast iron employ this hardness and compressive strength and they are used in rolling and crushing equipment, and other castings where wear resistance is necessary.

If a small percentage (about 0.05%) of magnesium, sodium, cerium, calcium, or lithium is added to the cast iron and the metal slowly cooled from the liquid state, the excess carbon will go into the form of little balls or spheres (called *spherulites*) in the iron instead of being in flat plates as in ordinary cast iron. This process removes the sharp points or stress risers which initiate the cracks in the metal. This *spheroidized iron*, named *nodular* or *ductile cast iron*, still has close to 4% carbon and 2.5% silicon in it, but the ductility is greatly improved and its tensile strength can be up to 120.000 lb/in^2. Ductile cast iron was developed about 1948.

The American Society for Testing and Materials (ASTM) has developed a method of naming these ductile irons. Sample designations for ductile cast irons might be 60:40:18, 65:45:12, 80:55:06, 100:70:03 and 120:90:02. The first number indicates the tensile strength of the metal in thousands of pounds per square inch,

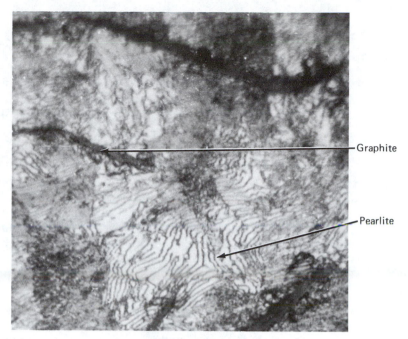

Graphite

Pearlite

FIGURE 5-13   Gray cast iron.

the second number denotes the yield strength, and the last number gives the percent ultimate elongation of the metal. The 60:40:18 ductile cast iron has a tensile strength of 60,000 lb/in^2, a yield point of 40,000 lb/in^2, and an ultimate percent elongation of 18%.

Ductile cast iron has the same melting point as gray cast iron, which is 400°F lower than steel. It is cheaper and easier to cast than steel. For this reason it is being used in internal combustion engines for engine blocks, pistons, crankshafts, gears, and other parts. It is more expensive than gray cast iron, but its improved properties in terms of strength and ductility make it a well-used material.

Malleable irons are produced by heat treatments of the white cast iron. If a cast iron with 2 to 3% carbon is cast, then heated to about 1750°F, the iron carbide or cementite (Fe_3C) is spheroidized. This process produces a material similar to ductile cast iron. There are two basic types of malleable iron: pearlitic malleable iron and ferritic malleable iron. *Pearlitic malleable iron* (also called *cupola malleable iron*) has more carbon and lower silicon than the ferritic malleable iron. The tensile strength of the cupola iron is about 40,000 lb/in^2 but can take only 5% elongation before breaking. The ASTM has two listings of pearlitic malleable iron: 35018 and 32510. Both of these materials have a minimum tensile strength of around 50,000 lb/in^2 and a minimum percent elongation of 10 to 18%. The maximum Brinell hardness number of these metals is about 250.

The production of malleable iron involves several steps. The casting is heated to 925°C (about 1700°F) for several hours. The *ferritic malleable iron* is cooled to just above the eutectoid temperature (about 1350°F) and cooled at the rate of 10 to 25°F per hour through the eutectoid temperature range. This step takes about 24 hours. This allows the excess carbon to diffuse into the graphite nodules. The ferritic malleable casting is then air cooled to room temperature. Pearlitic malleable iron is also heated to about 1700°F but is then quenched (rapidly cooled) in oil or air to room temperature. The part is then reheated to about 850 to 900°F for several hours to draw the metal and spheroidize the pearlite.

STAINLESS STEELS

One major problem with steels is that they corrode rather rapidly. At temperatures of 1800°F (red hot), as much as half of a steel bar will be oxidized or corroded in a single day. Although the rate of decay of steel is much lower at room temperatures, corrosion is still a problem with steel. Unless painted or otherwise surface treated, steel is not particularly a pretty material either. The addition of chromium and sometimes nickel to the steel greatly improves the resistance to corrosion. These high-chromium-content metals have been designated *stainless steels*.

Stainless steels have been divided into three major classifications. *Ferritic stainless steels* contain from 11 to 27% chromium and are always low-carbon steels. Usually, the carbon content is about 0.1% but can run as high as 0.35%. These steels give good corrosion resistance and are used in applications not requiring high strength. Automotive trim and even jewelry are sometimes made of ferritic stainless steels. These steels are relatively soft since their carbon content is not sufficiently high for hardening by heat treatments.

The addition of chromium and nickel can force the steel into a face-centered cubic structure even at low temperatures. The face-centered cubic steel is defined as austenite. The stainless steels, which are face-centered cubic, are therefore called *austenitic stainless steels*. Austenitic stainless steels are nonmagnetic, which gives them an easy method of identification. Simply place a magnet on the surface and see if it is attracted. Austenitic stainless steels are also very low carbon steels, but contain from 16 to 26% chromium and 6 to 22% nickel. Austenitic stainless steels are used for general low-strength structural applications. They are weldable, and to a certain extent, machinable. They cannot be hardened by heat treatments.

Martensitic stainless steels can be higher carbon steels with less chromium and nickel. The carbon content can be as low as 0.15% or as high as 1.2%. The chromium will vary from 4 to 18%. Most of the martensitic stainless steels do not have any nickel, but a few may have up to 2.5% nickel. These steels can be hardened by heat treatments. Martensitic stainless steels can achieve very high hardnesses with the upper limit near a Rockwell C hardness (HRc) of 65. This property, plus their corrosion resistance, makes them ideal for knives and other cutlery. As a rule, these high hardnesses prevent them from being very machinable. They are magnetic. Martensitic stainless steels are not easily welded.

ULTRAHIGH-STRENGTH STEELS

By adding other alloying elements and by cold working, high-strength stainless steels can also be formed. These are sometimes called *superalloys*. *Maraged steel* starts with a low-carbon steel with a composition of 18 to 25% nickel along with 7% cobalt, 5% molybdenum, 0.4% titanium, and trace amounts of zirconium, aluminum, boron, silicon, and manganese. The steels are then heated to 1500°F for 1 hour per inch of thickness, air cooled to room temperature, then reheated to 900°F for about 3 hours. These steels can be cold worked to achieve tensile strengths of close to 300,000 lb/in². They can be welded, but must be carefully heat treated following the weld. This heat treatment is difficult to do in large structures such as bridges and buildings.

If a chromium–molybdenum–vanadium steel is heated into the austenitic range, quenched or cooled very rapidly to a temperature of about 900°F, rolled or forged at this temperature, then rapidly quenched to room temperature, an *aus-formed steel* is the product. Experimental ausformed steels have been made with tensile strengths up to 370,000 lb/in².

FIGURE 5-14   Stainless steels.

Number	Composition	Tensile Strength (1000 lb/in²)	Uses
		Austenitic	
201	0.15 C, 16–18 Cr, 3.5–5.5 Ni, 7.5 Mn		
202	0.15 C, 10 Mn, 17–19 Cr, 4–6 Ni		General purpose
301	0.15 C, 2 Mn, 16–18 Cr, 6–8 Ni		Trim, general structural
302	0.15 C, 2 Mn, 17–19 Cr, 8–10 Ni	80–90 hot rolled, 100–150 cold rolled	General purpose
303	0.15 C, 2 Mn, 17–19 Cr, 8–10 Ni, 0.15 S	80–90	Free machining
304	0.08 C, 2 Mn, 18–20 Cr, 8–12 Ni		Weldable
308	0.08 C, 2 Mn, 19–21 Cr, 10–12 Ni		High corrosion resistance
316	0.08 C, 2 Mn, 16–18 Cr, 10–14 Ni, 2–3 Mo		Resists chemical attack
		Ferritic	
405	0.08 C, 11.5–14.5 Cr, 0.1–0.3 Al		Welding applications
430	0.12 C, 14–18 Cr		General purpose
445	0.2 C, 23–27 Cr		High temperature
		Martensitic	
403	0.15 C, 11.5–13 Cr, 0.5 Si	65–85 annealed, 170–220 cold rolled	Turbine blades
410	0.15 C, 11.5–13.5 Cr, 1.0 Si	65–88 annealed, 170–220 cold rolled	General purpose
420	0.15 C, 12–14 Cr	90–110 annealed, 150–250 cold rolled	Heat treatable

The American Iron and Steel Institute (AISI) has developed a three-number nomenclature system for stainless steels. The 200 and 300 series stainless steels are austenitic steels; the ferritic and martensitic steels are the 400 series. Figure 5-14 shows some of the compositions and properties of the most popular stainless steels. The properties of the iron–carbon alloys are summarized in Figure 5-15.

FIGURE 5-15   Properties of iron–carbon alloys.

Property	Plain Carbon	Alloyed Steel	Stainless Steel	Wrought Iron	Ductile Cast Iron	Gray Cast Iron
Density			7.86 g/cm³ or 490 lb/ft³			
Tensile strength (1000 lb/in²)	50–110	290	75–300	40	60–120	20–40
Yield strength (1000 lb/in²)	40–60	240	30–250	24	45–60	5–10
Young's modulus (lb/in²)	29×10^6		30×10^6	30×10^6		20×10^6
Ultimate percent elongation	27		60	2–20	3–10	—

PROBLEM SET 5-3

1. Match the material at the right with its best application on the left.
   ___ 1. A cutting tool for a plane   a. Cast iron
   ___ 2. A welded steel frame for a drill press   b. Ductile cast iron
   ___ 3. A decorative rail for a porch   c. Low-carbon steel
   ___ 4. A tank to hold seawater   d. Alloyed steel
   ___ 5. A shock-absorbing casting   e. Stainless steel
   ___ 6. An axle for an automobile   f. High-carbon steel
   ___ 7. A melting pot for lead   g. Wrought iron

2. What is the tensile strength of wrought iron in megapascal? (Use Figure 5-15.)

3. What is the maximum yield strength of ductile cast iron in megapascal?

4. What is the Young's modulus of gray cast iron in gigapascal?

5. What is the Young's modulus of stainless steel in gigapascal?

6. A stainless steel bar has an original length of 10 in. What would be its length if it were stretched to its maximum length just prior to breaking?

7. How many inches could a 5-ft-long bar of plain carbon steel be stretched before it would be expected to break?

8. A customer brings you a 4-ft-long stainless steel bar from an exercise set. He wants it stretched 4 in. Could this be done?

9. Would a plain carbon steel be a good choice of material for cans for packing tomatoes? (Explain your answer.)

10. Would a plain carbon steel be a good choice of material for a hunting knife? Why?

11. Would a plain carbon steel be a good choice of a material for a spring? Why?

12. Would a plain carbon steel be a good choice of a material for a wheelbarrow axle? Why?

13. Would cast iron be a good choice of material to be welded to other metals? Why?

14. Would gray cast iron make a good hammer? Why?

15. List by number a stainless steel that is not ferromagnetic.

16. List by number a martensitic stainless steel.

17. List by number a ferritic stainless steel.

18. State the proper number notation of a ductile cast iron which had a yield strength of 45,000 lb/in^2, a tensile strength of 60,000 lb/in^2, and a 12% ultimate elongation.

19. What is the difference between a maraged and an ausformed steel?

20. If a casting was to be made and both ductile cast iron and a low-carbon steel would work equally well, which would be the best choice? Why?

CORROSION

Iron and steel react with oxygen to form iron oxide to form rust. The mechanism of *corrosion* is related to electrolysis. When the surface of a metal comes in contact with water containing oxygen or another negative element, an electrolytic cell is set up. The grain boundaries of the metal have higher energy than the center of the grain. The grain boundaries therefore become the anode of the cell and will go into solution. (See Figure 5-16 to follow this explanation.)

FIGURE 5-16   Corrosion.

As the grain boundaries are eroded by the action of the electrolyte, their energy level is lowered. Eventually, the grain boundaries are eaten away to the point where their energy level is less than that of the middle of the grain and the polarity is reversed. Now the middle of the grain becomes anodic and it starts to dissolve. Soon the energy level of the middle of the grain once again becomes lower than the grain boundary and the polarity switches back to the original situation. This cyclic reversal of polarities continues until the entire piece has reacted with the electrolyte.

The oxide of steel forms a larger crystal than that of steel alone. The larger size of the oxide causes it to buckle away from the surface of the steel. The oxide will not stick to the surface, so it flakes off, exposing the next layer of metal to the attack of the electrolyte. In turn, each layer falls off and the steel will corrode through.

Thinker 5-2:

Will steel immersed in oil corrode? Why or why not?

In contrast to steel, aluminum and magnesium form oxides which have the same crystal size as the parent metal. As a result, these oxides stick to the metal, protecting it from further corrosion. Metals such as sodium form oxides whose crystal structure is significantly less than that of the parent metal. This causes the oxide to shrink and crack, allowing the electrolyte to penetrate to the surface of the metal and corrode it further. Finally, these metals will corrode completely through.

Corrosion can also be caused by placing the metal in an atmosphere that will leach out some of the alloying elements. Brass and bronze are especially susceptible to this mechanism. The copper of brass will react with chlorine, oxygen, or sulfur to form a green or black coating of copper chloride, copper sulfate, or copper oxide on the surface. Usually, these coatings stick to the surface of the metal and protect it against further attack. Some artists prefer this green color and artificially corrode their bronze statues for aesthetic purposes.

Thinker 5-3:

Would it be wise to run copper pipe through concrete next to the steel reinforcing bars? Explain your reasoning.

Heat Treatment of Steels

INTRODUCTION

Steel has a unique quality among metals. The hardness of steel can be significantly changed by heat treatments. A medium-carbon steel with a Brinell hardness number of 200 can easily be hardened to 400 by heat treatment. High-carbon steels can be made much harder than this. Copper, aluminum, and magnesium will change their hardness only slightly by cold working or annealing. The reason for this ability of steel to change its hardness radically lies in the fact that steel changes its crystal structure upon heating. Most other metals do not.

At low temperatures, steel is a mixture of ferrite and pearlite. Both ferrite and pearlite have body-centered cubic (BCC) structures. If the steel is heated above the Ac_3 line (see Figure 6-1), that is, into the cherry-red temperature range, the steel will transform into a face-centered cubic (FCC) structure, austenite.

If a steel is heated into the austenite range and slowly cooled, it will return to the body-centered cubic structure of pearlite and ferrite. The rearrangement of the atoms from a face-centered cubic to a body-centered cubic structure takes a certain amount of time. If the steel is heated into the austenitic range and cooled rapidly, the movement of the atoms is drastically slowed and there is not sufficient time for the face-centered cubic structure to go back to the body-centered cubic system. The result will be a "strained" crystal, believed to be body-centered tetragonal, called *martensite*. Martensite is the hardest structure formed in steel and has a Rockwell C (HRc) hardness of about 68.

One reason for the difficulty in changing from a face-centered cubic crystal to a body-centered cubic crystal is illustrated in Figure 6-2. Using the same atomic radius for the iron atoms in both structures of 1.7×10^{-8} angstrom or 0.17 nanometer*, the lattice constant (the length of the edge of the cube) of the body-

---

*An angstrom unit (Å) is equal to 1×10^{-8} centimeter or 1×10^{-10} meter. A nanometer (nm) is equal to 1×10^{-9} meter.

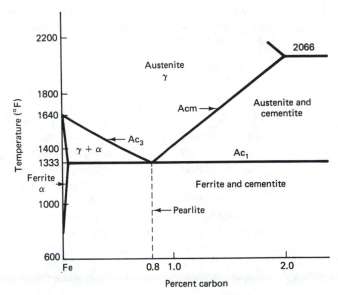

FIGURE 6-1   Iron–carbon phase diagram (steel range).

centered cubic unit cell is 3.93 Å or 0.393 nm, while the lattice constant of the face-centered cubic cell is 4.81 Å or 0.481 nm. The volume of the body-centered cube, which is the lattice constant cubed, is 60.5 Å^3 or 6.05 × 10^{-29} m^3 meters, but the volume of the face-centered cube is 111.2 Å^3 or 1.112 × 10^{-29} m^3. This represents a change of about −45% in going from the face-centered cubic austenite to the body-centered cubic pearlite and ferrite.

Thinker 6-1:

Why is martensite not found on the iron–carbon phase diagram?

Since martensite is the hardest structure in a heat-treated steel, it stands to reason that the higher the percentage of martensite, the harder the steel becomes. Martensite is formed only from the part of the steel that is pearlite. Pearlite has a composition of 0.8% carbon (eutectoid composition). A steel that has 100% pearlite, such as a 1080 steel, can be heated into austenite then cooled to 100% mar-

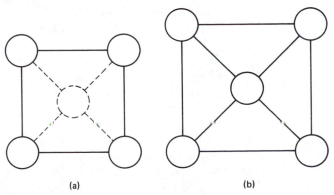

(a)   (b)

FIGURE 6-2   (a) Body-centered cubic and (b) face-centered cubic sizes.

tensite. A low-carbon steel such as a 1020 steel which has 25% pearlite could only form 25% martensite upon heat treatment. The higher the carbon content, the higher the hardness into which a steel can be heat treated.

TTT CURVE

Consider the following experiment. Several identical pieces of steel are heated to a temperature just above the Ac_3 line, into austenite. One piece of steel is removed from the oven and dunked immediately into cold water. A second sample of the steel is removed from the oven and allowed to cool below the austenitic temperature for 1 second, then plunged into the water. A third piece of steel is withdrawn from the oven and allowed to air cool for 2 seconds before it is dunked in the water. Subsequent pieces of the steel are withdrawn from the oven but allowed to age for 4, 8, 16, and 32 seconds, respectively, before they are doused in the cold water.

If Rockwell hardness tests are now run on each of the steel pieces listed above, the one that was placed in the water immediately after removal from the oven will have the highest hardness. The second piece would be slightly lower in hardness. The successive pieces would get progressively softer until a point is reached where the hardnesses remain constant.

If a grain of austenite is cooled, it will go to either pearlite or martensite. It cannot form both. If the grain is cooled slowly enough to go entirely to pearlite, no martensite will form. If martensite is formed, it will have to be reheated into austenite to be allowed to be transformed into the softer pearlite. The reason the immediately quenched steel of the previously described experiment would be the hardest is that that steel would be converted to the maximum amount martensite and no pearlite would be formed. Those samples allowed to air cool for a bit before quenching would already start to form some pearlite, precluding the formation of martensite in that grain.

A graph of the temperature and times a steel is allowed to cool in order to get pearlite or martensite has been developed. These graphs, such as the one shown in Figure 6-3, are known by several names. The *time–temperature transformation curve*, or *TTT curve*, is the most descriptive of the names. These graphs are also known as *C curves*, *Bain-S curves*, *Isothermal transformation curves*, or *I-T curves*. They are all the same. They all depict the times and temperatures at which the change from austenite to martensite or pearlite will occur. They are the basis for all heat treatments of steels.

As seen from Figure 6-3, temperature is plotted on the y axis while the log of time forms the x axis. A line drawn from the austenitic temperature or just above the Ac_3 line angling down to the right would be the cooling rate at which a steel could be cooled. The Ps line is the "pearlite start" line, and Pf is the "pearlite finish" line. The Ms means martensite start and the Mf is the martensite finish line.

If a steel has a cooling rate represented by line A in Figure 6-3 it is cooled rapidly enough to miss the "knee" of the Ps line and will be converted entirely to martensite. No pearlite will be formed and the steel will be at its hardest. If a steel is cooled very slowly from the austenitic region (above the Ac_3 line), represented by line C on the graph, when the temperature reaches the pearlite start (Ps) line, pearlite will begin to form. Once the temperature and time pass the pearlite finish (Pf) line, no further transformation can occur and the only change possible in the steel is for the grains to grow. The steel will be in its softest condition.

If a cooling rate shown by line B on the graph in Figure 6-3 is used, a mixture of pearlite and martensite grains will be formed. As the temperature hits the Ps

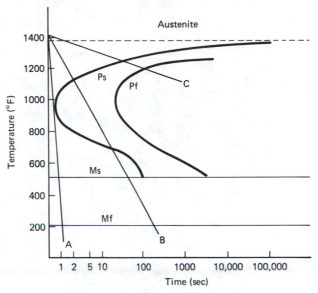

FIGURE 6-3 Typical TTT curve.

line, pearlite will start to form but since the line does not cross the Pf line, the transformation is not complete. At the temperature where line B crosses the Ms line, martensite will start to form in the grains that have not already been transformed to pearlite. The transformation of the remaining grains will have been completed as the temperature dips below the Mf line. A steel cooled at this rate will have some martensite with some pearlite. It will have a hardness between the fully martensitic and fully pearlitic steels.

The TTT curve can also be used to predict the type of grains and the grain sizes resulting from different cooling rates. A very slow cooling rate that hits only the top of the C curve will produce very coarse grained pearlite. A cooling rate that intersects the "C" just above the knee of the curve will produce fine-grained pearlite. If the cooling rate intersects the knee of the curve, nodular pearlite will be formed. Nodular pearlite is a structure in which the laminar or layered structure of the pearlite is rolled up, producing a jellyroll-looking grain. The region just below the knee of the I-T curve between the Ps and the Pf lines produces *bainite*, which is a structure of the steel that can be identified by its feathery edges of ferrite at the grain boundaries. Figure 6-4 shows some of these structures.

The TTT curve shown in Figure 6-3 would be for a plain carbon (no alloying elements present) steel. From this graph it can be seen that for a plain carbon steel to be transformed completely into martensite, a cooling rate that would miss the knee of the pearlite start line would have to cool the steel from the austenitic temperature (about 1400°F) to below 900°F in less than 1 second. Although this cooling rate can be obtained on the surface of a steel, it cannot be cooled this rapidly very deeply into the steel. The result would be that the maximum percent martensite would form only on the surface of the steel. Below the surface the percent martensite declines rapidly. A plain carbon steel would be hard only on the surface but not very deeply into the part.

If the part in question is one that must withstand a lot of wear or abrasion, such as a gear tooth, camshaft, or plow blade, the depth of hardness becomes important. As soon as the outer layer of steel is worn off, the relatively soft steel would be exposed and could not prevent further wear. It would be advantageous if a steel could be hardened more deeply.

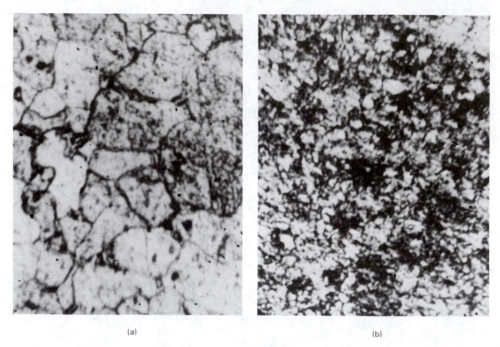

(a)   (b)

FIGURE 6-4   Grain structures produced at various cooling rates (1018 steel): (a) slow cooling (annealed); (b) fast cooling (quenched).

The addition of alloying elements has a tendency to shift the TTT curve to the right. Figure 6-5 shows a curve for an 8660 steel which contains 0.64% chromium and 0.53% nickel. This steel allows a full 5 seconds to cool from the austenitic temperature (1400°F) to below 800°F and still miss the knee of the curve to get 100% martensite. The 8660 steel will harden significantly to a much greater depth than will the plain carbon counterpart.

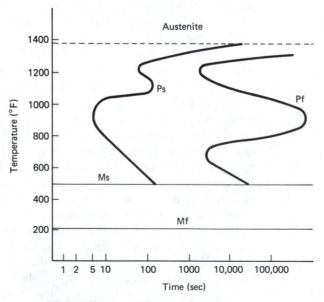

FIGURE 6-5   TTT curve for 8660 steel.

JOMINY TEST

The standard test to determine the *hardenability* of a steel is the *Jominy test*. Jominy hardenability is defined as the depth to which 50% martensite will be formed. To determine this depth a standard Jominy bar (shown in Figure 6-6) is used. This bar is 1 in. in diameter 3.5 to 4 in. long and has a cap on one end. The bar is heated into the austenitic region, left at that temperature for 1 hour to ensure that it is uniformly heated throughout, then quickly transferred from the oven to the Jominy test bucket (shown in Figure 6-7). Water is forced against the bottom of the bar and it is allowed to cool to room temperature.

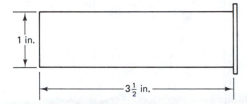

FIGURE 6-6   Standard Jominy bar.

After the bar is cooled, it is removed from the Jominy bucket and a flat area about $\frac{1}{4}$ in. wide is sanded on one side to remove the scale. Rockwell C hardness tests are made every $\frac{1}{16}$ in. for 1 in., then every $\frac{1}{8}$ in. for the second inch, every $\frac{1}{4}$ in. for the third inch, and a reading is taken at 3.5 in. from the quenched end. A graph that has the Rockwell C hardness plotted on the y axis and the distance from the quenched end on the x axis. The line obtained on this graph is called the Jominy hardenability curve. Figure 6-8 is the Jominy curve for a plain carbon steel.

If one imagines that the quenched end of the bar lies on the surface of a steel part and the length of the bar being the depth into the steel, the Jominy test simulates the depth to which a steel can be hardened. The most accurate method of finding the depth to which 50% martensite is obtained would be to highly polish the length of the bar and examine it with a metallurgical microscope. This involves a lengthy procedure. The Jominy depth can be approximated by assuming that the

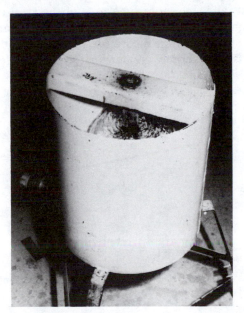

FIGURE 6-7   Standard Jominy test bucket.

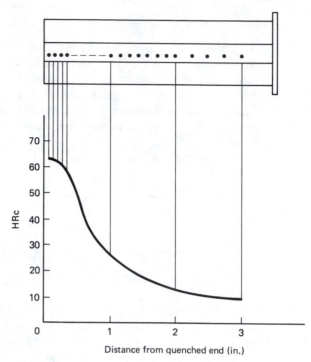

FIGURE 6-8   Jominy curve for a plain carbon steel.

quenched end of the bar is 100% martensite, the other end having 0% martensite. The middle of this range hardness would be approximately the point at which 50% martensite is formed. A horizontal line from this midhardness point carried over to the Jominy curve, then vertically to the x axis, would give the Jominy hardenability depth for that steel. There is a fallacy of this technique in that the slowly cooled end of the bar may not be at 0% martensite. The results are close for most steels.

One of the purposes of adding alloying elements to a steel is to shift the TTT curve to the right, allowing more time to cool the steel and still get martensite. Figure 6-9 is a Jominy curve for an alloyed steel. Comparison of this graph with the plain carbon steel shown in Figure 6-8 shows that the hardenability depth of the alloyed steel is considerably greater for the alloyed steel.

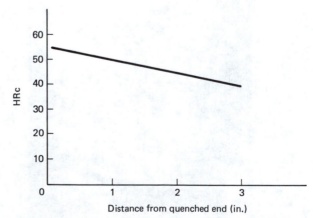

FIGURE 6-9   Jominy curve for an alloyed steel.

METHODS OF SOFTENING STEELS

By using the iron–carbon phase diagram, the TTT curve, and the Jominy hardenability curve, methods of changing the hardness of steels can be derived. All the heat treatments involve heating a steel and cooling it down. The critical differences in the heat treatments involve how hot the steel is heated and how fast it is cooled.

A steel is put in its softest condition by a process called *full annealing*. The steel is heated to a temperature about 50°F above the Ac_3 line (see Figure 6-1) into the austenitic range. The exact temperature will depend on the carbon content for steel. For a steel with 0.8% carbon, a temperature of 1390°F would be sufficient, while a steel with only 0.4% carbon would require a temperature close to 1500°F. The steel is held at this temperature for 1 hour per inch of thickness to saturate the part thoroughly or obtain an even temperature throughout. The temperature is then lowered very, very slowly (20 to 40°F per hour) until it has passed the pearlite finish line on the TTT curve. It is then slowly air cooled to room temperature. The process may take 48 to 72 hours to cool the part. Not only will the part have the softest condition possible, but the grain size will be very large and uniform.

Fully annealing a steel requires sophisticated equipment and considerable time. A simpler technique often used is *box annealing*. In this method the steel part is placed in an oven or box furnace and heated to 50°F into the austenitic range as was explained for full annealing. The part is held at the high temperature for 1 hour per inch of thickness and the oven turned off. A well-insulated oven heated to 1600°F, with the door left closed, will require at least 24 hours to cool to room temperature. A box-annealed part will not be quite as soft as a full anneal, but it will be fairly close. This process is quicker and cheaper than a full anneal, which makes it attractive to some industries.

One of the problems with annealing a steel is that the surface of the steel becomes severely corroded. The oxygen of the air reacts very quickly with the iron at these high temperatures. To combat this problem, the process of *bright annealing* has been developed. Bright annealing is done in a *muffle furnace*. A muffle furnace is one that can be sealed to prevent the gas inside it from escaping or the air from the outside from entering. The part is placed in the muffle furnace, and the air inside is purged or replaced with an inert gas such as helium or nitrogen. The part is heated to the required 50°F into the austenitic region and the temperature lowered gradually for the annealing to occur. The result is an annealed part that has no surface corrosion.

Steels that have been cold rolled or forged may not have a even grain size throughout the part. The part will have a nonuniform hardness which makes machining and other manufacturing processes difficult. To overcome this problem the parts are often *normalized*. Normalizing a steel involves heating it 100°F above the Ac_3–Acm line (well into austenite), holding the part at that temperature for 1 hour per inch of thickness, then removing it from the oven and allowing it to cool in still air to room temperature. The steel will not be nearly as soft as a full anneal, but it will have an even grain size. Normalizing is used before other heat treatments and to give the steel good machineability.

Some of the heat treatments used in industry do not involve heating the steel into the austenitic range. *Process annealing* is one such technique used to give a steel good malleability. The steel is heated to 1000 to 1300°F, which is just below the Ac_1 line. It is then air cooled to provide the finished product.

A technique used to give a steel its maximum ductility so that it can be drawn into wire is *patenting*. Patenting involves heating the steel into the austenitic range, then quenching it in molten lead which is held at a temperature of 800 to 1000°F, depending on the carbon content of the steel.

Spheroidizing the steel reduces the cementite to spheres, which produces a much tougher steel. High-carbon parts are spheroidized by heating them to about 1200°F for several hours, then slowly cooling them. Tool steels having alloying elements are heated slightly above the Ac_1 line, from 1380 to 1480°F, and hold at that temperature from 1 to 4 hours. The steel is then slowly cooled in the furnace.

METHODS OF HARDENING

Theoretically, all that need be done to harden a steel is to heat it into the austenitic temperature range and rapidly cool it. A medium- to high-carbon steel can be hardened by heating it above the Ac_3 line, then plunging it in cold water. This is called *quenching* the steel. There are a few problems that must be considered. The heat must be removed as rapidly as possible. For a plain carbon steel a cooling rate of 600°F per second is needed to miss the knee of the TTT curve. A cooling medium that can take the heat away the quickest is needed.

Two other problems in quenching still exist. Once the part is plunged into the cooling solution, the red-hot steel tends to boil the water forming a pocket of steam about the metal. This gas pocket insulates the metal from the cold water and slows the cooling rate. The use of a *brine* or saltwater solution which has a higher boiling point is helpful in easing this problem. Rapid agitation of the part in solution or playing a cold stream of water or brine from a hose on the part is also done.

The second problem involves the geometry of the part. If the part to be quenched is reasonably uniform in thickness and shape, a water or brine quench can be used. If the part is a little delicate, a water or brine quench may crack it due to the different rates of shrinkage within the part. The thinner places will cool and shrink faster than the thicker ones, causing them to pull away from the thicker sections and form a crack. In this case, an *oil quench*, which does not put as severe a strain on the part as the water, would have to be used. The oil quench does not harden the steel as well as the water, but it does not break it either.

If the part is extremely delicate or has large differences in thicknesses of the material, even an oil quench could be too severe. In this case all that can be done is to quench the part in a blast of *cold air*. Again, this is a trade-off between the maximum hardness and the chance of fracturing the material.

A glance at a TTT curve (Figure 6-5) shows the martensite finish (Mf) temperature at about 200°F. Some of the steels have Mf lines at even lower temperatures. In some cases it is necessary to place the steel in a deep freeze after quenching and to cool it to well below 0°F to ensure that the steel goes entirely to martensite, and that no pearlite will form.

Thinker 6-2:

If the maximum cooling rate is desired in quenching a steel, why isn't some cryogenic liquid such as liquid nitrogen (boiling point = −195.8°C or −320.4°F) used?

The rapid drop in temperature required in quenching a steel introduces nonuniform or *local strains* in the metal. These local strains give the structure poor machinability, making it more subject to impact failure and fatigue in the metal. To even out the strains in the steel it must be *tempered* or *drawn* after quenching. After quenching, the steel is reheated to about 700°F and held at that temperature

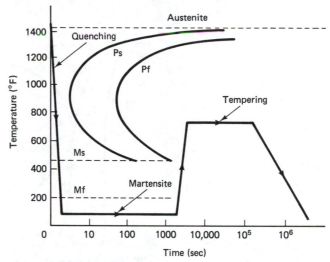

FIGURE 6-10 Quenching–tempering process.

for 1 hour per inch of thickness, then air cooled to room temperature. This is well below the temperature, which would convert the martensite back into austenite. Tempering greatly increases the ductility of the steel, has little effect on the tensile strength or yield strength of the steel [unless the temperatures get above 750°F (400°C)], but will decrease the hardness of the steel a little. The reduction in hardness is a function of the tempering temperature. Tempering greatly increases the toughness or impact strength of the quenched steel. Figure 6-10 shows the time–temperature curve for the complete quenching–tempering process. Figure 6-11 relates tensile strength, yield strength, Rockwell hardness, and percent elongation to the tempering temperature for a typical steel.

Thinker 6-3:

A good knife blade requires a high-carbon steel which can be hardened to keep an edge. Files and large hacksaw blades have the carbon content required. However, files are as hard as most drills and other cutting tools, making it almost impossible to cut, drill, or shape the file into the shape of a knife blade. Outline a procedure by which you could take a file, make it into a knife, and still have the needed hardness and toughness.

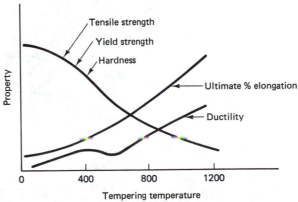

FIGURE 6-11 Properties of steel versus tempering temperature graph.

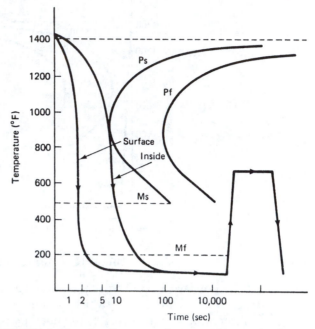

FIGURE 6-12   Cooling rates in normal quenched steel.

    In cooling the steel from the austenitic temperature to room temperature, the outside of the piece will cool at a faster rate than that steel below the surface. Figure 6-12 shows the relative cooling rates. The outer skin of the steel will have finer grains than the inside grains. It would be desirable for the steel to be uniform throughout as it enters the martensitic transformation. Two techniques have been developed to accomplish this end. *Marquenching* or *martempering* is a technique in which the part is quenched to a temperature just above the Ms line (about

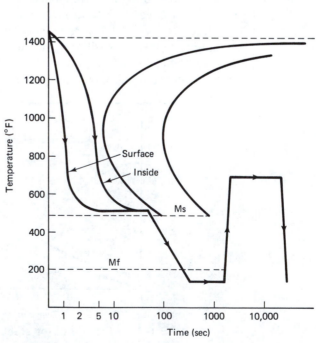

FIGURE 6-13   Cooling rates in martempered steel.

500°F), then removed from the quenching medium for a few seconds to allow the temperature to become uniform through the steel. It is then quenched through the martensitic range to room temperature. Once quenched, the steel must be tempered to remove local strains. Figure 6-13 illustrates the cooling rates for martempering.

A second way of getting the uniform grain structure prior to going through the martensitic region is called *austempering*. In austempering the quench is interrupted at a temperature well below the knee of the TTT curve but above the Ms line and above the temperature at which martempered steels are interrupted (about 700°F). The steel is held at this temperature while it passes through the Ps and Pf lines. In this region of the TTT curve, bainite will be formed. This may require holding the temperature of the steel at the 700°F from 20 minutes to 3 hours, depending on the type of steel. Once the steel has completed the bainite formation, it is cooled to room temperature. Although austempering requires no further tempering or drawing to remove local strains, it is much slower than the entire martempering-drawing method. Austempering is also limited to small parts, usually not over $\frac{1}{2}$ in. in diameter. Figure 6-14 shows the cooling rates for austempered steel.

Case Hardening

Parts such as gear teeth or cutting tools need considerable toughness to withstand impact and good surface hardness to resist wear or abrasion. If a high-carbon steel was used to get the high hardness, the part would be brittle and would not take impact. A low-carbon steel would be tough enough to take the impact but could not get the surface hardness needed to resist the wear. Certainly a dilemma exists. The solution lies in *case hardening*. Case hardening can be done by at least four different methods: *carburizing*, *nitriding*, *cyaniding*, or *carbonitriding*.

Carburizing is done by placing a low-carbon steel in a high-carbon atmosphere and heating it into the red-heat range (1400 to 1600°F). The atmosphere can be a carburizing gas such as natural gas or a solid containing carbon and barium carbonate. The carbon from the surroundings diffuses through the surface of the steel and goes into the interstices of the crystal structure. The depth to which the carbon will penetrate is dependent on the time it is left in the carbon-rich compound.

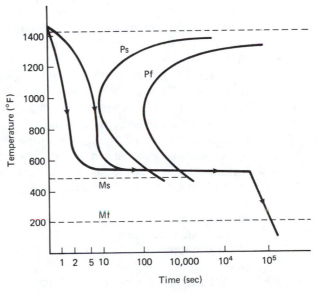

FIGURE 6-14   Cooling rates for austempering.

FIGURE 6-15   Case-hardened wrist pin.

Several hours would be required for a depth of $\frac{1}{16}$ in. of high-carbon steel to be formed. Wrist pins in automobile engines are traditionally hardened by this technique. Figure 6-15 is a photograph of the cross section of a case-hardened cam shaft.

Thinker 6-4:

It is said that ancient soldiers heated their swords red hot and plunged them into a log to harden them. Is this a believable story or not? Why?

Nitrogen and carbon have similar affects in the steel. Both of them can be in the steel interstitially and harden it. Further, nitrogen can react with the chromium, molybdenum, and certain other elements in alloyed steels to form very hard brittle compounds. Nitriding involves placing the heated steel in an atmosphere containing high quantities of nitrogen and allowing the nitrogen to diffuse into the outer layers of the steel. The nitriding agent is usually either ammonia gas (NH_3) or nitrogen gas.

The cyaniding and carbonitriding processes supply both carbon and nitrogen to the steel. The cyanide radical $(CN)^-$ is obtained by using sodium cyanide (NaCN) in the molten form. The hot steel is left buried in the molten sodium cyanide for several hours while the carbon and nitrogen diffuse into the surface of the metal. One precaution must be taken with the cyaniding process; it must not be allowed to become acidic or be around any acids. Sodium cyanide reacts with any acid to form hydrogen cyanide gas, which is extremely deadly when breathed.

Carbonitriding produces the same results as cyaniding. In carbonitriding the hot steel is immersed in an atmosphere of hot gases containing both the carbon and nitrogen. Nitrogen taken into the austenite will also slow down the critical cooling rate that is needed to miss the knee of the TTT curve. This allows the steel to form martensite at a greater depth than would be expected for a plain carbon steel.

If the steel does not need the impact resistance of a low-carbon steel, a medium- or high-carbon steel can be surface hardened two ways. *Flame hardening*

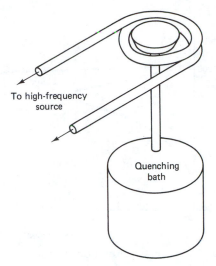

To high-frequency
source

Quenching
bath

FIGURE 6-16   Induction hardening.

involves heating the surface of the steel with a torch or flame, then immediately quenching it. The heat does not penetrate deeply enough to convert the inside of the steel to austenite; therefore, martensite will be produced only on the surface. Good-quality lathes, milling machines, shapers, and similar machine tools have beds that are flame hardened to prevent wear.

Induction hardening uses electricity to heat the steel. An electrical current flowing through a coil will create lines of magnetic flux about that coil. As the current increases, the magnetic field of these lines of flux expands. An alternating current creates and destroys magnetic fields about a coil many times a second (120 times a second for electric power lines in the United States). If a metal is crossed by a magnetic field and cuts its lines of magnetic flux, an electric current is generated in that metal (see Figure 6-16). Since the magnetic field expands and collapses many times a second about the metal, an alternating current is induced in the metal. The electrical resistance of the metal causes it to heat up when an excessive amount of electrical current flows through it. If the frequency of the alternating current is sufficiently high, the electric current will be generated only in the surface of the metal. This is known as the skin effect. With the current being induced only in the surface of a metal, only the surface will be heated. This heat will raise the surface temperature of the medium- to high-carbon steel into the austenitic range in a few seconds. The metal can be quenched, producing a hard martensitic surface with a softer core.

Age Hardening

Aluminum and other alloys sometimes have a problem in that they become harder, stronger, and more brittle as they get older. Aluminum rivets, for example, become brittle in a few months. If not used while they are relatively fresh, they will crack when peened into place. The problem is *age hardening* or *precipitation hardening*.

To illustrate the age-hardening process, let us consider the aluminum–copper system. An intermetallic alloy $CuAl_2$ is formed in this system. Up to 5.6% of the copper will dissolve in the aluminum at its eutectic temperature of 548°C (1018°F) (refer to Figures 6-17 and 6-18). If an aluminum alloy with less than 2% copper cools slowly from the melt, large grains will be formed with the $CuAl_2$ lying along the grain boundaries. The result is a rather soft material. By heating the material into the alpha (α) range at about 510°C (950°F) the $CuAl_2$ will dissolve into the copper and finely disperse in the solid solution. Quickly quenching the material so

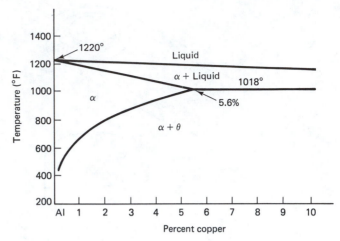

FIGURE 6-17   Copper–aluminum phase diagram (aluminum end).

that equilibrium is not reestablished, the $CuAl_2$ will stay evenly dispersed in the matrix. Reheating the alloy to a temperature of 200 to 400°F will cause the $CuAl_2$ molecules to precipitate together and effectively harden the alloy. This alloy will age harden naturally over a few months but can be accelerated by the *solution heat treatment* just explained.

Alloys such as copper–silver and others that have large alpha regions can also be age hardened. Figure 6-19 shows sketches of the three steps of age hardening $CuAl_2$.

Thinker 6-5:

If aluminum rivets age harden and become brittle in a few months, what steps can a large user of the rivets (such as the aircraft industry) take to keep from having brittle rivets?

Grain Size

One of the factors affecting the hardness, ductility, and malleability is the size of the grains in a metal. Other factors being ignored, the larger the grain, the softer the metal. Grains can be seen through an optical metallurgical microscope. There

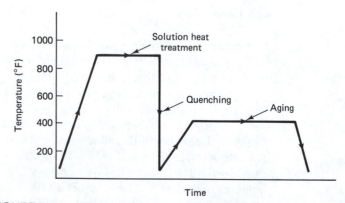

FIGURE 6-18   Time versus temperature chart for age hardening.

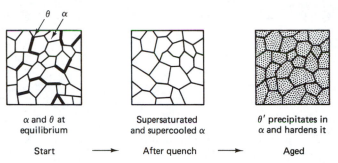

α and θ at
equilibrium

Supersaturated
and supercooled α

θ' precipitates in
α and hardens it

Start ⟶ After quench ⟶ Aged

FIGURE 6-19   Steps in age hardening.

is a standardized method of calculating the size of the grain sizes. It does not mean much to measure the diameter of a grain on a picture taken of the surface of a metal. The grains may not be cut through the center and the grains may not be spherical. The American Society for Testing and Materials (ASTM) has developed a formula for determining the average size of the grains. It is

$$N = 2^{n-1}$$

Where N is the number of grains seen in a photograph 1 in. square which has been magnified 100 diameters and n is the grain size number.

Suppose that a $100\times$ photograph shows 34 grains in a 1-in. square. The grain size number would be

$$34 = 2^{n-1}$$

$$\log 34 = (n - 1) \log 2$$

$$n = \frac{\log 34}{\log 2} + 1$$

$$= 6$$

The grain size number is always rounded to an integer. If the magnification of the photograph is not $100\times$ the number of grains that would be in the square inch under $100\times$, magnification must first be calculated. One square inch equals 645 square millimeters:

$$(1 \text{ in.})^2 (25.4 \text{ mm/in.})^2 = 645 \text{ mm}^2$$

Under $100\times$ magnification this 645 mm^2 would be an actual area of 0.0645 mm^2:

$$\frac{645}{100^2} = 0.0645$$

Suppose that a photograph measuring 35 mm by 45 mm showed 55 grains under a magnification of 250 diameters. The grain size number could be calculated as follows.

$$\text{area of the photograph} = 35 \text{ mm} \times 45 \text{ mm} = 1575 \text{ mm}^2$$

$$\text{actual area} = \frac{1575 \text{ mm}^2}{250^2} = 0.0252 \text{ mm}^2$$

The equivalent number of grains in 0.0645 mm² is

$$\frac{N}{0.0645} = \frac{55}{0.0252}$$

$$N = \frac{55(0.0645)}{0.0252} = 141$$

The grain size number is then

$$n = \frac{\log N}{\log 2} + 1$$

$$= \frac{\log 141}{\log 2} + 1 = 8.14 \quad \text{or} \quad 8+$$

The area of the photograph does not have to be square. Suppose that a $500\times$ photograph of a metal showed 24 grains in a circle having a diameter of 2 cm. What would be the grain size number?

$$2 \text{ cm} \times 10 \text{ mm/cm} = 20 \text{ mm diameter}$$

$$\text{area on the photograph} = \pi r^2 = 3.1416 \times (10 \text{ mm})^2 = 314.16 \text{ mm}^2$$

$$\text{actual area} = \frac{314.16 \text{ mm}^2}{500^2} = 0.00126 \text{ mm}^2$$

$$N = \frac{24(0.0645)}{0.00126} = 1232$$

$$n = \frac{\log 1232}{\log 2} + 1 = 11.3 = 11+$$

PROBLEM SET 6-1

1. A photograph taken through a $100\times$ microscope showed 15 grains in a 1-in. square. What is the grain size number?
2. A photograph taken through a $250\times$ microscope showed 45 grains in a circle having a 1-in. radius. What is the grain size number?
3. A photograph taken through a $500\times$ microscope showed 37 grains in a square 20 mm on a side. Find the grain size number.
4. A photograph taken through a $500\times$ microscope showed 18 grains in a circle having a radius of 20 mm. Find the grain size number.
5. Which grains are actually larger, those with a grain size number of 8 or those with a grain size number of 5?
6. How many grains would you expect to find in a 1-in. square on a photograph taken through a $250\times$ microscope if the grain size number was 5?
7. How many grains would you expect to find in a 1-in.-radius circle on a photograph taken through a $100\times$ microscope if the grain size number was 7?
8. The photograph shown in Figure 6-20 was taken with a magnification of $500\times$. Calculate the grain size number.

FIGURE 6-20   (Photo for Problem 6-1, number 10).

9. What is the grain size number of the metal in Figure 6-21 if it was taken with a magnification of 500×?

10. What is the grain size number for Figure 6-20 if the magnification was only 100×?

11. What is the grain size number for Figure 6-21 if the magnification was only 200×?

12. If 24 grains were found in a picture measuring 50 mm × 50 mm taken through a 250× microscope, what was the grain size number?

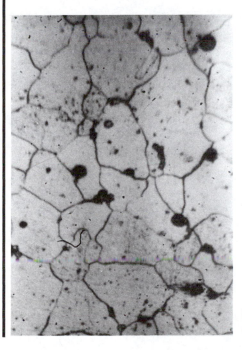

FIGURE 6-21   (Photo for Problem 6-1, number 11).

13. If a photograph measuring 56 mm by 56 mm had 36 grains under a magnification of $200\times$, what was the grain size number?

14. A photograph measuring 50 mm $\times$ 50 mm had 30 grains and a grain size number of 8. What was the magnification of the metal in the photograph?

15. A photograph showed 25 grains in a circle of 1 in. diameter and had a grain size number of 11. What was the magnification?

16. Which heat treatment would you use to soften a file without corroding it?

17. Which heat treatment is used to get an even grain size in the metal?

18. Which heat treatment is used prior to drawing a steel into a wire?

19. Which heat treatment would put a steel in the hardest condition?

20. Which heat treatment could be used to harden only the surface of the steel?

Nonferrous Metals

INTRODUCTION

Although steel is a widely used metal in engineering, many other metals have their place in industry. Those that contain no iron are called the *nonferrous metals*. A few of the more important and most used of these metals are discussed in this chapter.

COPPER

Copper was one of the first metals to be used. Entire civilizations were based on the use of *copper*. Copper probably derived its Latin name, *cuprum*, from one of the earliest recorded source of the metal, the island of Cyprus or Kypros. Copper jewelry has been found in Yugoslavia dating from 5400 B.C. Chinese copper artifacts from as early as 1030 B.C. are recorded. Copper has been essential to cultures from the early Egyptians to the North American Indians.

Early copper was found in its native nugget or pure form and could be cold worked or hammered into the desired shape or melted and cast without refining. Smelters for refining copper from its ores have been dated as early as 5000 B.C. in the Mesopotamian region of the Tigris and Euphrates valley. (This area is now part of Iraq, Syria, and eastern Turkey.)

There are over a dozen copper ores, but not all of them have much commercial importance. The ores of copper are predominately oxides, sulfides, or basic carbonates of copper. The main oxide of copper is cuprite. The principal sulfide ores are chalcopyrites, bornite (referred to as horseflesh ore, peacock ore, or variegated ore by miners), chalconite, and covellite. Basic copper carbonates are found in the form of malachite and azurite. The silicate of copper, chrysocolla, is also a usable ore. Although not used extensively as a source of copper, turquoise is a copper silicate. Figure 7-1 lists the predominate coppers and their characteristics.

FIGURE 7-1 Ores of copper.

Name	Formula	Density (g/cm^3)	Mohs Hardness	Color
Azurite	$Cu(OH)_2(CO_3)_2$	3.8	3.5	Blue
Bornite	Cu_5FeS_4	5.1	3	Red-brown
Chalcocite	Cu_2S	5.5	2.5	Gray
Chalcopyrites	$CuFeS_2$	4.1	3.5	Brass-yellow
Chrysocolla	$CuSiO_3 \cdot 2H_2O$	2.4	2	Green
Covellite	CuS	4.6	1.5	Blue
Cuprite	Cu_2O	6.14	1.5	Red
Malachite	$Cu_2(OH)_2(CO_3)$	4	3.5	Green
Turquoise	$Cu(Al_3Fe)_6(PO_4)_4(OH)_8 \cdot 4H_2O$	2.6	4.5	Blue-green

Chalcopyrites is the chief copper ore found in Utah, Michigan, and Nevada as well as Chile, Canada, Africa, England, and Spain. Chalcocite is found in Montana, Arizona, Alaska, Peru, Mexico, and Bolivia. Cuprite is also found in Arizona and South America. Chrysocolla is found in New Mexico and Arizona. The Lake Superior region of Michigan is rich in copper, and even native copper is found there. The 3-ton Ontonagon boulder of native copper came from Michigan. The largest piece of native copper ever found was discovered in 1847 and weighed 6 tons.

The copper oxides and basic carbonates are found, along with some native copper, rather close to the surface of the earth. These are the simplest to refine and were probably the first to be used. It was a major breakthrough when human beings learned to refine copper from the ore. No longer were they confined to using stones, wood, and native metals for their utensils. Perhaps the greatest significance of this advancement was that they learned that the rocks and soils had useful metals and other materials hidden in them.

Many of the present-day ores may have only 5 to 6% copper in them. The traditional method of smelting copper ores is a multi-step process. Ores that contain silicas (sand), aluminas (clays), or other undesirable materials must first be concentrated by a *flotation* process. Flotation involves grinding the ore to a powder, then putting it in water. A foaming material such as detergent is added to the solution. Air is bubbled through the slurry, creating a froth that carries the ore to the surface. The foam is skimmed from the surface and the ore recovered through a filtration process.

Most of the ores contain some iron in the form of iron sulfide. To convert the iron sulfides to iron oxides, the concentrated copper ores must be roasted. The ancient method of roasting an ore was to leave it in the sun for several days. Now the ore is roasted in an oven. After roasting the ore, which consists of the copper sulfides, copper oxides, iron sulfides, iron sulfates, some silicates, and other impurities, the ores are placed in a smelting furnace, where it is heated to 2600°F and melted. The product of this step is called *matte copper*, which is about 30% copper and may contain other metals, including gold and silver as impurities. The matte is put into a converter with a silica (sand) flux where air is blown through the melt. The process is similar in theory to that of converting iron into steel. The oxygen of the air reacts with the sulfur of the matte to produce reasonably pure copper. The chemical reaction for this step is

$$Cu_2S + O_2 \rightarrow 2Cu + SO_2$$

The molten copper is cast into cakes or ingots for further refining. Since there is a gas of sulfur dioxide and trapped air in the melt, as the copper cools it forms

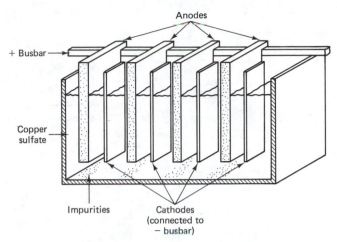

FIGURE 7-2   Copper plating cell.

bubbles on the surface, which gives the product its name: *blister copper*. Blister copper is approximately 98 to 99% copper. The slag from this process is refined further to extract the precious metals (gold and silver) and other valuable by-products.

Since a very large portion of the copper produced in the world is used in electrical wire and other electrical products, it must be extremely pure. A very small percentage of impurities in copper increases its electrical resistance drastically. The impurities in copper are removed by electroplating. The blister copper is remelted and cast into *anodes*. These anodes are placed in a plating cell containing a solution of copper sulfate. Refined copper *cathodes* are placed on the other side of the plating cell. The anode is connected to the positive pole of a direct current and the negative pole is connected to the cathode. The electric current causes the anode to go into solution and the pure copper to plate out on the cathode. The impurities will be left behind in the plating solution and the completed cathode will be in excess of 99.9% pure copper. This process is illustrated in Figure 7-2.

The completely plated cathodes may weigh up to 300 lb and may measure about 3 ft by 3 ft by about $\frac{3}{4}$ in. thick. These cathodes may be melted and used in castings directly. If the metal is to be used in electrical wire, it must be remelted in a reverberatory furnace with an oxidizing flame to prevent sulfur from being reabsorbed from the fuel and poled to keep the oxygen content at less than 0.04%. The old method of poling involved keeping a green log in the melting pot. The carbon from the wood reacted with the oxygen, keeping it from reacting or getting into the copper. The copper produced in this manner is called *electrolytic tough pitch (ETP) copper*. Figure 7-3 is the flowchart for refining copper.

Other methods of refining copper have been tried. Electrolytic leaching, a method similar to electroplating, of the copper directly from the ores has been used. In an effort to control the oxygen in the copper the addition of about 0.02% phosphorus has been added. This type of copper is called phosphorus deoxidized (DHP) copper.

The properties of copper make it a very useful metal. Its electrical conductivity is second only to silver. Its thermal conductivity is also extremely good. It has good corrosion resistance, although it will react with sulfur, oxygen, and chlorine when heated. Its tensile strength and yield strength are considerably less than steel and its modulus of elasticity is roughly half that of steel. The properties of copper depend somewhat on the grain size of heat treatment of the metal. Figure 7-4 summarizes the properties of copper.

Copper has long been used for its electrical conductivity. At temperatures

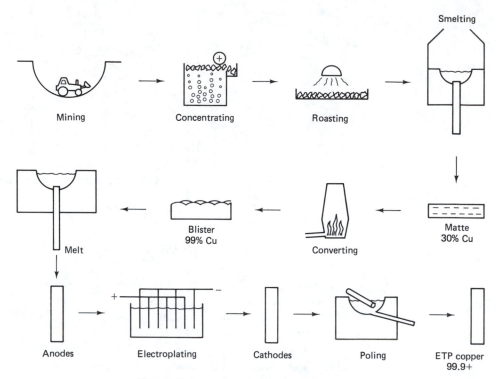

FIGURE 7-3   Copper refining flowchart.

close to absolute zero it becomes a *superconductor.* Superconductors have almost no resistance to electrical flow. A current started in a superconducting circuit will continue to flow indefinitely. Impurities in the metal were found to increase the resistance sharply. Recently, certain compounds of copper were found to be superconductors at temperatures as high as 90K ($-183°C$ or $-297°F$). These temperatures can be obtained by submerging the circuit in liquid nitrogen, whose boiling point is $-196°C$ (77K). The rare earth ceramic compounds of copper oxides $YBa_2Cu_3O_7$ and $La_{1.8}Sr_{0.2}CuO_4$ have been found to be superconductors at temperatures of 90K. A material containing yttrium, strontium, barium, and copper oxides has been reported to be a superconductor at a temperature of $-9°F$. The goal of the scientists working in this field is to find materials that will be superconductors at temperatures that can be maintained in atmospheric transmission

FIGURE 7-4   Properties of copper.

Property	*Hard Copper*	*Soft Copper*
Density	8.96 g/cm³ (559 lb/ft³)	
Melting point	1083°C (1981°F)	
Tensile strength ($\times$1000 psi)	44–55	30–38
Yield strength ($\times$1000 psi)	45	10
Modulus of elasticity	17×106	15×10^6
Maximum percent elongation	1.5–16	35–55
Poisson's ratio	0.33	0.33
Rockwell hardness	50 (HRb)	40 (HRf)
Electrical resistivity	1.673×10^{-6} Ω/cm³ at 20°C	
Thermal conductivity	226 Btu/hr/ft²/°F	
Coefficient of thermal expansion	9.2×10^{-6} in./in./°F	
Heat capacity	0.092 Btu/lb/°F	
Tensile strength/density	155,000 in.	

lines. This would make possible electrical transmission over long distances with very little loss of energy or electrical power.

BRONZE

Early in the civilizations that used copper, it was found that some metalsmiths produced harder and stronger coppers than did others. The impurities in the ores also presented problems. In some ores there is a significant amount of arsenic, which no doubt had a detrimental effect on the health of the workers. These impurities could change the properties of the copper product. Mixing other elements with the copper also produced better properties in the metal. Clay tablets recovered from the Babylonian empire (2500 B.C.) even had the formulas for mixing elements. One such formula specified "one part by weight of tin to nine parts of copper." Today these alloys are known as *bronzes*.

Bronze is defined as copper alloyed with any other metal. There are many different formulas for bronzes, depending on the percent of the alloying elements present. Traditionally, copper and tin were combined to form tin bronzes. Bronzes with 4.5, 8, and 10% tin have been known as A, C, and D bronzes, respectively. Small amounts of phosphorus are sometimes added to the tin bronzes to improve the ductility. These bronzes are erroneously called phosphor-bronzes, but phosphorus is not the primary alloying element.

Naval bronze or hard bearing bronze contains 85% copper, 13% tin, 1.5% zinc, and the rest nickel and phosphorus. Bronzes that have more than 90% copper are red in color. Architectural bronze has 97% copper, 2% tin, and 1% zinc. With more than 10% alloying elements added to the copper, the bronzes become yellow and even white. Gold bronze contains 89.5% copper, 2% tin, 5.5% zinc, and 3% lead. This type of bronze is resistant to acids and is used in many castings.

Aluminum bronzes are mixtures of copper and from 4 to 11% aluminum. These aluminum bronzes form the highest-strength alloys of copper, with tensile strengths over 100,000 lb/in². Aluminum bronzes also have outstanding corrosion resistance, which makes them well suited for structural purposes. Silicon bronzes have properties similar to mild steel in workability, yet have the corrosion resistance of copper. The typical silicon bronze will have about 5% silicon dissolved in the copper. Silicon bronzes find considerable use in marine construction, where both strength and corrosion resistance are important.

The addition of about 2% beryllium to the copper creates a beryllium bronze. Beryllium in its pure state is an extremely toxic metal. Combined with the copper it forms a stable, nontoxic metal that will not create sparks when struck with another metal. This quality alone makes beryllium bronze the choice of metal for mixing implements for explosives, rocket fuels, and other flammable materials.

Copper–nickel alloys or nickel bronzes are often white in color and do not look like traditional bronzes. *Nickel silver* and *German silver* are names often used for nickel bronzes. Present-day U.S. clad coinage is an example of nickel silver. The outer layers of U.S. dimes, quarters, and half dollars are made of 75% copper and 25% nickel bonded to an inner core of copper.

Intermetallic Compounds

Not all mixtures of metals are solutions. Some combinations form what are known as *intermetallic compounds*. Intermetallic compounds are true compounds just as are sodium chloride and potassium nitrate. They react in definite ratios with each other and form a new material unlike either of the parent materials. Epsilon (ε) bronze has the formula Cu_3Sn, while eta (η) bronze has the composition Cu_6Sn_{11}.

The percent composition of these intermetallic compounds can be figured from their formulas. There are three atoms of copper and one atom of tin in the molecule, so the molecular weight of epsilon bronze would be

$$3 \times 63 = 189$$
$$1 \times 119 = \underline{119}$$
$$\text{mol. wt.} = 308$$

Since three of the atoms are copper, the percent copper would be

$$\frac{189}{308} (100) = 61.4\%$$

This coincides with the percent copper of epsilon brass in the phase diagram.

BRASS

Copper and any other element makes bronze. A specific type of bronze, copper and zinc, is *brass*. As with bronze, there are many formulas for brass. Brass may also contain other elements besides the copper and zinc. Admiralty brass has 70% copper, 29% zinc, and 1% tin. Its tensile strength is about 100,000 lb/in². Cartridge brass is a straight 70–30 brass (70% copper, 30% zinc). Muntz metal, invented in 1832 by George F. Muntz, also called malleable brass, has a composition of 60% copper and 40% zinc. Its annealed tensile strength is 52,000 lb/in², it has an ultimate elongation of 50%, and it has a Rockwell B hardness (HRb) of 30. The addition of 0.75% tin to the Muntz metal makes it into naval brass. The further addition of 0.25% lead adds considerable ductility to the metal and it becomes known as *free machining brass*.

Red brass or rich low brass has 85% copper and 15% zinc. It has a tensile strength in the annealed state of 38,000 lb/in², but can be hardened by cold rolling to 70,000 lb/in². It is one of the most ductile and malleable of the brasses. Its reddish color, corrosion resistance, and ability to take a polish make it ideal for castings. Low brass has a composition of 80% copper and 20% zinc. Figure 7-5 shows the compositions of several brasses.

Brass and bronze are two-component systems for which phase diagrams exist. Figure 7-6 is the phase diagram for a copper–zinc system. The copper–tin phase diagram is shown in Figure 7-7. Many of the properties of brass and bronze can

FIGURE 7-5   Compositions of brass.

Type	Copper (%)	Zinc (%)	Other (%)
Red	85	15	
Low	80	20	
Yellow	65	35	
Muntz metal	60	40	
Lead brass	65	32	3 lead
Admiralty brass	70	29	1 tin
Cartridge brass	70	30	
Naval brass	60	39.25	0.75 tin
Free machining	60	39	0.75 tin 0.25 lead

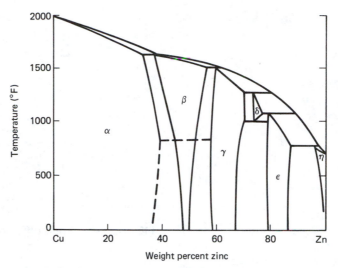

FIGURE 7-6   Copper–zinc phase diagram.

be explained from their phase diagrams. As seen in the copper–zinc phase diagram, up to 40% by weight of zinc can dissolve into copper without forming a second solid phase. Brasses that fall in this region can be referred to as alpha (α) brasses. They would tend to be more ductile than their two-phase counterparts. Intermetallic compounds are also formed at 59 to 68% zinc and 80 to 87% zinc. These compositions are not used significantly.

In the copper–zinc phase diagram (Figure 7-6) it can be seen that the melting point of zinc is the lowest-melting composition on the graph. When a pure element has the lowest melting point on the phase diagram, the system is termed a *monotectic* system. Monotectic systems occur frequently in systems of two or more components in which the melting points of the components differ greatly. The iron–zinc system is another example of a monotectic. Iron melts at 2795°F and zinc melts at 787°F, which is the monotectic point on the phase diagram.

In the phase diagram for bronze (copper–tin) we see that although tin will dissolve into copper up to 20% at temperatures of 950°F, at room temperatures (70°F) a second phase will form with the addition of very small amounts of tin. An intermetallic epsilon (ε) bronze will form at about 38% tin. A second intermetallic

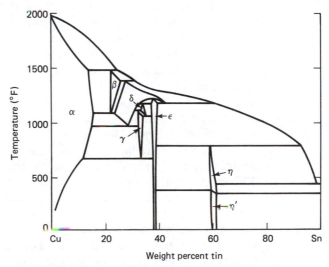

FIGURE 7-7   Copper–tin phase diagram.

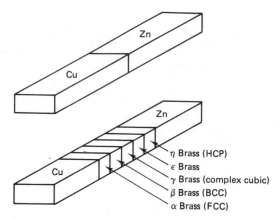

FIGURE 7-8   Solid diffusion in copper and zinc.

compound, the eta (η) phase, is formed at about 62% tin, but this composition, being more tin than copper, is not considered a bronze, nor is it used. In both brass and tin bronzes, adding an alloying element to the copper lowers the melting point, which makes it easier to cast.

An interesting experiment to illustrate the intermetallic compounds of brass is as follows. Polish a surface on both a piece of copper and a piece of zinc. Clamp them together and raise the temperature to about 700°F and hold them at that temperature for several hours to a day or two. The metals will have been welded together by the end of this time. Polishing and etching the side of this new block and examining it under a microscope will reveal at least five different intermetallic compounds. Through the process of diffusion, discussed in Chapter 4, the atoms have migrated from one metal into the other. The sketch in Figure 7-8 illustrates the different brasses obtained.

ALUMINUM

The history of aluminum is one of the most dramatic in science. Aluminum was discovered in the year 1825 by Hans Christen Oersted and isolated in 1827 by Friedrich Wöhler, a German professor of chemistry at Heidelberg University. By 1854, a French scientist, Henri Deville, and Wöhler were able to produce aluminum from the ore by reacting it with sodium metal. Emperor Napoleon III was greatly impressed with this new metal and set about to equip his armies with aluminum helmets and investigated the possibility of lightweight aluminum weapons. Since the price of aluminum in 1852 was $545 per pound, the honored guests at French state dinners had forks and spoons made of aluminum while the lesser dignitaries had to use the less expensive gold and silver utensils. The year 1884 saw the first use of aluminum in architecture, when about 6 lb of aluminum was placed on the tip of the Washington Monument.

Attempts to reduce aluminum oxide to aluminum metal met with many difficulties. A general rule of chemistry is that the easier a metal oxidizes, the more difficult it is to decompose it. Aluminum is a very active metal, being only slightly less active than magnesium. It will not react with carbon to reduce the oxide as did iron and copper. Aluminum oxide is so thermally stable that it is used in ceramics and therefore will not break down by heat. Some metals can be reduced by converting their ores into the hydroxide, then heating them. Efforts to reduce aluminum hydroxide to aluminum by heating only sent the aluminum hydroxide back to aluminum oxide. To reduce the oxide of aluminum by electrolysis, it must be put in solution. Aluminum oxide is insoluble in water.

In the early 1880s a young chemistry student, Charles M. Hall, accepted the challenge issued by one of his professors at Oberlin College (Ohio) and sought a commercial method of producing aluminum. At the same time, working independently in France, a metallurgist, Paul Heroult, started working on the same project. After graduation, Hall continued working on the project in a shed behind his home. He even made his own batteries to supply the electric current needed for his experiments. By 1886, both Hall and Heroult had succeeded in producing aluminum with basically the same technique. After many years of law suits and court battles, Hall succeeded in establishing his claim to the U.S. patent as the first inventor of the process. His troubles were not over, however. He founded the Aluminum Company of America (ALCOA) and served as its vice-president, but unscrupulous business interests practically stole his process from him. Charles Hall, the man who invented the process that brought the price of aluminum down to as low as 15 cents a pound at one time, died in 1914 at the age of 51 years.

Although it is the most abundant metal in the earth's crust (8.1%) and the third most abundant element in the crust behind oxygen and silicon, it is never found in the free state. Aluminum is found in clay, bauxite, mica, feldspars, cryolite, corundum, sapphires, and rubies in the form of alumina (aluminum oxide or Al_2O_3). Its richest ore is bauxite (also known as gibbsite, boehmite, and diaspore). Bauxite was originally found near the French town of Le Baux and named after it. The main U.S. source of bauxite comes from Arkansas, but the majority of the aluminum ore used in the United States is imported from Surinam, Guyana, and Jamaica.

Hall found that bauxite having the chemical formula $AlO(OH)$ could be dissolved in molten cryolite, which is sodium aluminum fluoride (Na_3AlF_6). Once dissolved, the aluminum could be separated by electrolysis. Cryolite is a mineral that was once mined for this purpose. Now an artificially made sodium aluminum fluoride is used.

The modern refining technique called the Bayer process requires several steps. The ore is usually found near the surface, so it is open-pit or strip mined in a method similar to that of coal and copper. The ore is then crushed, washed to remove undesirable residues, and dried at temperatures from 200 to 250°F. The dried powder is mixed with soda ash (Na_2CO_3), lime (CaO), and water and placed in a digester to form sodium aluminate ($Na_2Al_2O_4$). After filtering to remove impurities (called red mud) the sodium aluminate is placed in a precipitator where the sodium aluminate decomposes into aluminum hydrate [$AlO(OH)$]. The aluminum hydrate is heated to 2000°F to convert it to alumina (Al_2O_3). The rest of the refining is the Hall process. The alumina is put into a carbon-lined pot containing the molten cryolite at 1800°F. Large carbon electrodes are lowered into the molten solution and an extremely high direct current (on the order of 100,000 A) is passed through the solution. The electrodes are positively charged, while the carbon lining of the pot is the negative pole. The product is metallic aluminum, which is drained periodically from the bottom of the cell and cast into ingots measuring some 28 by 6 by 5 in. and weighing about 80 lb each.

The reduction process is made possible by the availability of large amounts of cheap electricity. The electrolytic cells or pots are capable of producing about 2000 lb of aluminum per 24-hour day. To produce 1 ton of aluminum around 20,000 kW of electricity are required. This is equivalent to 10,000 W per pound of aluminum. Aluminum refineries are usually located near a power dam or some other source of electricity and contract for most of its output. Figure 7-9 shows a flow diagram for the production of aluminum from its ore.

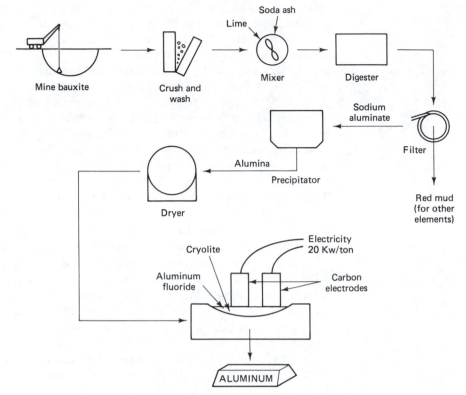

FIGURE 7-9 Aluminum production flowchart.

The properties of aluminum when combined with its availability and low cost make it one of the most useful materials in the metals inventory. Although the tensile strength, ductility, and other properties can be changed somewhat by heat treatments and alloying the average properties can be listed. The density of aluminum is 2.7 g/cm³, or only 168 lb/ft³. The only metal lighter in density is magnesium, at 1.74 g/cm³. The tensile strength of aluminum is about 10,000 lb/in², but it can be alloyed to a tensile strength of 80,000 lb/in². The aluminum industry is quick to mention that on a weight basis aluminum is as strong as some steels. The *tensile strength-to-density ratio* of aluminum varies from about 100,000 to 800,000. The tensile strength-to-density ratio for steels is about 229,000 for low-carbon steels to over 1,000,000 for the superalloys.

To get the tensile strength-to-density ratio, divide the tensile strength in lb/in² by the density of the metal in pounds per cubic inch. The density of a metal in the English system can be found by multiplying the density of the metal in grams per cubic centimeter by

$$\frac{(\text{g/cm}^3)(2.54 \text{ cm/in.})^3}{(454 \text{ g/lb})} = 0.036 \text{ lb/in}^3$$

The tensile strength-to-density ratio for an aluminum having a tensile strength of 10,000 lb/in² and a density of 2.7 g/cm³ would be

$$\frac{(10,000 \text{ lb/in}^3)}{(2.7 \text{ g/cm}^3)(0.0360)} = 103,000 \text{ in.}$$

The electrical conductivity (which is the reciprocal of the resistivity) of alu-

FIGURE 7-10   Properties of aluminum.

Property	Value
Density	2.7 g/cm³
Melting point	660°C (1220°F)
Tensile strength	10,000–80,000 psi
Yield strength	5000–68,000 psi
Modulus of elasticity	10.6×10^6 psi
Maximum percent elongation	14–15%
Poisson's ratio	0.33
Electrical resistivity	3×10^{-6} Ω/cm³
Thermal conductivity	130 Btu/hr/ft²/°F
Coefficient of thermal expansion	13×10^{-6} in./in./°F
Heat capacity	0.23 Btu/lb/°F
Tensile strength/density	100,000–800,000 in.

minum is approximately 60% that of copper, but on a weight basis the electrical conductivity per pound is up to twice that of copper, depending on the alloy of aluminum used. For this reason, some electrical transmission lines that have to cover long spans between supports are made of aluminum. Figure 7-10 summarizes the properties of aluminum.

An apparent contradiction occurs with aluminum. It is a very chemically active metal, yet it never seems to corrode. Aluminum window sashes, chalkboard trays, and other fixtures always have the dull-gray finish but do not look oxidized. The fact is that aluminum oxidizes very rapidly, within seconds. Unlike iron or steel, for which the oxide falls off the surface, the oxidation on aluminum sticks to the surface of the metal and forms a protective coating which prevents any further corrosion. The fact that many compounds of aluminum will adhere to the metal surface forms the basis for *anodizing*. If an aluminum part is placed on the anode (positive pole) of an electrolytic (electroplating) cell with oxalic acid, sulfuric acid, or chromic acid as the electrolyte (plating solution), the aluminum will form an oxide, sulfate or chromate coating, which is a hard, wear-resistant, electrical insulating material. This coating, which will protect the surface against abrasion and further attack by oxygen, gives the material an attractive appearance. These coatings are on the order of 0.002 to 0.008 in. thick. The surfaces of aluminum can be given different colors by the addition of other metals to the electrolyte or by dying.

Thinker 7-1:

Aluminum castings and extrusions usually have a dull-gray finish. Should these materials be polished to improve their appearance?

The Aluminum Association has devised a system for designating the types of wrought aluminum alloys. The naming of the alloys is based on a four-digit numbering system. Aluminum alloys may have such numbers as 1075, 6016, 7075, or 1130. In this system the first digit indicates the alloying elements present in the aluminum. The number "1" indicates that no major alloying elements are in the metal and the aluminum would be 99% or more pure aluminum. A 2––– series aluminum would have copper as the alloying element. Figure 7-11 lists the other alloys of aluminum.

The second digit indicates the alloy modifications or degree of control of the

FIGURE 7-11   Aluminum alloy designation system.

Designation	Major Alloying Element
1 – – –	None (99 + % pure aluminum)
2 – – –	Copper
3 – – –	Manganese
4 – – –	Silicon
5 – – –	Magnesium
6 – – –	Magnesium and silicon
7 – – –	Zinc
8 – – –	Other
9 – – –	Unused as yet

impurities. The number "0" in this slot indicates that the original alloy had no modifications or no special control of the impurities. A "1" in the second digit indicates a maximum of 0.15% silicon and iron. The second digit is also used to indicate the succession of the design modification of the formula for the alloy. A 20– – indicates that the original alloy had 4% copper, and a 21– – is the first modification of the alloy, having only 2.5% copper. The last two digits are arbitrary numbers to identify a specific alloy within the series or to indicate the purity of the aluminum over 99%.

The internal structure of aluminum metal can also be modified somewhat by heat treatments. The four-digit numbering system of aluminum may be followed with a heat treatment designation. Thus an aluminum designated xxxx-T6 would be solution heat treated, then artificially aged. The heat treatment designations are shown in Figure 7-12. Figure 7-13 shows a few of the more popular aluminum alloys with their compositions.

FIGURE 7-12   Aluminum heat treatment designations.

Designation	Meaning
– F	As fabricated
– O	Annealed and recrystallized
– H	Strain hardened
– H1 (+ a digit)[a]	Strain hardened only
– H2 (+ a digit)[a]	Strain hardened and partially annealed
– H3 (+ a digit)[a]	Strain hardened, then stabilized
– W	Solution heat treated, unstable temper
– T	Treated to produce stable tempers other than – F, – O, or – H
– T2	Annealed
– T3	Solution heat treated, then cold worked
– T4	Solution heat treated and naturally aged to a stable condition
– T5	Artificially aged only
– T6	Solution heat treated, artificially aged
– T7	Solution heat treated, then stabilized
– T8	Solution heat treated, cold worked, then artificially aged
– T9	Solution heat treated, artificially aged, then cold worked
– T10	Artificially aged, then cold worked

[a]The digits 1 to 8 may follow, indicating the degree of cold working or strain hardening. The number 2 indicates $\frac{2}{8}$ ($\frac{1}{4}$) hard, 4 indicates $\frac{1}{2}$ hard, and 8 means fully hardened.

FIGURE 7-13   Wrought aluminum alloys.

Alloy Number	Composition
1100	99.00%+ aluminum 0.2 copper, 0.05 manganese, 0.1 zinc, 0.65 silicon, iron
2011	93%+ aluminum 5–6% copper, 0.4 silicon, 0.7 iron, 0.3 zinc
3003	96%+ aluminum 1–1.5 manganese, 0.6 silicon, 0.7 iron, 0.2 copper, 0.1 zinc
4043	92%+ aluminum 4.5–6 silicon, 0.8 iron, 0.3 copper, 0.05 manganese, 0.05 magnesium, 0.1 zinc, 0.2 titanium
5050	96%+ aluminum 1–1.8 magnesium, 0.4 silicon, 0.7 iron, 0.2 copper, 0.1 manganese, 0.1 chromium, 0.25 zinc
6061	96%+ aluminum 0.4–0.8 silicon, 0.8–1.2 magnesium, 0.7 iron, 0.15–0.4 copper, 0.25 zinc, 0.15 manganese, 0.15 titanium, 0.15–0.35 chromium
7075	91%+ aluminum 5.1–6.1 zinc, 2.1–2.9 magnesium, 1.2–2.0 copper, 0.5 silicon, 0.7 iron, 0.18–0.4 chromium, 0.2 titanium

PROBLEM SET 7-1

Completely describe the alloys in Problems 1 through 5, giving their composition and heat treatment.

1. 2011-T3

2. 1100-H32

3. 7075-T6

4. 6061-T2

5. 4043-T8

6. What is the proper designation for the following aluminum alloy? 99.05% Al, 0.2% Cu, 0.1% Zn, 0.65% Si, and Fe, artificially aged only.

7. What is the proper designation for the following aluminum alloy? 93% Al, 5.6% Cu, 0.4% Si, 0.7% Fe, and 0.3% Zn, solution heat treated and artificially aged.

8. A nonsparking alloy is needed for use in a fireworks plant. What metal could be used?

9. What is the percent copper in the eta (η) bronze (Cu_6Sn_{11})?

10. What is the percent tin in delta (δ) bronze ($Cu_{31}Sn_8$)?

11. Beta brass has the formula CuZn. What is the percent zinc?

12. What is the difference between brass and bronze?

13. Are all brasses a form of bronze?

14. From the copper–tin phase diagram, how many intermetallic compounds are there?

15. What is the eutectic composition and temperature in the copper–tin phase diagram?

16. In the copper–zinc phase diagram, how many intermetallic compounds are there?

17. Is Muntz metal (60% copper and 40% zinc) an intermetallic compound or a mixture of two phases?

18. How many peritectics are there in the copper–zinc system?

19. How many eutectoids are there in the copper–zinc system?

20. What phases would be present at 70% zinc at room temperature in the copper–zinc system?

Each of the alloying elements performs a specific function in the aluminum alloy. Copper generally increases the strength of the aluminum. Silicon improves the corrosion resistance. Magnesium also gives excellent corrosion resistance, high strength, and lets the alloy take a good finish. Zinc improves the machinability of the metal. Nickel increases the high-temperature tensile strength. Iron helps reduce shrinkage but must be controlled to obtain maximum strength.

The 1000 series aluminum is used for electrical and electronics application and many other uses where strength is not the primary concern. The 2000 (copper) series is the strongest aluminum alloy. It has good machinability but has poor corrosion resistance and weldability. The 3000 (manganese) alloys have good workability and are used for many architectural and civil engineering applications. If wear resistance is important, the 4000 (silicon) alloys are good. They have low coefficients of thermal expansion, which makes them good for castings, forgings, and uses that must withstand heat. They are used in boats and other marine equipment and automobile engine pistons. The magnesium (5000) series of alloys have good weldability and ductility. They are used for extrusions and automotive body parts. The 6000 series alloys, containing magnesium and silicon, have good weldability and corrosion resistance, which makes them suitable for piping, sheet, and extrusions. The 7000 aluminum alloys (zinc) have very high strength (over 85,000 lb/in^2) and are easily heat treated. This combination makes the 7000 aluminums the choice for many aircraft applications.

MAGNESIUM

Magnesium, the lightest structural metal, was discovered by the English chemist Sir Humphry Davy in 1808. It is the eighth most abundant in the earth's crust. It is found in such minerals as magnesite ($MgCO_3$) and dolemite [$CaMg(CO_3)_2$], but its primary source is seawater. Magnesium is the third most abundant element in seawater. Only sodium and chlorine have higher concentrations. Seawater has 1272 g of magnesium per metric ton, which converts to about 6,000,000 tons per cubic mile or almost 0.73 lb/yd^3. A pound of magnesium can be obtained from every 100 gallons of seawater. (A foot of water in the average bath tub is about 60 gallons.) Other common chemicals in which magnesium is found are epsom salts ($MgSO_4$) and milk of magnesia [$Mg(OH)_2$]. It is present in the minerals asbestos and talc. Magnesium is also found in plants and animals. It is involved in the structure of chlorophyll. Adult humans need about 300 mg per day of magnesium for good health.

Magnesium is a very lightweight metal having a density of only 1.738 g/cm^3

FIGURE 7-14 Properties of magnesium.

Property	Value
Density	1.738 g/cm³
Melting point	651°C (1204°F)
Tensile strength	37,000 lb/in²
Yield strength	22,000–29,000 lb/in²
Modulus of elasticity	6.5×10^6 lb/in²
Maximum percent elongation	4–15%
Poisson's ratio	0.35
Electrical resistivity	—
Thermal conductivity	99 Btu/hr/ft²/°F
Coefficient of thermal expansion	14.5×10^{-6} in./in./°F
Heat capacity	0.25 Btu/lb/°F
Tensile strength/density	590,000 in.

or 108 lb/ft³. Its melting point is about 1204°F or 651°C. Its tensile strength is only about 37,000 lb/in³, but compared to its low density, its tensile strength-to-density ratio is almost 590,000. The hexagonal close-packed structure of magnesium does not give it very high ductility (only 8 to 15%). One of its unique properties is the ease with which it will react with oxygen. A magnesium ribbon or wire can easily be ignited at relatively low temperatures and will burn with an extremely hot flame and emit an intense light. This property of magnesium is put to use in flash bulbs and incendiary bombs. The properties of magnesium are summarized in Figure 7-14.

The production of magnesium from seawater is relatively uncomplicated. The seawater is filtered through a bed of lime [Ca(OH)₂] and oyster shells to ensure that the magnesium is converted to magnesium hydroxide, which is insoluble in water and precipitates to the bottom of a settling tank. Hydrochloric acid is added to the filtered solution to convert the magnesium hydroxide to magnesium chloride:

$$Mg(OH)_2 + HCl \rightarrow MgCl_2 + H_2O$$

Once the magnesium solution is thickened by evaporators and dried, it is sent through electrolytic cells, where the molten magnesium chloride is decomposed into magnesium metal and chlorine gas. The chlorine gas is recycled to the hydrochloric acid plant and reused. The chemical equation for this process is

$$MgCl_2 \rightarrow Mg + Cl_2$$

As with aluminum, the production of magnesium uses a large amount of electricity per pound of metal. Figure 7-15 is the schematic diagram of magnesium production process.

About half of the magnesium used is in the alloyed form. The American Society for Testing and Materials (ASTM) and the Magnesium Association have issued a designation system for magnesium alloys which uses one or two capital letters followed by one to three numbers, then another letter. An alloy might be designated AZ81A, EK41A, or AZ91C. The first two letters represent the two major alloying elements in descending order of percentages. The two or three numbers are the percent of the alloying elements rounded off to the nearest percent. The last letter is the modification of the alloy. Thus the AZ81A means that the magnesium alloy has aluminum and zinc in it. The 8 means that there is 8% aluminum, the 1 indicates 1% zinc, and the last A shows that this is the first alloy qualifying for the AZ81 designation. Figure 7-16 shows the letters representing the

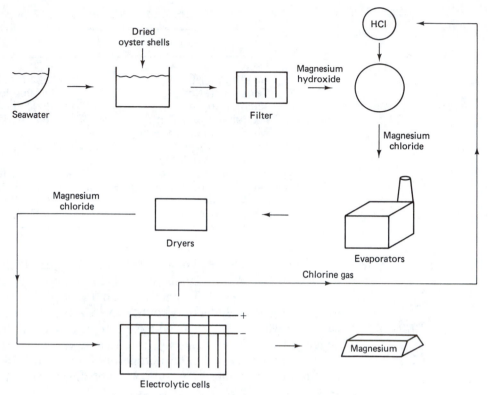

FIGURE 7-15   Magnesium production flowchart.

alloying elements. The designation system for heat treatments is the same as that used for aluminum.

The aluminum–zinc (AZ) and the aluminum–magnesium (AM), zinc–zirconium (ZK), zinc–thorium (ZH), and some of the rare-earth (EK and EZ) magnesium alloys are used for casting. The AZ91A is best suited for *die casting*. The M1A is a cheap, low-strength alloy that can be welded and has good corrosion

FIGURE 7-16   Alloying elements of magnesium.

Letter	Element
A	Aluminum
B	Bismuth
C	Copper
D	Cadmium
E	Rare earths
F	Iron
G	Magnesium
H	Thorium
K	Zirconium
L	Beryllium
M	Manganese
N	Nickel
P	Lead
Q	Silver
R	Chromium
S	Silicon
T	Tin
Y	Antimony
Z	Zinc

resistance. The ZK51A and ZK61A have very good tensile strength and ductility but tend to crack during casting. HK31A is used as a sheet alloy for high-temperature applications (to 700°F).

There are problems in joining magnesium parts. Although it can be welded (with care), most of the joining of magnesium parts is by riveting or by the use of the new adhesives or glues. Caution must be used in machining magnesium, since fine chips and dust are easily ignited and burn fiercely. It is the lightweight applications for which magnesium and magnesium alloys are most used. Much of the use of magnesium and its alloys is in the production of aircraft.

NICKEL

The U.S. five-cent coin is not called a nickel without reason. Ever since it was authorized by Congress in 1866, it has been made of an alloy of 75% copper and 25% nickel. The Canadian nickel is 99.4% nickel. Although nickel is found in many of the ancient alloys and is a predominant component of meteorites, it was officially discovered by Baron Axel Fredric Cronstedt in 1751. Cronstedt discovered nickel in the ore niccolite (nickel arsenide) ($NiAs$), but it is also found in millerite (NiS), rammelsbergite ($NiAs_2$), and garnierite, a nickel-bearing serpentine [Ni + $Mg_3Si_2O_5(OH)_4$]. The major commercial ores of nickel are pentlandite [$(FeNi)_9S_8$] and pyrrhotite (also known as magnetic pyrites) (FeS with nickel). Over 80% of the world's supply of these ores comes from the Sudbury region of Ontario, Canada (just north of Lake Huron) and the Lind Lake region of Manitoba, Canada (in northwest Manitoba). Most of the remainder of the commercial nickel ores come from New Caledonia, the Soviet Union, and Australia.

Nickel is refined from the sulfide ores by two methods. One of the processes is very similar to the process by which copper is extracted from the ore. After the ore is mined, crushed, and ground, it is concentrated by the flotation process. From here the ore is roasted and smelted in an electric furnace to produce the furnace matte. The matte is sent to the converter, where air is blown through the molten metal to produce the blister nickel. The blister is then remelted and cast into anodes for electrorefining in a plating cell. The nickel is plated out on the cathodes and sent on for further fabrication or use as an alloying element.

The *Mond process*, although simpler, will not work for all ores of nickel. In this process carbon monoxide gas is fed over the heated ores to convert the nickel to nickel carbonyl [$Ni(CO)_4$]. Nickel carbonyl is a volatile compound which sublimes (goes directly from a solid to a gas) at 973°C (1783°F) and decomposes into metallic nickel dust and carbon monoxide shortly thereafter:

$$Ni(CO)_4 \rightarrow Ni + 4CO$$

Many ores, including nickel, copper, lead, tin, and others, are found in association with other minerals. When refining these ores the slags are often treated and re-refined to recover the other elements. The slags and "muds" from the copper and nickel refining operations are further treated to recover gold, silver, selenium, tellurium, cobalt, platinum, palladium, rhodium, ruthenium, iridium, and others as the economy dictates. Sometimes the metals are in the slags but cost more to recover than can be gained by their sale.

Nickel has many properties that make it a valuable industrial metal. It is critical to many of today's alloys. In World War II the nickel in the five-cent coins was replaced by silver in order to free the nickel for alloys used in armaments.*

*The 1943, 1944, and 1945 nickels were composed of 56% copper, 35% silver, and 9% manganese.

FIGURE 7-17 Properties of nickel.

Property	Value
Density	8.84 g/cm³
Melting point	2647°F (1453°C)
Tensile strength	
Hard	115,000 lb/in²
Annealed	70,000 lb/in²
Yield strength	
Hard	90,000 lb/in²
Annealed	20,000 lb/in²
Modulus of elasticity	32×10^6 lb/in²
Maximum percent elongation	5–45%
Electrical resistivity	8.54×10^{-6} Ω/cm³
Thermal conductivity	53 Btu/hr/ft²/°F
Coefficient of thermal expansion	7.4×10^{-6} in./in./°F
Heat capacity	0.11 Btu/lb/°F
Tensile strength/density	360,000 in.

Nickel is hard, relatively noncorroding, magnetic up to 680°F, and takes a high polish. Its density is 8.84 g/cm³. It has a tensile strength of 70,000 lb/in². but can be heat treated to obtain a tensile strength up to 115,000 lb/in². In its hardened condition, it has a modulus of elasticity slightly higher than steel, 32×10^6 lb/in². Its electrical conductivity is only 16% that of copper, so it is not used as an electrical conductor. It is alloyed with other metals to make electrical resistance wires and electrical heating elements. Its tensile strength-to-density ratio is a maximum of 360,000. Other properties of nickel are listed in Figure 7-17.

The uses of nickel are varied. Since it has such good resistance to corrosion and takes a good polish, it is used to electroplate parts for decorative purposes and protection. About 10% of all nickel produced is used in electroplating. Nickel does not stick or adhere well when electroplated on steel. It is therefore necessary to copper plate the steel before it is nickel plated to get a stable finish. The nickel plating is sometimes covered with a thin coating of chrome plating to give the ultimate in shine to the part. The electroplating is sometimes done with tin–nickel (65% tin and 35% nickel) using nickel chloride or nickel sulfate as the electrolyte solution.

The most common use of nickel is in alloying with other metals. Many companies produce these alloys under their own trade names. Very powerful permanent magnets are made from Alnico. Alumel and chromel are often used to make thermocouples used in temperature control devices. Several Hastelloy products are used for their high strength and corrosion resistance. Monel is a nickel–copper product of the Canadian nickel refineries and comes directly from the matte furnaces. Nichrome is a resistance wire used in heating elements. Nickel is now used in the Nicad battery. This electric cell uses a nickel hydroxide [$Ni(OH)_2$] positive plate and a cadmium negative plate with an electrolyte potassium hydroxide (KOH). The advantage of this electric cell is that it is rechargeable. Nickel–silver, better known as German silver, is an alloy of copper, zinc, nickel, and cobalt used in coinage and jewelry. Figure 7-18 lists a few of the more common alloys and their compositions.

Superalloys

The term "superalloys" has been applied to some of the nickel alloys in as much as they can have tensile strengths around 200,000 lb/in². Beryllium–nickel alloys used in springs have 97.6% nickel, 1.9% beryllium, and 0.5% manganese. Several

FIGURE 7-18 Nickel alloys.

Alloy	Composition
Alnico I	20% nickel, 67% iron, 12% aluminum, 5% cobalt
Alnico VIII	15% nickel, 34% iron, 35% cobalt, 7% aluminum, 5% titanium, 4% copper
Alumel	94% nickel, 2.5% manganese, 0.5% iron
Beryllium–nickel	97.6% nickel, 1.9% beryllium, 0.5% manganese
Chromel A	80% nickel, 20% chromium
Chromel C	60% nickel, 16% chromium, 24% iron
Chromel D	35% nickel, 18.5% chromium, 46.5% iron
Duramet	23% nickel, 20% chromium, 3% iron
Duranickel	93.7% nickel, 4.4% aluminum
Hastelloy A	60% nickel, 20% molybdenum, 20% iron
Hastelloy B	64% nickel, 28% molybdenum, 5% iron, 1% chromium, 1% manganese
Hastelloy C	55% nickel, 17% molybdenum, 6% iron, 4% tungsten, 1% manganese
Hastelloy D	85% nickel, 4% copper, 9% silicon, 1% manganese, 1% iron
Hastelloy G	20% nickel, 20% chromium, 7% iron, with niobium and tantalum
Hastelloy X	45% nickel, 22% chromium, 24% iron, 9% molybdenum
Inconel 600	15.5% nickel, 8% chromium, 76% iron
Inconel X	72% nickel, 15% chromium, 5% iron, 2.5% titanium, with niobium, copper, and aluminum
H Monel	63% nickel, 31% copper
K Monel	66% nickel, 29% copper, 2.75% aluminum, with iron, silicon, manganese, and titanium
R Monel	67% nickel, 30% copper
Z Monel	98% nickel, 2% copper
Nichrome S	25% nickel, 17% chromium, 2.5% silicon
Nichrome V	80% nickel, 20% chromium
Nickel silver	57% copper, 25% zinc, 15% nickel, 3% cobalt

of these alloys have been trademarked by companies. Hastelloy is a registered trademark of the Union Carbide Corporation. The name Inconel belongs to the International Nickel Company. Hoskins Manufacturing Company uses the trade mark Chromel. Durimet is the property of the Duriron Company.

CHROMIUM

Chromos is the Greek work for "color." The compounds of chromium have many brilliant colors, so it is aptly named. Chromic oxide (Cr_2O_3) is dark green. Potassium chromate ($KCrO_4$) is bright yellow, potassium dichromate ($K_2Cr_2O_7$) is orange, chromium trioxide (CrO_3) is red, and lead chromate ($PbCrO_4$) is yellow. Chromium was discovered in 1797 by Louis Nicolas Vauguelin, a professor of chemistry at the College of France.

Chromium is the third hardest element, with a Mohs hardness of 9. Only

FIGURE 7-19 Properties of chromium.

Property	Value
Density	7.2 g/cm³ (450 lb/ft³)
Melting point	1890°C (3434°F)
Modulus of elasticity	36×10^6 lb/in²
Coefficient of thermal expansion	3.4×10^{-6} in./in./°F
Heat capacity	0.11 Btu/lb/°F

boron and diamond are harder. It is extremely resistant to corrosion even in the presence of nitric acid. It can be dissolved in hydrochloric acid and sulfuric acid. The melting point of chromium is 3434°F (1890°C). The density of chromium is 7.2 g/cm³ or 450 lb/ft³. Figure 7-19 summarizes the properties of chromium metal.

The primary ore of chromium is chromite ($FeOCr_2O_3$), which is found in Rhodesia, Turkey, the Soviet Union, Iran, Albania, Finland, Malagasy, and the Philippines. Some low grades of chromite are found in California, Maryland, and Oregon in the United States. The reduction process is theoretically simple but expensive. The chromite is crushed and powdered. It is then reacted with powdered aluminum metal, which releases a mixture of iron and chromium. The chemical reaction is

$$(FeOCr_2O_3) + Al \rightarrow Cr + Fe + Al_2O_3$$

The chromium can be purified by electroplating, but most of it is used in alloys, chrome-containing chemicals, or electroplating baths. Pure chrome is not always needed. Chromic oxide and chromite ore are also used in making refractory or firebricks.

There are some problems in chrome plating steel. Chromium will not stick to steel very well. Chromium will adhere to nickel. A process called triple plating is used. The steel is first stripped or washed in a degreasing compound to ensure that it is clean. It is then pickled or etched with nitric acid to erode the grain boundaries and give the plated metal a roughened surface to hold onto. The steel then has a very thin layer of copper plated on the surface. After washing, the copper plate is then covered with a thin coating of nickel plating. This is followed by washing again, then a plate of chromium is deposited on the metal. A chromium plate as thin as 0.0002 in. provides a decorative, shiny surface but coatings up to 0.05 in. are used to provide wear resistance. Chromic acid, which is chromic oxide dissolved in water, is used as the plating solution. See Figure 4-61 for the electroplating apparatus.

The use of chromium as an alloying element has already been discussed in the section on stainless steels and nickel alloys. Chromium is also alloyed with copper. Chromium bronze, which contains 1% chromium, 1% iron, 2 to 10% tin, and the remainder copper, is used for bearings.

TITANIUM

Named after the Greek gods, titanium is one of the most interesting and useful metals now being produced. Discovered in 1791 by the English clergyman William Gregor it remained for the German chemist Martin Heinrich Klaproth to name it in 1795. Titanium is the ninth most abundant element on the earth and is found in the minerals rutile (TiO_2), ilmenite ($FeTiO_3$) and sphene ($CuTiSiO_4$) as well as many iron ores and other minerals. The *Apollo* missions to the moon brought back samples of rock containing from 7 to 12% titanium. The major source of titanium ore in the United States, near Jacksonville, Florida, produces about 5% of the world's supply. Titanium is also mined in Australia, Canada, Sierra Leone on the west coast of Africa, and the Soviet Union.

The metal was not purified until 1910 by Hunter, and only since World War II has it been produced in quantity. The refining process is not simple. The finely crushed and powdered ores are run through electrostatic and gravitational separators to remove the titanium-bearing crystals. The titanium oxides are then treated with chlorine gas to convert the oxides to titanium tetrachloride:

$$TiO_2 + 2Cl_2 \rightarrow TiCl_4 + O_2$$

Titanium tetrachloride is itself an interesting compound. When combined with moist air it forms a dense white cloud of smoke and is used in skywriting and in tracer bullets. A stainless steel tank, or "bomb" as it is called, is filled with magnesium or sodium metal and welded shut. The titanium tetrachloride is then run through the tank at about 800°F for 2 days. The titanium metal collects as a sponge metal on the steel. The reaction for this process is

$$TiCl_4 + Mg \text{ or } Na \rightarrow Ti + MgCl_2 \text{ or } NaCl$$

The tank is then cut open and the titanium chipped out of the tank. The sponge is then crushed and remelted in a vacuum arc furnace and cast into ingots.

Titanium metal is as strong as steel having a tensile strength in excess of 95,000 lb/in^2. Its density is 4.54 g/cm^3 or 283 lb/ft^3, roughly half that of steel. This gives a tensile strength-to-density ratio of 1,220,000. It is the only metal that will burn in nitrogen as well as oxygen. Titanium has a hexagonal close-packed crystal structure at low temperatures but transforms to a cubic system at 880°C (1616°F). A thin transparent oxide coating will form on the surface of titanium, making it very resistant to attack of acids and corrosion. The other properties of titanium are listed in Figure 7-20.

Titanium has some exotic uses. Its light weight, high strength, and ability to withstand high temperatures makes it the metal of choice for high-speed aircraft. The SR-71 *Blackbird*, which can fly at 2195 miles per hour, over three times the speed of sound, depends on a titanium skin to withstand the heat. The supersonic transport (SST) *Concorde* uses over 600,000 lb of titanium metal and alloys. Its ability to withstand corrosion combined with the fact that it is nonmagnetic makes

FIGURE 7-20   Properties of titanium.

Property	Value
Density	4.54 g/cm^3 (283 lb/ft^3)
Melting point	1675°C (3100°F)
Tensile strength	95,000 lb/in^2
Yield strength	80,000 lb/in^2
Modulus of elasticity	14.8 × 10^6 lb/in^2
Maximum percent elongation	0%
Poisson's ratio	0.3
Thermal conductivity	9 Btu/ft^2/°F/hr
Coefficient of thermal expansion	5.2 × 10^{-6} in./in./°F
Heat capacity	0.13 Btu/lb/°F
Tensile strength/density	1.22 × 10^6 in.

it ideal for submarines and other marine applications. The nonmagnetic quality helps guard submarines against detection by magnetic anomaly detectors (MADs). It is ironic that the titanium-hulled deep-diving submarine *Alvin* was used to explore the sunken ship *Titanic* on the floor of the ocean at a depth of 12,500 ft. (Both titanium and the *Titanic* were named after the Greek gods, the Titans). The *Alvin* has gone even deeper in exploring ocean trenches.

Titanium is also used in the medical profession. Surgeons prefer titanium instruments over stainless steel because of their light weight. Titanium replacement joints from the ankle to the shoulder are in common usage. Like aluminum, titanium will also form a thin oxide layer on its surface, which adheres and protects it against further corrosion. Jewelers have found that by connecting the metal to an electric current, this surface can be colored by painting the surface with an electrolyte such as ammonium sulfate. The colors obtained depend on the voltage and the depth of the plating. Everything from watches, pens, and cameras to sunglasses is now being made from titanium.

Titanium is also used in combined form and as alloys. Titanium oxide is used to produce a brilliant white paint. Steel alloys containing titanium offer high strength at high temperatures. Many parts of jet engines are made of titanium steels and titanium metal. About 75% of today's use of titanium is in the aircraft industry.

One of the most interesting alloys of titanium is nitinol, a mixture of titanium and nickel. Nitinol, developed by the U.S. Naval Ordnance Laboratory, is an intermetallic compound (NiTi) having 54.5% nickel and 45.5% titanium. It has a tensile strength of 110,000 lb/in^2. The remarkable property of nitinol is that it can be bent into a particular shape, then heated to about 1000°F and it will retain that shape. If deformed from that shape and slightly heated it will try to return to the desired shape with a pressure of up to 100,000 lb/in^2. This is the "metal with a memory," for which new uses are being found daily.

The manufacture of titanium parts is not easy. Most of the parts must be made by the "lost wax" technique. The lost wax process is thousands of years old, yet it is used on one of the newest materials. In this process, an exact model of the part to be produced is made from wax. The wax can be formed directly by carving or it can be poured into a mold. The wax pattern is then placed in a flask and a refractory material such as crystobalite (plaster of paris and clay) or ceramics are poured or sprayed around it. When the refractory material has set, the flask is placed in an oven and the wax is burned out, leaving a cavity in the exact shape of the wax. Molten metal can then be poured or forced into the cavity. Centrifugal and vacuum casting methods are often used to drive the molten metal into the hole. In the case of titanium, the casting process must be carried out in a vacuum and is done in the furnace by robotics or remote control. Once cooled, the mold is stripped away from the metal and the part is ready for further processing.

Titanium tends to alloy easily with other elements, especially when heated. Titanium parts have been damaged merely by coming in contact with cadmium or other elements. Titanium tends to adhere to cutting tools and alloy itself with the tool steel. The combination of the cutting steel with titanium often embrittles the cutting tool making it worthless. The cutting of titanium at cryogenic temperatures produced by liquid nitrogen has been tried to improve the machinability. Pure titanium parts are often very brittle. This property usually requires that it be alloyed with other materials. Further, all aircraft and other critical parts must be X-rayed for flaws. It is little wonder that titanium parts are very expensive, but manufacturers are willing to use it to get the desirable properties of what has been called today's "miracle metal."

WHITE METALS

Lead

The low-melting elements antimony, bismuth, cadmium, lead, tin, and zinc are called the *white metals*. Lead has been known since the days of the Pharaohs and is mentioned in the Old Testament Books of Exodus, Numbers, Job, and Ezekiel.* It was used by the ancient Romans for water pipes. The lead pipes in the Roman baths in Bath, England, are still in use. The Latin name for lead is *plumbum*. The men who worked with plumbum in making water pipes became known as "plumbers."

There are several minerals that contain lead, including anglesite ($PbSO_4O$), cerrusite ($PbCO_3$), minum (Pb_3O_4), and galena (PbS). Galena is the most common ore used. The ores are easily refined by roasting to convert the sulfide to oxides, then smelting with carbon to produce lead and carbon dioxide. The chemical reactions for this process are

$$2PbS + 3O_2 \rightarrow 2PbO + 2SO_2$$

$$2PbO + C \rightarrow 2Pb + CO_2$$

Lead is a very soft, heavy metal with a tensile strength of only 2600 lb/in². Its melting point is 621°F (327°C). It is a poor conductor of electricity, having only 8.1% of the electrical conductivity of copper and 7.7% that of silver. Lead has been found to be toxic to all animal life. The dosage is cumulative and it tends to build up in the system. Lead poisoning has been partially blamed for the collapse of the Roman empire since it was used in the piping for their water supply. Other properties of lead are shown in Figure 7-21.

FIGURE 7-21   Properties of lead.

Property	Value
Density	11.34 g/cm³ (708 lb/ft³)
Melting point	326°C (621°F)
Tensile strength	2600 lb/in²
Yield point	1300 lb/in²
Modulus of elasticity	2,000,000 lb/in²
Poisson's ratio	0.4
Maximum percent elongation	20–50%
Electrical resistivity	22×10^{-6} Ω/cm³
Thermal conductivity	20.3 Btu/ft²/°F/hr
Coefficient of thermal expansion	18.3×10^{-6} in./in./°F
Heat capacity	0.031 Btu/lb/°F
Tensile strength/density	63,500 in.

The greatest use of lead is in storage batteries. Battery-plate lead contains 7 to 12% antimony, with small amounts of tin, arsenic, and copper. The lead recovered from these batteries after use is remelted and marketed as secondary lead. Lead was used for many years in paints as white, yellow, and red pigments. Since lead has been found to be hazardous to the health of animals and birds as well as people, it has been phased out in the United States for these purposes. Similarly, the use of lead in the compound lead tetraethyl [$Pb(C_2H_5)_4$] as an additive for

*Exodus 15:10, Numbers 31:22, Job 19:24, and Ezekiel 27:12.

FIGURE 7-22   Low melting alloys of lead.

Melting Temperature		Composition (%)				
°F	°C	Bismuth	Lead	Tin	Cadmium	Other
478	248[a]	—	82	—	18	—
361	187[a]	—	38	62	—	—
350	176[a]	—	—	68	32	—
291	144[a]	60	—	—	40	—
281	138[a]	57	—	43	—	—
255	124[a]	55	45	—	—	—
205	96 (Rose's metal)	50	28	22	—	—
203	95[a]	52	32	16	—	—
160	71 (Wood's metal)	50	25	12.5	12.5	—
158	70 (Lipowitz metal)	50	27	13	10	—
136	58[a]	49	18	12	—	21 In
117	47[a]	45	23	8	5	19 In

[a]Eutectic alloy.

gasoline (Ethyl gasoline) is being curtailed by the Environmental Protection Agency. Since lead is one of the densest metals available, it makes excellent shielding for X-rays and other radioactive materials.

Lead is also used in many alloys. Added to bronze and brass it gives them good machinability. It was mixed with antimony for type metal in the days of the mechanically set type for newspapers and books. Tin–lead alloys are used in solder for both plumbing and electrical applications. Babbitt, from which crankshaft bearings is made, is an alloy of lead composed of 80% lead, 15% antimony, and 5% tin.

Several very low melting alloys of lead are used as thermal fuses in sprinkling systems and hot water heaters. Many compositions of these low-melting alloys have been given names such as rose metal, Wood's alloy, or Lipowitz metal. Many of these named alloys have differing compositions with the different producers; there are many "Wood's metals," and so on. The compositions of a few of these low-melting alloys are shown in Figure 7-22. The last two mixtures in Figure 7-22 melt at temperatures that could be held in the hand.

Tin

A second of the white metals that has long been known is *tin*. Known as "stannum" by the Romans, tin is the major element in pewter. Pewter, used mainly for decorative purposes, once contained enough lead to cause it to darken with age. Modern pewter, also known as Britannia metal, contains 91% tin, 7% antimony, and 2% copper. The predominant use of tin is still as an alloying element with other metals.

The principal ore of tin is cassiterite (SnO_2) found in the most part in Malaya, Thailand, Bolivia, Indonesia, Nigeria, and the Republic of Congo. Some poor grades of tin ore are found in California and Alaska. The oxide ore is reduced to tin by smelting in a reverberatory furnace in the presence of coke (carbon). The reaction is

$$SnO_2 + C \rightarrow Sn + CO_2$$

Pure tin can take several different crystal structures. Below 13.2°C (56°F) tin is in the alpha (α) or "gray tin" form, which has a cubic structure. Above 56°F the

FIGURE 7-23   Properties of tin.

Property	Value
Density	7.3 g/cm³ (455 lb/ft³)
Melting point	232°C (449°F)
Tensile strength	3100 lb/in²
Modulus of elasticity	6×10^6 lb/in²
Maximum percent elongation	55%
Poisson's ratio	0.33
Electrical resistivity	11.6×10^{-6} Ω/cm³
Tensile strength/density	12,000 in.

tetragonal "white tin" or beta (β) form exists. Alpha tin has a density of 5.8 crystals into a gray powder can actually be heard at low temperatures (called the g/cm³, while the beta structure has a density of 7.3 g/cm³. The gray form of tin is very brittle, which causes it to crack very easily when bent. The cracking of the "tin cry") and is referred to as the "tin pest" by those in the industry. This property keeps pure tin from being a useful metal at low temperatures. Pure beta tin takes a good shine, is rather weak in tensile strength, and has a low modulus of elasticity. The properties of pure tin are shown in Figure 7-23.

Tin is used as a plating for steel, in solders when mixed with lead, and in tin babbitts when mixed with antimony and copper. Several formulas for tin babbitts exist which contain from 83 to 91% tin, 4 to 8% antimony, and 3 to 8% copper. They are used for bearings and die castings. A tin foil is made which has 8% zinc and 92% tin. Stannic oxide is also used as a glaze for ceramics. Some toothpastes use stannous fluoride to provide a fluoride treatment for teeth.

Zinc

Zinc is one of the white metals that finds its major use in coating steel plates in a process known as *galvanizing*. Since zinc is more active than iron in the activity series, it will be attacked by acids and *oxidizing agents* easier than the iron. The zinc acts as an anode in the electrolytic cell which is sacrificed to save the iron. As long as there is any zinc on the iron, it will go into solution, and in doing so, gives up electrons to the iron which protects it from corrosion. This is known as *cathodic protection* of the steel. Ocean-going ships use this principle to protect the hulls. The bronze bushings around the propeller shafts are in contact with the steel of the shaft and hull. Since the iron in the steel is a more active metal than the copper in the bronze, an electrolytic cell is formed with the seawater as the electrolyte. This action would cause the hull to corrode very rapidly. Large zinc bars are bolted to the hull next to the bronze bushings. Zincs can also be strapped around the shafts. The zinc anode is dissolved quickly and they must be replaced every few months, but the hull is protected.

A second use of zinc is in the production of dry cells for flashlight and other batteries. Zinc also finds considerable use in brasses and as an alloying element for other metals. Zinc oxide is also used in paints, glass, matches, medicines, and dental cements. It can be die cast and machined easily when alloyed with slight amounts of aluminum and magnesium.

Zinc is found in several minerals, including sphalerite (zincblende) (ZnS), smithsonite ($ZnCO_3$), gahnite ($ZnAl_2O_4$), and wurtzite (ZnS), but it is refined chiefly from the minerals franklinite ($ZnFe_2O_4$) and zincite (ZnO). Zinc ores are found throughout the world. Australia, the United States, the Soviet Union, South

FIGURE 7-24   Properties of zinc.

Property	Value
Density	7.13 g/cm³ (445 lb/ft³)
Melting point	419°C (787°F)
Tensile strength	16,000 lb/in²
Modulus of elasticity	12×10^6 lb/in²
Maximum percent elongation	5–65%
Thermal conductivity	784 Btu/ft²/in./hr/°F
Coefficient thermal expansion	$9–22 \times 10^{-6}$ in./in./°F
Heat capacity	0.092 Btu/lb/°F

America, and South Africa all produce zinc. The zinc is extracted from the ore by roasting the sulfides to oxides, then heating the zinc oxide to 1200°C (2200°F) in the presence of coke (carbon). Since zinc boils at 907°C (1635°F), the zinc emerges as a vapor that must be condensed. Zinc can also be extracted from the ore by leaching the oxide with sulfuric acid, then using electrolysis to produce the pure metal. Zinc has a hexagonal close-packed crystal structure. The major properties of zinc are shown in Figure 7-24.

Antimony, Bismuth, and Cadmium

Antimony is found in lead ores and minerals such as senarmonite (Sb_2O_3), cervantite (Sb_2O_4), and kermensite ($Sb_2O_4 \cdot H_{20}$), but the primary ore is stibnite (Sb_2S_3). Stibnite is found in China, Mexico, Japan, West Germany, Boliva, Alaska, and the western United States. Antimony has a rhombohedral crystal structure.

Bismuth is found in the ores bismuthinite (Bi_2S_3) and bismutite ($Bi_2O_3CO_3$) found in Bolivia, Peru, Europe, Australia, and the United States. Bismuth is also a rhombohedral crystal structure. Bismuth is *diamagnetic* and is one of the few metals that expands upon solidifying. Bismuth also has the lowest thermal conductivity of any metal except mercury.

Cadmium's only commercial mineral is greenockite (CdS), found in Missouri. It is found in other minerals, such as sphalarite. One unique use of cadmium is as a neutron absorber in the control rods of nuclear power plants. The cadmium will absorb the neutrons emitted by the splitting of an atom to prevent a chain reaction from getting to the explosive stage. The properties of antimony, bismuth, and cadmium are summarized in Figure 7-25.

FIGURE 7-25   Properties of antimony, bismuth, and cadmium.

Property	Value		
	Antimony	Bismuth	Cadmium
Density (g/cm³)	6.62	9.8	8.6
Melting point (°F)	1167	520	610
Tensile strength (lb/in²)	1560	—	10,300
Modulus of elasticity (lb/in²)	11.3×10^6	4.6×10^6	5×10^6
Maximum percent elongation	—	—	50%
Thermal conductivity (Btu)	131	58	639
Coefficient of thermal expansion (in./in./°F)	8.5×10^{-6}	7.4×10^{-6}	16.6×10^{-6}
Heat capacity (Btu/lb/°F)	0.049	0.034	0.055

PRECIOUS METALS

Gold, silver, and platinum have traditionally been known as the precious metals because of their value and their use in jewelry and coinage. For most of recorded history, the value of the coins was the worth of the metal in the coins. A $20 gold piece had $20 worth of gold in it. Now the weight of gold in the coin is worth far more than the value of the coin. For this reason, coins are made of the cheaper metals. Now the noble metals of iridium, rhodium, osmium, ruthenium, and palladium are being added to this group.

Gold

Gold has long been called the "king of metals." It is found as native metal in the form of nuggets, gold dust, and in quartz rock. Although gold is sought the world over, about two-thirds of today's production comes from South Africa. Gold is also found in seawater in a concentration of about 0.1 to 2 mg per ton of water. Before starting to seek a fortune by extracting gold from seawater, one should consider that this amounts to only 2.7×10^{-7} oz per gallon and that over 3.7 million gallons or about 500,000 ft^3 of seawater would have to be processed to recover an ounce of gold.

The properties of gold that make it a valuable industrial metal is its corrosion resistance, electrical conductivity, and malleability. Gold is not attacked by air but can be dissolved by *aqua regia* or hot chlorine gas. Aqua regia, so named because of its ability to attack the "king of metals," is a solution of one-fourth nitric acid and three-fourths hydrochloric acid. Gold will also form an *amalgam* with mercury and can be dissolved in cyanide solutions. Reacting it with mercury or cyanide have been the traditional methods of extracting it from the gold-bearing quartz ores.

Gold can form a protective coating on a metal by being electroplated from the gold chlorine solution. Gold is the most malleable and ductile of metals. One ounce of pure gold can be beaten out into a sheet of 300 ft^2. Gold is a heavy metal having a density of 19.32 g/cm^3 or 1205 lb/ft^3 (0.697 lb/in^3). Gold is too soft to be used in the pure form, so it is alloyed with other metals. Twenty-four-carat gold is pure gold. Twelve-carat gold would have 50% gold and 50% other metals. Copper, nickel, and platinum are common alloying elements. Gold is face-centered cubic in crystal structure and has a melting point of 1064°C or 1947°F.

Gold has many practical uses. Besides being used in jewelry and coinage, it is used in dentistry for caps, crowns and fillings of teeth. One dental alloy contains 64 to 70% gold, 0 to 5% platinum, 0 to 5% palladium, 14 to 25% silver, 11 to 18% copper, 0 to 3% nickel, and 0 to 1% zinc. The properties of gold are listed in Figure 7-26.

Silver

Named *argentum* in Latin, from which the country Argentina is named, silver has been known since ancient times. Silver occurs as a free metal in nature and in such ores as argentite (Ag_2S) and horn silver ($AgCl$). It is also found in copper, lead, zinc, nickel, and other ores. Silver is the best electrical and heat conductor of all the metals. It is face-centered cubic and is very malleable and ductile. Silver was used in coins for centuries. The British "sterling" was composed of 92.5% silver and 7.5% copper. Until 1964 silver was used in the coins of the United States. Since the price of the silver in the coins has exceeded the face value of the coins,

FIGURE 7-26   Properties of gold.

Property	Value
Density	19.32 g/cm³ (1205 lb/ft³)
Melting point	1064°C (1947°F)
Tensile strength	19,000 lb/in²
Modulus of elasticity	11×10^6 lb/in²

it was replaced by nickel silver and copper. At one time the price of silver rose to more than $35 per troy ounce.

Besides being used to silver plate other metals, as a conductor of electricity and in jewelry, silver is also used to make the light-sensitive compounds in photographic films and papers. About 30% of the annual production of silver is used for photographic purposes. Silver nitrates, chlorides, bromides, and other salts are light sensitive and decompose rapidly when exposed to light. The black areas on black-and-white photographic negatives and prints are just a layer of finely deposited silver. Silver is also used in making mirrors. Silver iodide is used to seed clouds to cause rain. Silver is also used in making photochromic (light-sensitive) glasses which darken when exposed to ultraviolet rays or sunlight. Photochromic glass is discussed in Chapter 8.

Other uses of silver include brazing alloys and silver–cadmium batteries. Certain salts of silver are known for their germ-killing properties, so they are used in salves, ointments, and other medicinal preparations. Silver fulminate ($Ag_2C_2N_2O_2$) is an explosive. Some of the properties of silver are shown in Figure 7-27.

FIGURE 7-27   Properties of silver.

Property	Value
Density	10.5 g/cm³
Melting point	960°C (1761°F)
Tensile strength	18,200 lb/in²
Yield strength	7900 lb/in²
Modulus of elasticity	10.3×10^6 lb/in²
Electrical resistivity	1.59 μΩ-cm²/cm

Platinum

Although used by the Indians of South America for some time before the arrival of the explorers, platinum was discovered in Columbia for the Europeans by Antonio de Ulloa in 1735. Platinum is found as the free metal in nature and in the mineral sperrylite ($PtAs_2$). Besides being found in South America, platinum is found in the Soviet Union and a few western American states. Sperrylite occurs with the nickel-bearing ores of Ontario. Platinum is also accompanied by the other noble elements iridium, osmium, palladium, ruthenium, and rhodium in many ores. Platinum is malleable and does not readily corrode, but like gold, it will dissolve in aqua regia.

Besides being used for corrosion-resistant coatings, platinum is used as a catalyst for many chemical reactions and in the petroleum industry. High-resistance electrical wires used for heating of furnaces are sometimes alloys of platinum. Finely ground platinum will give off a considerable amount of heat in the presence

FIGURE 7-28   Properties of platinum.

Property	Value
Density	21.45 g/cm^3
Melting point	1773°C (3224°F)
Tensile strength	17,000–19,000 lb/in^2
Modulus of elasticity	21.3 × 10^6 lb/in^2
Poisson's ratio	0.39

of methyl alcohol and other hydrocarbons, which makes it useful in hand warmers. It is also used in the catalytic converters used with internal combustion engines to convert unburned hydrocarbons and carbon monoxide to carbon dioxide and water. Because of its corrosion resistance, platinum is used as a coating for electrical contacts and for anodes for some electroplating operations. Considerable platinum is used in laboratory equipment, medical instruments, and fine jewelry. Platinum is a face-centered cubic crystal structure.

The main drawback of platinum as an industrial metal is its cost: It is more expensive than gold. Figure 7-28 gives some of the properties of platinum.

Iridium, Osmium, Palladium, Rhodium, and Ruthenium

The metals iridium, osmium, palladium, rhodium, and ruthenium have many things in common. All are found in the same ores as platinum and nickel. All are high-melting metals with good corrosion resistance and electrical and thermal conductivity. They all are used primarily as alloying agents with other metals.

Osmium and iridium were discovered in 1803 by Smithson Tennant, an English chemistry professor at Cambridge University. Palladium and rhodium were discovered in 1804 and are credited to the British chemist William Hyde Wollaston. Although a German chemist, Gottfried Wilhelm Osann, named the element ruthenium in 1828, the credit for its isolation is given to K. K. Klaus in 1844.

Some of the unique properties of these metals include the density of iridium and osmium. At 22.6 g/cm^3, which would equal 1410 lb for a cube 1 ft on a side, these elements have the highest density of all metals and are roughly twice as heavy as an equal volume of lead. Osmium has the highest melting point of any metal at 2700°C or 4890°F. Osmium has the highest modulus of elasticity of any element at 81 × 10^6 lb/in^2. The modulus of elasticity for iridium, osmium, rhodium, and ruthenium all exceed that of steel. Like platinum, palladium is an excellent catalyst for many chemical reactions. Ruthenium is added to platinum and palladium as a hardener for the alloy. It also becomes a superconductor at about 11 K. Figure 7-29 relates some of the properties of these elements.

FIGURE 7-29   Properties of iridium, osmium, palladium, rhodium, and ruthenium.

Property	Value				
	Iridium	Osmium	Palladium	Rhodium	Ruthenium
Density (g/cm^3)	22.0	22.0	12.02	12.44	12.2
Melting point (°C)	2454	2700	1554	1966	2500
Melting point (°F)	4450	4890	2829	3571	4550
Tensile strength (lb/in^2)	—	—	21,000	73,000	—
Modulus of elasticity (lb/in^2 × 10^6)	76	81	16	42.5	60

PROBLEM SET 7-2

Use the following list of metals to answer Problems 1 through 10:
    Copper
    Aluminum
    Magnesium
    Nickel
    Titanium
    Lead
    Tin
    Zinc

1. Which has the highest tensile strength?
2. Which has the highest density?
3. Which has the lowest density?
4. Which is the best thermal conductor?
5. Which is the best electrical conductor?
6. Which has the highest melting point?
7. Which has the best corrosion resistance?
8. Which is the most chemically active?
9. Which has the highest strength-to-density ratio?
10. Which would you select to give a nice-looking electroplated surface? Why?
11. Malachite $[Cu_2(OH)_2(CO_3)]$ is an ore of copper. What is its molecular weight?
12. What is the theoretical percent copper in malachite? (Use the formula from Problem 11.)
13. Which oxide would make the better firebrick, magnesium oxide or nickel oxide? Why?
14. What is the commercial source of magnesium?
15. How much silver is there in nickel silver?
16. Would placing a piece of lead sheet around a steel shaft of a power boat protect the steel from corrosion?
17. If copper and lead were tied together in an electrolytic solution, which would go into solution?
18. Given 100 lb of millerite (NiS), how many pounds of nickel could theoretically be obtained?
19. How many pounds of chromium are theoretically obtainable from a ton of chromite $(FeOCr_2O_3)$?
20. Rutile (TiO_2) is an ore of titanium. If 800 lb of titanium is obtained from a ton of rutile, what percent of the titanium in the ore is recovered?

Glass and Ceramics

GLASS

No one knows exactly when or how glass was discovered. Pliny, the Roman historian, credits the Phoenician sailors with discovering glass being formed around their campfires on the sandy beaches nearly 5000 years before the birth of Christ. Although this story is probably in error, glass beads have been found that date as far back as 4500 B.C. Mesopotamian glass bottles believed to have been made around 4000 B.C., and Egyptian glass artifacts from as early as 3000 B.C. are in museums. Most of the earliest glass was simply decorative glaze used on pottery. The invention of the blowing iron or pipe used for glass blowing represents a major technological innovation and is credited to the Babylonians about 200 years B.C. Fully transparent glasses date from about the time of Christ.

The production of flat glass for windows came much later. In theory, it would seem that flat glass could be made by pouring it out onto a flat surface. The problem with this technique is that in the large glass slabs the finish is so rough that light passing through it is severely distorted making the glass useless for windows. Pieces of 1-cm-thick colored glass slabs were made in Roman times and used in decorative mosaic or stained glass windows. Another early method of making flat glass that was used through the early nineteenth century consisted of blowing a large spherical globe of glass about 20 in. in diameter with a blowpipe. The glass was reheated and then attached to a long iron rod called a pontil (pronounced "punty" by the workers). The reheated sphere was then flattened or "marveled" against an iron slab. A small hole was punched through the soft glass disk and the pontil pierced through the center. After reheating the glass, the pontil was twirled and the centrifugal force would cause the glass to form a disk nearly a yard in diameter. The flattened sphere was spun into a disk in much the same way that a pizza is spun to make a thin disk of the dough. The glass was kept spinning until it was cool. Glass plates made in this manner were thicker in the middle than on the edges so that only the outer pieces could be cut for use in window panes. This accounts for the fact that most of the windows in the buildings built in the sixteenth and seventeenth centuries were made up of many small panes.

Another technique of making flat glass, developed in about the year 1688 in France, consisted of pouring molten glass onto flat iron tables and rolling it flat with iron rollers. This method is similar to rolling out biscuit dough with a rolling pin. A rough surface was produced, but this technique, which required hand polishing, was a very lengthy and laborious task. A technique similar to this is still used for plate glass today, but the rolling and polishing is now done with machines. Glass can be cast in continuous sheets in rolling mills similar to the continuous casting process in steel mills. After casting and rolling it must still be polished to prevent the rough uneven surfaces.

The definition of *glass* is difficult. Almost every definition has its exceptions. One dictionary definition of glass is: "a substance formed by the fusion of silica with two or more metallic oxides"; another is: "Glass is a hard substance, usually brittle and transparent, commonly composed chiefly of silicates and alkali fused at high temperatures." Glass is a liquid in that it does not have a definite crystal structure. It is often called an *amorphous* (without body) structure. Glass has no fixed melting point, nor does it break along slip planes as do solids. Most glasses are made from silica (sand or silicon dioxide), but a few, such as the rare earth glasses, have no silica. Not all glass is transparent. For purposes of definition let us say that glass is a superviscous liquid which is generally made from silica and usually contains one or more metallic oxides.

If glass is a liquid, it should flow. In fact, it does. If a glass rod is supported on both ends with a weight (not heavy enough to break the rod) applied in the middle, after a period of time (several months at least) the weight can be removed and the glass will not return to its original straight shape. Quartz is made of silica also, but it has a definite crystal structure. If a similar experiment is performed with a quartz rod, the quartz, being a true solid, would return to its original shape. The glass has flowed very slowly under the force of the weight and gravity. This is illustrated in Figure 8-1.

Glass does not have a melting point. Heat a glass rod to about 700°C (1300°F), and it will start to soften and can be bent. This temperature is called its softening point. At a higher temperature, it will bend even easier. The higher the temperature, the softer the glass gets. At high temperatures it can be poured as any liquid. A quartz rod, in contrast, will retain its rigidity to its melting point (1610°C or 2930°F), where it will turn into a liquid at a single temperature.

Any material that has a definite melting point must lose its heat of fusion when changing from a liquid to a solid. Further, the solidification process usually involves a change in volume. Quartz shows this change in volume upon solidifying, but glass does not. Figure 8-2 is a graph showing the change in volume of quartz and glass upon cooling. This additional evidence further proves that glass remains a liquid even though apparently quite rigid.

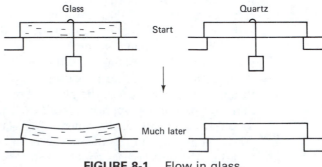

FIGURE 8-1   Flow in glass.

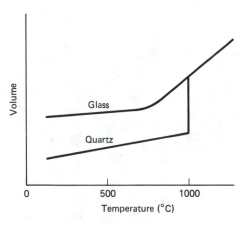

FIGURE 8-2 Change in volume on cooling for quartz and glass.

Sand, being silicon dioxide (SiO_2) with the quartz crystal structure (hexagonal), has a melting point of 2930°F. Early workers in glass had difficulty in obtaining a temperature high enough to melt the sand. They found that the addition of soda ash (Na_2CO_3) or lime (CaO from limestone) would cause the sand to fuse or melt at lower and readily attainable temperatures. The product was glass. We now know that it is these metallic oxides that keep the silica from forming a solid.

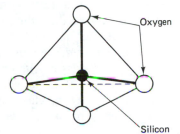

FIGURE 8-3 Silica tetrahedron.

Silica or silicon dioxide actually forms in small tetrahedra with a silicon atom at the center and four oxygen atoms at the corners (see Figure 8-3). These oxygen atoms are attached to silicons of other tetrahedra. In quartz, these tetrahedra go together in a definite arrangement as shown in Figure 8-4. The sodium or other metallic atoms in the structure prevent these tetrahedra from going into the quartz structure. The structure of glass is shown in Figure 8-5. The sodium atoms break up the regular structure of the silica tetrahedra, breaking some of the bonds and leaving a random structure. This random structure does not have even-length bonds,

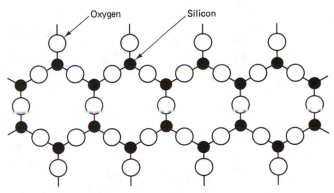

FIGURE 8-4 Quartz structure.

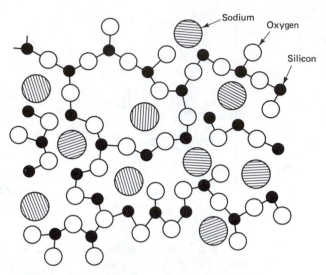

FIGURE 8-5   Glass structure.

which would all break at the same energy or temperature, therefore no fixed melting point.

Most glass is transparent or translucent. In transparent glass an image can be seen through it. In translucent glass the light comes through, but it is diffused and the image is blurred, as would be the case with frosted glass. When light enters a piece of glass, it is bent or *refracted*. The amount that it is refracted depends on the type of glass. Each type of glass has its own *index of refraction* or *refractive index*. The angle that a light ray enters the glass is its *angle of incidence*, labeled (θ_i) (see Figure 8-6). Upon entering the glass, the light ray will be bent toward the perpendicular. The angle the light ray makes with the perpendicular inside the glass is its *angle of refraction*, labeled θ_r, in Figure 8-6. The index of refraction (n)

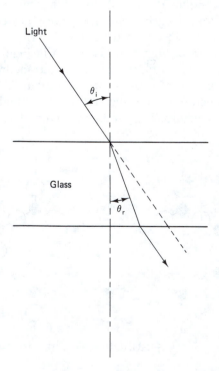

FIGURE 8-6   Index of refraction.

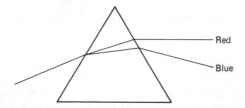

FIGURE 8-7   Effect of a prism on light.

is calculated by dividing the sine of the angle of incidence by the sine of the angle of refraction:

$$n = \frac{\sin \theta_i}{\sin \theta_r}$$

For example, if a light ray entered the glass at an angle of 15° from the perpendicular and was bent to an angle of 10° from the perpendicular, the index of refraction would be

$$n = \frac{\sin 15°}{\sin 10°} = \frac{0.2588}{0.1736} = 1.49$$

As the light ray leaves the glass it will be bent back away from the perpendicular to the surface it leaves. By shaping the surfaces of the glass, the light rays can be bent in the desired directions. This is the principle upon which lenses are made.

Not all frequencies of light are refracted the same amount. Blue light will be bent more toward the perpendicular than will red rays. White light contains all colors or wavelengths of light. By using a prism the white light can be broken into its component colors, as shown in Figure 8-7.

TYPES OF GLASS

There are well over 100,000 formulas for glass, but they can be divided into seven classes or categories. Each class has many variations. Only the average of each class will be presented here.

Soda-Lime Glass

The oldest glass is the common *soda-lime glass*. Soda-lime glass or soft glass contains 70% silica (SiO_2), 15% soda or sodium oxide (Na_2O), and 10% lime or calcium oxide (CaO). Soda-lime glass has a tensile strength of 10,000 lb/in^2, a compressive strength of 50,000 lb/in^2, and a modulus of elasticity of 10×10^6 lb/in^2. The other properties of soda-lime glass are summarized in Figure 8-8.

FIGURE 8-8   Properties of soda-lime glass.

Property	Value
Tensile strength	10,000 lb/in^2
Compressive strength	50,000 lb/in^2
Modulus of elasticity	10×10^6 lb/in^2
Density	2.3–2.6 g/cm^3 (143–156 lb/ft^3)
Softening point	700°C (1300°F)
Coefficient of thermal expansion	$40–60 \times 10^{-7}$ in./in./°F
Thermal conductivity	0.4–0.6 Btu/ft^2/ft/°F
Heat capacity	0.16–0.2 Btu/lb/°F
Index of refraction	1.51

FIGURE 8-9   Properties of borosilicate glass.

Property	Value
Modulus of elasticity	8.8×10^6 lb/in^2
Density	2.25 g/cm^3 (143 lb/ft^3)
Softening point	820°C (1508°F)
Coefficient of thermal expansion	18×10^{-7} in./in./°F
Thermal conductivity	0.21 Btu/ft^2/in./°F
Heat capacity	0.2 Btu/lb/°F
Refractive index	1.47

Soda-lime glasses are still the most common glass in use and the cheapest. Window panes, bottles, lamp bulbs, and a host of other items are made from this oldest form of glass. Window glass is manufactured in sheets up to 6 ft. wide in two thicknesses, single strength and double strength, which are $\frac{1}{16}$ in. and $\frac{1}{8}$ in. thick, respectively. Plate glass is similar to window glass but made up to 15 ft. wide and polished on both sides. It is sold in many thicknesses and grades.

Borosilicate Glass

A second glass which has become commonplace in the world is *borosilicate glass*. In borosilicate glasses, much of the sodium is replaced with the element boron. A typical formula for the borosilicate glasses averages 80.5% silica, 13% boron oxide (B$_2$O$_3$), 2% alumina (aluminum oxide or Al$_2$O$_3$), 4% sodium oxide, and 0.5% potassium oxide. This glass is marketed under the trade names Pyrex, Kimax, and others. Since its coefficient of thermal expansion is roughly one-third that of soda-lime glass, it does not shrink as much when cooled and it can withstand thermal shock much better. Borosilicate glasses find considerable use in cooking ware and in scientific and technical apparatus. Figure 8-9 outlines a few of the properties of an average borosilicate glass.

Aluminosilicate Glass

A third type of glass is *aluminosilicate glass*. These glasses average about 59% silicon dioxide (SiO$_2$), 20% alumina (Al$_2$O$_3$), 5% boron oxide (B$_2$O$_3$), 9% magnesia (MgO), 6% lime (CaO), and 1% soda (Na$_2$O). The aluminosilicates have a softening point of about 915°C (1650°F), which is much higher than that of the other glasses. For this reason it is used in high-temperature applications. Its coefficient of thermal expansion is 23×10^{-7} in./in./°F. This glass has an index of refraction of 1.53. The aluminosilicate glass is also used in applications where chemical resistance is important. The main disadvantage of aluminosilicate glass is its cost, which is up to three times more than the borosilicates.

Lead Glass

The *lead glasses* or lead alkali glasses come in several compositions. Optical lead glass, commonly called *flint glass*, has only about 35% silica, about 7% potassium oxide (K$_2$O), but 58% lead oxide. This glass has a very high index of refraction of 1.69. Its coefficient of thermal expansion is about the same as that of soda-lime glass, 50.5 in./in./°F, and it has one of the lowest softening temperatures of all glasses, 1046°F (580°C). Its primary uses are in lenses and prisms. Another lead glass formula has 57% silica, 1% aluminum oxide, 5% sodium oxide, 8% potassium oxide, and 29% lead oxide. This modification of lead glass is used in applications where high electrical resistance is needed, such as in stems for light bulbs and electrical seals. This low-lead glass has a refractive index of 1.56, a coefficient of thermal expansion of 49.4×10^{-7} in./in./°F, and a softening point of 1166°F (630°C). It is used for decorative cut-glass tableware, called crystal or English crystal glass. Because of its high density, lead glasses are also use to provide protection against X-rays and gamma rays in laboratories.

Fused Silicates

The fused silicates or quartz glass are of two main varieties. The 100% pure fused silica approximates quartz in its properties. It has a very high softening point, 2954°F (1640°C). Its index of refraction is 1.46, and its coefficient of expansion is about 3×10^{-7} in./in./°F. Its optical transparency is very high and, unlike most glasses, it will transmit ultraviolet wavelengths. Fused silicates are very expensive because of the high temperatures required in their manufacture and the purity of the sand from which they are made.

A second type of fused silicates are the 96% silica glasses. These contain 96.5% silica, 3% boron oxide, and 0.5% aluminum oxide. This is a compromise between the borosilicates and the 100% fused silica. It is made by using a borosilicate glass and leaching the nonsilicate ingredients out with an acid and treating it at high temperatures. This process gives the glass good thermal properties with a softening point of 1520°C (2768°F) and a coefficient of thermal expansion of 4.4×10^{-7} in./in./°F. The index of refraction is the same as that of pure fused silica, 1.46. This glass has been used in missile nose cones, windows of space vehicles, and heat-resistant laboratory glassware.

Rare-Earth Glass

The rare-earth glasses contain no silicates at all. One rare earth glass contains 21.5% boron oxide (B_2O_3), 3% barium oxide (BaO), 2% barium tungstate ($BaWO_4$), 28% lanthanum oxide (La_2O_3), 20% tantalum pentoxide (Ta_2O_5), and 25.5% thorium dioxide (ThO_2). The rare-earth glasses have the highest index of refraction of any class, 1.85. Their clarity and high refractive index make them used almost solely in lenses and other optical devices.

Phosphate Glass

Like the rare-earth glasses, the phosphate glasses have no silica content. They are composed of 72% phosphorus pentoxide (P_2O_5), 18% alumina, and 10% zinc oxide. These glasses have high transparency to the infrared region of the spectrum and find one use in the nose cones of heat-seeking missiles.

SPECIALIZED GLASS

Besides the commercially available glasses listed above, there are some specialized glasses that have specific applications. A silicon-free glass having a composition of 36% boron oxide (B_2O_3), 27% barium oxide (BaO), and 20% magnesium oxide (MgO) has been developed for use in sodium vapor lamps.

Since large metal atoms absorb nuclear radiation, they cannot be used in applications that must transmit very short wavelength radiation, such as X-rays. Lindemann glass, with atoms of very low atomic number, was developed for this purpose. It consists of 83% boron oxide, 2% beryllium oxide (BeO), and 15% calcium fluoride (CaF_2). Conversely, a neutron-absorbing glass having a high cad-

FIGURE 8-10   Photochromic eyeglasses.

mium content is used for radioactivity shielding. This glass has 26% silica, 2% alumina, 64% cadmium oxide, and 8% calcium fluoride.

One of the newer innovations in the old art of glassmaking is *photochromic glass*. These glasses have a composition of about 60% silica, 10% soda, 10% alumina, 20% boron oxide, 0.6% silver, 0.3% chloride ion, and 0.9% iodide ion. The silver ions in the glass become silver metal when struck by a photon of light. This causes the glass to darken when hit by rays of sunlight. When the light source is removed, the silver atoms become silver ions again and the glass clears up. Figure 8-10 shows the use of photochromic glass in eyeglasses.

Photosensitive glasses can be made to act like a photographic film in the presence of light. This glass has a composition of 72% silica, 17% soda, 11% lime (CaO), 0.02% gold, and 0.04% selenium. Besides gold, cerium or copper can be used as the sensitizing agent. Using photosensitive glasses, a pattern can be embedded in the glass which can then be chemically machined or eroded along the pattern. Microscopic circuit boards for electronics application can be made in this way.

The silicon-to-oxygen bond is very strong. Theoretically, glass should have a tensile strength of almost 1 million pounds per square inch (70,000 kg/cm^2). In practice the tensile strength is about 1% of that theoretical value. The reason for this loss of tensile strength is that the surface of glass develops very minute surface cracks during cooling. These cracks act as stress risers or points that concentrate the stress. When stress is applied, the cracks simply open up and break bond after bond until the crack goes all the way through the material (see Figure 8-11). Glass would have a much higher strength if these cracks could either be eliminated or the stress not be allowed to concentrate at them. This can be done by two techniques: chemical tempering or thermal tempering.

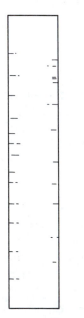

FIGURE 8-11   Cracks in glass.

TEMPERED GLASS

Some glasses can be *chemically toughened* or *chemtempered* by immersing them in a bath of molten potassium chloride. The larger potassium atoms will replace the sodium atoms in the surface of the glass, crowding the atoms together on the surface (Figure 8-12). This crowding produces a denser surface and keeps the cracks

Potassium ions Sodium ions

Molten salt

Glass →

Crack

FIGURE 8-12 Chemically tempering glass.

from opening up. Eyeglasses are strengthened by this technique and can withstand a considerable impact without breaking.

A second method of strengthening glass is *thermal tempering*. This glass is also known as *prestressed glass*. Glass, as with many materials, can be made more uniform throughout by annealing or slow cooling. Annealing also reduces internal stresses in the material. If glass is heated above its softening point, then plunged into an oil bath, the surface is immediately cooled without being given a chance to shrink significantly. The slower-cooling middle of the glass has a chance to shrink. This allows the center of the glass to pull the outside of the glass together, keeping it in compression. Since glass breaks in tension, the compressive forces on the outer layer of the glass must be exceeded before the surface will even go into tension. The glass becomes much stronger. Figure 8-13 illustrates this principle.

One problem exists with prestressed glass. If the surface of the glass is scratched or nicked, the internal stresses are released and the glass will explode into very small pieces. For this reason, glass must be cut to size and shaped before it is tempered. Trying to cut glass after it is tempered will destroy it. Tempered glass does have a safety advantage. Big pieces of broken glass can cause serious injury when falling on a person. Large glass doors and windows, basketball backboards, automobile glass windows, and glass tabletops are all tempered. Many states require eyeglasses to be tempered. If they should break, they will break into tiny pieces of glass that will not cut a person severely.

Viewing thermal tempered glass through a Polaroid screen will show the internal stresses. These stresses show up as dark lines or spots in the glass and can often be seen by using Polaroid sunglasses when looking through the automobile windshield. The stresses in a thermal tempered glass can be seen in Figure 8-14.

There are many other new innovations using glass. Glass can be blown into small fibers by hitting very small droplets of very fluid glass with a blast of high-velocity air. These fibers can be made into glass wool mats for insulating material or combined with a resin for use in fiberglass products (more about this application in Chapter 11). Very pure glass can also be drawn into continuous flexible filaments as thin as a human hair which will transmit light extremely well. Bundles of these fibers can be used for fiber optic light pipes, which allow one to see around corners or into places too small for normal vision. Surgeons use fiber optic instruments for visual inspection of the intestines and other parts of the body. Fiber optic com-

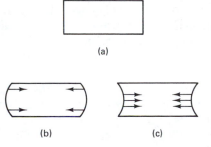

(a)

(b)   (c)

FIGURE 8-13 Tempered glass: (a) hot glass; (b) initial quench; (c) after cooling.

FIGURE 8-14   Stresses in prestressed glass.

munication lines used in conjunction with lasers and electronics can transmit 10,000 voice signals over a single fiber. This is 200 times more conversations than could be carried over a metal wire.

Automobile safety glass is made from two layers of plate glass laminated to an intermediate layer of plastic such as polyvinyl butyral. After the glass is cut and shaped, it is often thermal tempered. In case of an accident, safety glass will splinter into very small pieces and not let chunks of glass fly out to injure passengers. The glass is usually about $\frac{1}{4}$ in. thick, but other thickness can be ordered.

CERAMICS

Ceramics in the form of clay pottery are among the oldest objects made by the human race. The raw material, clay, is found everywhere. Early pottery was sun dried and not fired. Evidence of firing of pottery has been found in archeological finds dating from 2000 to 3000 B.C.

The word "ceramics" is derived from the Greek *keramos*, meaning "burnt stuff." Whereas glass is a noncrystalline structure, ceramics can have a crystal structure. Some experts list glass as a subcategory of ceramics. Ceramics are primarily alumina or clay products, while glass is made from silicas or sand. There is not a sharp dividing line between the two. Some ceramics have silicates in their formulas and some glasses contain aluminas. Clay is a hexagonal structure that breaks up into flat platelets. Each platelet can have a monomolecular (one molecule thick) layer of water adhering to it. This gives the clay good plasticity and workability. While this water is present, the clay can be formed into any shape desired. The invention of the potter's wheel allowed the formation of bowls, pitchers, cups, vases, bottles, and other cylindrical objects. Clay can also be extruded through a die for other shapes or it can be made into a slurry called *slip* and poured into a mold. Once the clay is formed into its final shape, the water can be driven off by heat, allowing the clay particles to come together in a rigid form. This product is porous and not very strong. Firing the clay product causes the clay particles to fuse together in a process called *sintering*. In sintering, only the edges of the individual particles are bonded together; the whole particle does not melt. The end product is a strong, rigid, brittle product that has good compressive strength, high melting points, and good heat resistance. Figure 8-15 illustrates the ceramic forming process.

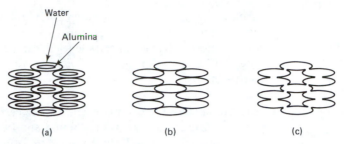

FIGURE 8-15 Forming of ceramics: (a) formed; (b) dried; (c) fired and sintered.

Besides the usual pots and vases commonly associated with ceramics, bricks, tile, terra-cotta statuary, and refractory materials also fall in the category of ceramics. The 7000 terra-cotta statues of the Ch'in dynasty in China (221–206 B.C.) found near the city of Xian are proof of the durability of that material. Common red brick has a compressive strength of 5000 to 8000 lb/in².

Thinker 8-1:

The uses of ceramics are limited only by one's imagination. Brick is usually thought of as a building material. Make a list of at least 10 other uses of brick besides its use as a structural material.

Refractories

Refractories are materials that will withstand high stresses at extremely high temperatures. They must be made of very pure components. Very often the best refractories are the oxides of very active metals. The materials from which refractories are made include the silicas (SiO_2), aluminas (Al_2O_3), magnesias (MgO), iron oxides (Fe_2O_3), and chromic oxides (Cr_2O_3). The principle involved here is that elements which react the easiest, in general, are the most difficult to disassociate. Magnesium, for instance, reacts very easily with oxygen to form magnesium oxide. In forming magnesium oxide a tremendous amount of energy is given off. At least that same amount of energy must be added to the magnesium oxide to disassociate the magnesium from the oxygen. For this reason, magnesium oxide will withstand temperatures of well over 4000°F. The melting point of magnesium oxide is 5072°F (2800°C). Aluminum oxide is very similar and has a melting point of 3713°F (2980°C).

Refractory materials are composed of coarse oxides called *grog* mixed with some finer refractory material which bonds the particles together. They are divided into three types: acidic, basic, and neutral. The acidic refractories are mostly silicas and aluminas. Acidic firebrick can have up to 80% aluminum oxide, with the rest silicon dioxide. The basic refractories are mostly magnesium oxides (up to 93%) with a little iron oxide mixed in. Neutral refractories contain 30 to 50% chromium oxide, up to 40% magnesium oxide, with the remainder being alumina, silicas, and iron oxide. The basic refractories are used in steelmaking and other applications, where they must be compatible with the smelted metals. Acidic and basic refractory bricks are not compatible. Where the cheaper acidic brick can be used in conjunction with basic brick, a layer of neutral refractory material must be used to separate them.

Other materials can also be used in refractories. Zirconia (ZrO_2), titanium carbide (TiC), zirconium carbide (ZrC), and some of the nitrides and borides have high enough melting temperatures to serve as refractories. Some ceramics can conduct electricity. Silicon carbide (SiC) can be used as heating elements for furnaces.

Cermets

A relatively new class of materials are the *cermets*. Cermets are made from ceramic particles dispersed in powdered metal and fused together. A common cermet uses aluminum oxide mixed chromium or chromium alloys. Cermet parts from this composition can have tensile strengths varying from 20,000 to 40,000 lb/in^2 with densities from 4.5 to 9 g/cm^3 (280 to 560 lb/ft^3). Their modulus of elasticity is higher than steel, ranging from 37 to 50 $\times$ 10^6 lb/in^2. Other cermets are the tungsten carbides, titanium carbides, chromium carbides, and barium carbonate–nickel. Uranium dioxide, chromium-alumina, nickel magnesia, and iron–zirconium carbide cermets have been used in nuclear reactors and related equipment.

The modern uses of ceramics include heat shielding for space shuttles, magnetic ceramics, and automobile engines. Silica–alumina fibers formed into insulating blankets that can take heats to 2000°F are now in use. Nose cones of sintered alumina are used on rockets. The recently discovered superconducting rare-earth copper oxide materials are sintered to form a ceramic. The material that the human race has known the longest is now becoming one of its most modern materials.

PROBLEM SET 8-1

Use the following list of materials to answer Problems 1 through 9:

Soda-lime glass
Borosilicate glass
Fused silicates
Chemically toughened glass
Lead glass
Phosphate glass
Aluminosilicate glass
Rare-earth glasses
Cermets

1. Which is the cheapest glass to purchase?
2. Which would be the best material for a nuclear radiation shield?
3. Which should be selected for use in a sliding glass door?
4. Which will take the highest temperature?
5. Which can be sintered easily?
6. Which can take thermal shock well?
7. Which would be the best choice to transmit ultraviolet light?
8. Which would be the best choice to transmit heat?
9. Which has the highest index of refraction?
10. The index of refraction of a glass is 1.51. If a ray enters this glass at an angle of incidence of 15°, what will be the angle of refraction?
11. If a ray of light enters a block of glass that has an index of refraction of 1.47, at an angle of 20° from the perpendicular to the surface, what will its angle with the perpendicular be inside the glass?
12. What is the difference between glass and ceramics?
13. If a ray of light enters a block of glass at 18° from the perpendicular to the

surface and is bent to an angle of 13° to the perpendicular, what is the index of refraction?

14. If a ray enters at an angle of 20° to the perpendicular of a glass that has an index of refraction of 1.5, by what angle will the glass be bent?

15. For a ray of light to be bent by an angle of 8° when it enters a block of glass at 25° from the perpendicular, what index of refraction must the glass have?

16. To determine the type of glass, an index of refraction can be determined. If an angle of incidence is 45° and the angle of refraction was 28°, which of the glasses listed in this chapter could it be?

17. A block of glass is 1.0 in. thick and has an index of refraction of 1.41. A ray of light enters at an angle of 30° with the perpendicular as shown in Figure 8-16. How much will it be offset when the ray leaves the block of glass?

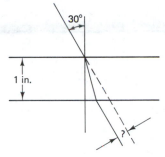

FIGURE 8-16   Diagram for problem 8-1, number 17.

18. If the coefficient of thermal expansion of a glass is 60×10^{-7} in./in./°F, how much will a 6-in. rod shrink in cooling it from 200°F to 70°F?

19. If a 6-in. rod of borosilicate glass was cooled from 200°F to 70°F, how much would it shrink?

20. A particular glass bar 8 in. long could shrink 0.0020 in. without breaking. If the coefficient of thermal expansion was 30×10^{-7} in./in./°F, would the glass bar break in cooling it from boiling water at 212°F to ice water at 32°F?

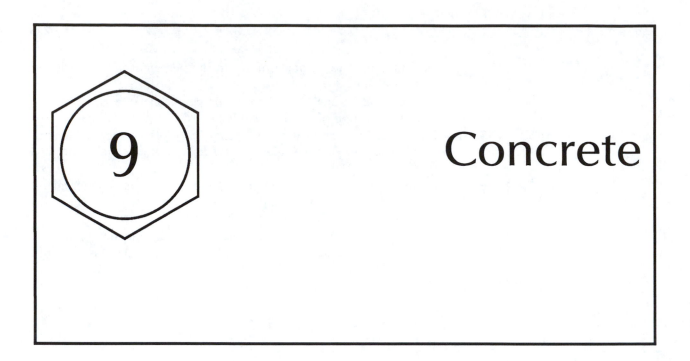

9 Concrete

INTRODUCTION

Concrete has been defined as any aggregate bonded together by a cementing agent that will harden into a solid mass. There are many cements, but rock and stone are usually the aggregate. Concrete of various types has been used since antiquity. The Romans in the second century B.C. used a volcanic product, found near the town of Pozzuoli on the bay of Naples, which when mixed with lime produced a cementing agent. This cement (called pozzuolana), mixed with an aggregate of stones, glass, and bricks made of clay, was used to build the baths of Caracalla (Figure 9-1) and the Pantheon in Rome.

The use of this type of concrete declined after the fall of the Roman Empire. Most of the building through the Middle Ages was of "stone on stone" or wood construction. In 1756, John Smeaton began experimenting with limestone. He found that limestone heated with a natural gas produced an intense white light—limelight. During these experiments with the heated limestone, he also found that certain limestones, after being thoroughly dried by the heating and pulverized, would form a monolithic type of stone when mixed with water. However, not all types of limestone would form this new cement. Only limestone from the quarries near Portland, England, seemed to work. This new type of bonding agent became known as portland cement. The widespread use of portland cement as a building material did not occur until the middle of the nineteenth century, and it was not introduced into the United States until after the Civil War, in 1871. The first portland cement plant in the United States was at Coplay, Pennsylvania. Since its introduction, portland cement has revolutionized the building industry, allowing types of construction not possible before its discovery.

Portland cement concrete is just one type of concrete. (Note that cement and concrete are two different things.) Rock and stone are often held together by other bonding agents. Asphalt, a petroleum product, is used as the bonding agent to

216

FIGURE 9-1 Baths of Caracalla, Rome.

produce asphaltic concrete. The primary use of this type of concrete is in the paving of roads. Currently, aggregates are also being held together by epoxies and other new adhesives to produce concretes with properties quite different from those produced by the portland cement concretes.

MANUFACTURE OF PORTLAND CEMENT

Portland cement is made from limestone, but not all limestone will make a good cement. It must have the proper composition to produce cement. To produce a good portland cement, the limestone must have the correct amounts of four ingredients:

1. Lime, which is calcium oxide (CaO)
2. Silica (sand), which is silicon dioxide (SiO_2)
3. Alumina (clay), which is aluminum oxide (Al_2O_3)
4. Iron oxide (rust) (Fe_2O_3)

Each of these compounds performs a specific function in the cement.

To convert the limestone to portland cement, it must first be removed from the ground by blasting it from the quarry. Then, in a cement plant (Figure 9-2), it is crushed and ground by a ball mill to a powder. If any of the four oxides are lacking or are not in the proper proportion, they are added at this point. The powder is then sent through a rotary kiln, where it is heated to about 2900°F. This process drives all the water from the mix.

The mix leaves the kiln in the form of a clinker, which must be rapidly cooled by an air blast to prevent it from regaining moisture. Here, too, about 2 to 4% of calcium sulfate ($CaSO_4$) in the form of gypsum is added to control the set up rate of the cement. Without the gypsum the concrete would set up so rapidly that it would be unworkable. The clinker with the added gypsum is then repulverized to

FIGURE 9-2   Cement plant.

about a 200 grit. (The 200 grit means that it will pass through a screen which has 200 openings per linear inch or 40,000 openings per square inch.) This product, which feels about like coarse face powder, is then bagged in 1-ft³ amounts which weigh 94 lb per sack.

During the drying process, the four oxides react to form four specific molecules that make up portland cement:

$$(CaO)_3(SiO_2) \longrightarrow C_3S$$

$$(CaO)_2(SiO_2) \longrightarrow C_2S$$

$$(CaO)_3(Al_2O_3) \longrightarrow C_3A$$

$$(CaO)_4(Al_2O_3)(Fe_2O_3) \longrightarrow C_4AF$$

The Portland Cement Association (PCA) has labeled these molecules C_3S, C_2S, C_3A, and C_4AF, respectively, as a shorthand method of identification.

In the finished cement, each tiny particle of the cement powder is in the form of a tetrahedron. It is these tetrahedra that are made up of the four molecules described above. When mixed with water, the C_3S, C_2S, C_3A, and the C_4AF that make up the tetrahedra react with the water (Figure 9-3) to form a gel that is made up of microscopic fibers. These fibers attach to the rock and form a matrix that holds the rock together. Electron microscope pictures have shown that these fibers are about 30 angstroms wide by about 250 angstroms long. (An angstrom is 0.00000001 cm or 1×10^{-8} cm.)

There are many common misconceptions about concrete. One of them is that water merely wets the cement and that the cement dries out on setting. Not so! Water is one of the chemicals needed to react with the cement to form the paste. If the water evaporates, the reaction stops and the concrete ceases to get stronger. Since the powdered cement is in the form of little tetrahedra (four-sided pyramids) and since only the surface of these tetrahedra can react with water, the reaction of the cement with the water is a very slow process. It takes days to months for the cement to react completely with the water. Even after a year some concretes

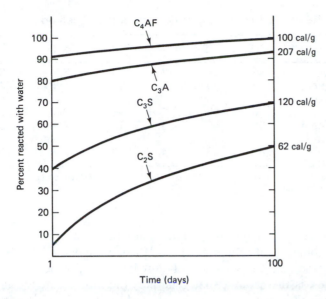

FIGURE 9-3   Hydration of portland cement.

are getting stronger. It is not uncommon to allow large concrete dams and spillways to cure for three years before the lakes behind them are allowed to fill to capacity. However, the engineering design strength of portland cement concrete has been standardized as the strength that the concrete attains at 28 days after mixing.

During the hydration process or reaction with water, all four molecules give off heat. This is known as an exothermic reaction and is called the heat of hydration. The C_3S gives off about 120 cal of heat per gram. A gram of C_2S emits 62 cal. The C_3A delivers a whopping 207 cal/g, while the C_4AF generates about 100 cal/g. On the average, a normal (type I) cement emits about 120 cal/g of cement during the curing of the concrete. The catalyst (C_3A) and the C_3S generate most of the heat. Most chemical reactions will proceed faster at higher temperatures. The reaction of portland cement with water is no exception. Therefore, if more C_3A and C_3S are added to the cement, it will generate more heat and cause the concrete to cure more quickly. Conversely, a lower percentage of the C_3A and a corresponding higher percentage of C_2S in the cement will result in a slower curing rate because it generates less heat. By varying the percentages of the four basic molecules in the cement, and by varying the particle size, different types of portland cement are manufactured.

TYPES OF PORTLAND CEMENT

There are five major types of portland cement. The types vary chiefly in the percentages of the four molecules present plus some additives (Figure 9-4). The standard types of portland cement are the following:

Type I   normal
Type II   modified
Type III   high early strength
Type IV   low heat
Type V   sulfate resistant

Type I cement is the most common and is used in general construction when the special properties of the other types are not required. Type II cement is used

FIGURE 9-4 Percent compositions of portland cements.

	C_3S	C_2S	C_3A	C_4AF	$CaSO_4$	Others[a]
Type I: normal	49	25	12	8	2.9	3.1
Type II: modified	46	29	6	12	2.8	4.2
Type III: high early strength	56	15	12	8	3.9	5.1
Type IV: low heat	30	46	5	13	2.9	3.1
Type V: sulfate resistant	43	36	4	12	2.7	2.3

[a]Included in "others" are free calcium oxide (CaO), magnesium oxide (MgO), and trace amounts of impurities from limestone.

where mild sulfate attack may occur. Sulfate attack occurs as a result of acid rain or pollution near some metal refineries. Type II cement has a relatively low C_3A content, which means that it has a relatively low heat of hydration.

Since the C_3A is the catalyst that generates the most heat, the high-early-strength cement will have more of this molecule and less C_3S. It will get its strength earlier, but the ultimate strength of the type III cement will be slightly less than that of the other cements. Another way of getting the high early strength is through the use of additives (called "admixtures" in the trade). The admixtures speed up the setting of the cement paste while eliminating the need for using different types of cement. Type III cement particles are also ground finer, resulting in a larger cement surface area. The finer particles react faster when mixed with the water, providing a higher strength in a shorter amount of time than the other cements.

Type IV cement was developed to reduce the thermal stress on concrete. In massive structures such as dams the heat generated by the reaction of the cement with the water causes the concrete to expand. Upon cooling, the concrete shrinks and cracks. Low-heat cement was designed to reduce excessive cracking. Type IV cement requires considerably more time for curing; at least 21 days must be allowed before attempting to build anything on top of it.

Type V cement is used in such structures as canal linings, culverts, and other applications where soils or groundwater with high sulfate contents could cause deterioration of the other types of portland cement concretes.

Thinker 9-1:

State which type of cement you would select for the following jobs:

a. A large hydroelectric power dam
b. A slab for a garage
c. Sidewalks
d. A foundation on which you need to build within a few days
e. A floor for a copper plating plant

SLUMP

One of the variables in mixing concrete is the thickness or viscosity of the mix. A thin, runny concrete would be easy to pour and would flow readily into complicated forms, but it would shrink considerably upon curing. Conversely, a thick concrete

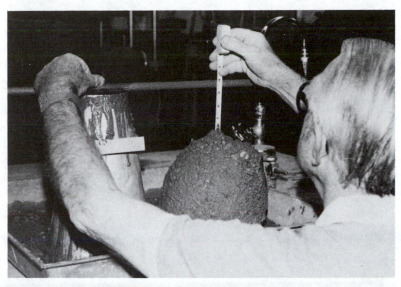

FIGURE 9-5   Standard slump test.

would be almost unworkable in casting. The viscosity of concrete is determined by the slump test (Figure 9-5). The standard slump cone used in the test is a truncated metal cone made of 18-gauge sheet steel, which is 12 in. high, 4 in. in diameter at the top, and 8 in. in diameter at the bottom. To conduct the slump test, the cone is placed on a level surface and held down firmly. It is then filled with the wet concrete mix to one-third of its height, then rodded or punched down with a $\frac{1}{2}$-in. steel rod 25 times to cause the concrete to compact completely into the bottom of the cone. It is filled another one-third and again rodded 25 times. The cone is then filled to the top, rodded another 25 times. The top is smoothed off (screeded) and the edges wiped free of any excess concrete. The slump cone is gently lifted straight up, clear of the concrete, and set beside the pile of concrete. When the cone is removed from around the concrete, the concrete will sag or slump. The rod is placed across the top of the slump cone over the concrete pile and the distance from the bottom of this rod to the topmost point on the pile of concrete is measured. This distance is the slump for that batch of concrete. For most construction, a normal slump would be from 2 to 4 in. There are times when a thinner (high slump) concrete or a thicker (low slump) mix would be needed.

MIXING CONCRETE

Imagine two pieces of material being glued together. If there is not sufficient glue to cover the glued surface completely, the bond will not be very strong. If there is enough glue to cover the joined surfaces completely, the joint will be quite strong. In concrete, rocks are glued together. To have a strong bond, each rock must be covered thoroughly with the cement paste. If the paste is thin with not much cement per unit of water, it will take a lot of aggregate to make the concrete have a slump of from 2 to 4 in. Therefore, there will not be much cement covering each rock particle and the concrete will be very weak. If the cement paste is thick with less water per unit of cement, it will take less aggregate to get the desired slump and the resulting concrete will be quite strong.

It has been proved by many tests that the *primary factor controlling the ultimate strength of concrete is the amount of water used per sack of portland cement*. This is called the *water/cement ratio* and is measured in gallons per sack. It takes about

4 gallons of water per sack of cement (minimum) to wet or mix a sack of portland cement completely. A paste of 4 gallons per sack would be very thick and not much aggregate would be added before the 2- to 4- in. slump was reached. The resulting yield or amount of concrete obtained from a sack of cement would be very low, but the concrete would be very strong. A mix of 6 gallons per sack would need more aggregate to get the same slump and would produce more concrete per sack of cement. This concrete would be weaker than the mix of 4 gallons per sack.

In mixing a trial batch of concrete, less than a bag of cement is used, so the amount of water would be measured in ounces. Suppose that $\frac{1}{8}$ sack of cement (11.75 lb) was to be mixed at a water/cement ratio of 6 gallons per sack. One-eighth of 6 gallons of water would be needed. One gallon of water is 128 oz. The amount of water for this trial batch would be 96 oz:

$$\frac{6(128 \text{ oz})}{8} = 96 \text{ oz}$$

Once the desired water/cement ratio is determined, the necessary weights of sand and aggregate are found to give the correct slump. The engineers will design the concrete on a weight basis. The specifications for a concrete will state the following if required by the job:

1. Water/cement ratio
2. Slump
3. Required compressive strength in pounds per square inch or megapascal
4. Any additives necessary
5. Aggregate size
6. Aggregate gradation
7. Handling specifications
8. Conditions of the mixing water
9. Curing specifications
10. Required tests on the batch or job
11. Delivery conditions

The small job does not require all of the specifications above. Many small jobs are done by the wheelbarrow load and can turn out well. A common backyard mix is 4–4–2: four shovels of sand, four shovels of aggregate, and two shovels of cement. Then enough water is added to get the desired slump. This type of mixing of concrete is sufficient for mowing strips around the house, but is not recommended for construction purposes.

Two terms commonly used in the concrete industry are *yield* and *cement factor*. The yield is defined as the cubic feet of concrete that is obtained per sack of cement. The cement factor is the number of sacks of cement used per cubic yard of concrete. If the yield is known, the cement factor is equal to 27 ft³/yd³ divided by the yield:

$$\text{cement factor} = \frac{(27 \text{ cubic feet/cubic yard})}{(\text{yield cubic feet/sack})}$$

For example, if a test batch of concrete used 80 lb of aggregate, 15 lb of cement, and 120 oz of water, what would be the yield and the cement factor of the batch? Concrete has a density of 150 lb/ft³.

$$\text{Total weight of batch} = 80 \text{ lb} + 15 \text{ lb} + \frac{120 \text{ oz}}{16 \text{ oz/lb}} = 102.5 \text{ lb}$$

$$\text{Volume of concrete produced} = \frac{102.5 \text{ lb}}{150 \text{ lb/ft}^3} = 0.68 \text{ ft}^3$$

$$\text{Volume per sack} = \text{yield} = \frac{(0.68 \text{ ft}^3)\,(94 \text{ lb/sack})}{15 \text{ lb}} = 4.28 \text{ ft}^3/\text{sack}$$

$$\text{Cement factor} = \frac{27 \text{ ft}^3/\text{yd}^3}{4.28 \text{ ft}^3/\text{sack}} = 6.3 \text{ sack/yd}^3$$

PROBLEM SET 9-1

1. If a job required 47 bags of cement, how many pounds of cement would that be?

2. If a job called for 473 ft³ of portland cement, how many bags should be ordered?

3. Concrete with a yield of 4.8 ft³/yd was used on a job. What was the cement factor on that batch?

4. A trial batch of concrete was mixed using 150 lb of aggregate, $\frac{1}{3}$ bag of cement, and 256 oz of water. If the density of this concrete was 148 lb/ft³, what was the yield of the batch?

5. What is the cement factor of the batch in Problem 4?

6. What is the water/cement ratio in gallons per sack for the batch in Problem 4?

7. Concrete with a cement factor of 5.5 sacks per cubic yard was used on a job that required 42.5 yd³ of concrete. How many sacks of cement would be used?

8. Concrete with a yield of 5.2 sacks per cubic yard was used on a certain project that required 67 yd³ of concrete. How many sacks of cement were used?

9. A backyard mix of concrete used 2 shovels of cement to 6 shovels of aggregate to 2 qt of water. Suppose that a shovel of cement weighed 5 lb and a shovel of aggregate weighed 8 lb and the density of the concrete was 150 lb/ft³.
   (a) What was the water/cement ratio in gallons per sack?
   (b) What was the yield?
   (c) What was the cement factor?

10. Cement costs $6.00 per sack, and aggregate costs $14.00 per cubic yard delivered to the site. A batch required 12 yd³ of concrete which had a cement factor of 5 sacks per cubic yard. Which would cost less for the materials:
    (a) the 12 yards of concrete delivered to the site at a price of $45 per cubic yard, or
    (b) buying the cement and aggregate and mixing it yourself?

11. It is desired to mix 7 yd³ of concrete having a cement factor of 5 sacks per

cubic yard. If the density of the concrete is 150 lb/ft³, how many sacks of cement are needed?

12. How much will a slab using all the concrete in Problem 11 weigh?

13. A test batch of concrete used 180 lb of aggregate, 22 lb of cement, 240 oz of water, and had a density of 145 lb/ft³. What was the yield?

14. What was the cement factor for the batch in Problem 13?

15. If the mixture of the test batch in Problem 13 was used to produce 50 yd³ of concrete, how many bags of cement would be needed?

16. What is the water/cement ratio in the batch described in Problem 13?

17. A trial batch of concrete used 11.75 lb of cement, 75 lb of aggregate, and 96 oz of water. One-tenth cubic foot of the concrete weighed 14.8 lb. What was the water/cement ratio of the batch?

18. What was the yield of the batch in Problem 17?

19. What was the cement factor of the batch in Problem 17?

20. If 10 yd³ of concrete of the mix in Problem 17 was needed, with the cost of cement at $5.25 per sack, what would be the cost of cement?

In industry, the sand, aggregate, cement, and water are mixed at the concrete mixing plant and loaded into 6- to 9-yd³ concrete hauling trucks. These trucks have a rotary drum which is kept turning during the trip from the mixing plant to the construction site. Without this constant agitation and mixing, the larger rocks would settle to the bottom of the mix while the cement rose to the surface. Once on the site, the concrete is poured into the forms with as little handling as possible. Where convenient, the concrete is poured directly into the forms from the truck. If this is not possible, the concrete is pumped to the forms by means of a concrete pumping crane or placed into large skip buckets and hoisted to the pouring site by means of a crane. On occasion, the concrete must be hauled to the forms by a wheelbarrow.

For commercial buildings, roads, dams, and other construction for public use, state or local laws may require that test samples be taken from every truckload of concrete. If this is the case, the inspector must be on the site when the concrete is delivered. The inspector will make three to six, 6- by 12- in. test cylinders from each truckload of concrete delivered. These samples must be cast in accordance with ASTM or state or local building codes, in much the same manner as the slump test. Each cylinder is filled one-third, rodded 25 times, filled another third, rodded 25 times, then filled to the top, rodded another 25 times, screeded off, and capped. The samples are marked with the necessary information to identify where the concrete is placed in the structure. The samples are then taken to the laboratory and cured for 28 days or for the time specified by the engineer or the architect. They are then tested to determine if the concrete has reached the compressive strength required by the architect or structural engineer.

Concrete has one major disadvantage—it has no strength initially when it is put in place. A steel beam has its maximum strength when it leaves the rolling mill. The strength of concrete increases after it is poured. If test samples are taken from a batch of concrete and a few broken at 1 day, 7 days, 14 days, 21 days, and 28 days, a graph of the strength versus the curing time can be plotted. Figure 9-6 shows the approximate strength of a normal concrete as a function of time. This graph shows that the strength of the concrete rises rapidly for the first few days but continues to get stronger even after 28 days. A normal concrete will attain approximately one-fourth of its 28-day strength in about 24 to 36 hours after mixing. By the end of a week it will have about half of its design strength. Many engineers require at least 14 days before any external load is placed on a normal concrete. The high-early-strength concretes can have loads placed on them much earlier.

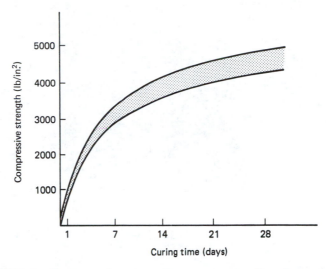

FIGURE 9-6   Compressive strength versus curing time for concrete.

A second useful method of plotting the same information is to plot the strength against the water/cement ratio. Curves can be drawn for 1-day to 28-day tests. This procedure allows the engineer to determine a water/cement ratio that will produce the required strength at the time it is needed. Figure 9-7 is such a graph.

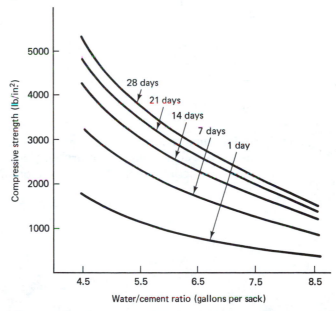

FIGURE 9-7   Compressive strength versus water/cement ratio for concrete.

AGGREGATES

If the aggregate consisted of perfect spheres of only one size, the spheres would occupy a maximum of 74% of the total volume. The rest of the space would have to be filled by the cement. A concrete made of this uniform-size aggregate would have a very poor yield, be very coarse or lumpy, have considerable shrinkage, be relatively weak with considerable cracking, and have very poor workability. An

(a)   (b)

FIGURE 9-8   (a) Well-sorted and (b) well-graded aggregates.

aggregate of uniform size is known as a *well-sorted aggregate* and is undesirable for use in concrete (Figure 9-8a).

The type of aggregate best suited for concrete would have rocks of reasonably large size ($\frac{3}{4}$ to $1\frac{1}{4}$ in. in diameter for an average type of concrete), then some smaller rocks to fill in the voids between the larger rocks. Still smaller rocks would fill in the smaller voids, and so on. Finally, the smallest voids would be filled with sand. This type of aggregate is called a *well-graded aggregate* (Figure 9-8b). An aggregate with sufficient sand to provide a smoothly troweled surface is called a well-sanded aggregate. The more sand that is in the aggregate, the lower will be the yield or amount of concrete per sack, but the resulting concrete will be more workable. In general, it is better to slightly oversand a concrete than to have too little sand in the aggregate.

Sieve Analysis of Aggregates

To determine the gradation of an aggregate, a standard sieve analysis is run. For this test a weighed amount of dry aggregate is placed on the screen of a standard Tyler sieve series (Figure 9-9). This standard sieve series has seven trays or screens consisting of a

    No. 4 screen
    No. 8 screen
    No. 16 screen
    No. 30 screen
    No. 50 screen
    No. 100 screen
    Pan

stacked in that order. Remember that the No. 4 screen has four openings per linear inch, the No. 8 screen has eight openings, and so on.

The stack of screens with the weighed amount of aggregate on the top screen is placed in a sand shaker which shakes the series of screens vigorously for about 3 minutes. The coarsest rock will remain on the top screen while the smaller rocks will fall through the screens until each rock or particle reaches a screen through which it cannot pass. Only the finest dust-size particles will reach the pan. The

FIGURE 9-9 Tyler sieve series.

screens are then removed from the shaker and the amount of aggregate retained on each screen is carefully weighed on a triple-beam balance.

One way of stating the gradation of an aggregate is by the use of the *fineness modulus*. Once the weight of aggregate retained on each screen is determined, the percent retained on each screen is calculated. The cumulative percent on or above each tray is computed. The sum of these cumulative percents, excluding the pan, divided by 100, is the fineness modulus of that aggregate. For the fineness modulus example, shown in Figure 9-10, the fineness modulus is 339.4/100 = 3.39.

A quick analysis of these calculations would show that if the aggregate were very coarse, with all of it staying on the top screen, a fineness modulus of 6 would result. If the aggregate was all dust which passed entirely through all of the screens and fell to the pan, the fineness modulus would be 0. A fineness modulus of 0 or 6 would indicate a well-sorted aggregate. A fineness modulus between 2.5 and 4.5 would generally be a well-graded aggregate.

A second method of showing the gradation of the aggregate is through use of the gradation curve. The aggregate is run through the sieve series as shown in Figure 9-10. The weight of aggregate retained on each screen is determined and

FIGURE 9-10 Calculation of the fineness modulus of an aggregate.

Tray	Weight Retained (g)	Percent Retained	Cumulative Percent Retained
No. 4	35	12.6	12.6
No. 8	44	15.8	28.4
No. 16	63	22.7	51.1
No. 30	52	18.7	69.8
No. 50	39	14.0	83.8
No. 100	27	9.7	93.5
Pan	18	—	—
	278		339.4

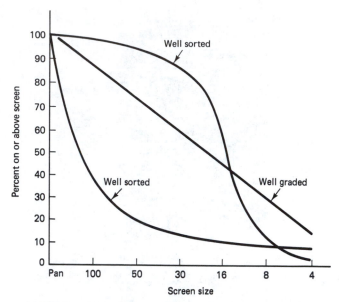

FIGURE 9-11   Gradation curve of an aggregate.

the percent of the aggregate retained on each screen is calculated. The graph of the cumulative percent of the aggregate on or above each screen is plotted on the *y* axis, with the size of screen, starting with the pan, on the *x* axis. Figure 9-11 shows a typical gradation curve for an aggregate sieve analysis.

Mortar used to join bricks or rock slabs would use a well-sorted fine aggregate in the form of sand. A sand that would all be retained on the No. 16 screen would have a fineness modulus of 4.0, while a sand that would be retained on the No. 30 screen would have a fineness modulus of 3.0. This illustrates the fact that the fineness modulus alone can be misleading. A fineness modulus combined with a gradation curve provides an excellent analysis of the aggregate.

PROBLEM SET 9-2

Use the following sieve analysis of an aggregate to answer Problems 1 through 12.

Sieve	Weight (g)
No. 4	15
No. 8	22
No. 16	30
No. 30	30
No. 50	26
No. 100	16
Pan	15

1. What is the fineness modulus of the aggregate?
2. Construct the gradation curve for these data.
3. Is this sample well graded or well sorted?
4. If a cubic yard of aggregate weighs 2000 lb, what weight of the aggregate with the gradation shown above will be a No. 4 grit or coarser?

5. At a density of 2000 lb/yd^3, what weight/ton of the gradation above will be a grit No. 50 or finer?

6. At a density of 2000 lb/yd^3, what weight/ton of the gradation above will be a grit No. 16 or coarser?

7. In a cubic yard of aggregate weighing 2000 lb and having the sieve analysis shown above, what weight/cubic yard would fall through to the pan?

8. Show by your calculations that the maximum fineness modulus would be 6.

9. What is the minimum possible fineness modulus? Show your calculations.

10. What would be the fineness modulus if exactly the same weight of material was found to be on each tray and the pan?

11. What would be the fineness modulus if all the aggregate stayed on the No. 30 pan? Would that aggregate be well sorted or well graded?

12. If 100 lb of a well-sorted aggregate stayed on the No. 4 screen and was added to 100 lb of sand that stayed on the No. 50 screen, what would be the new fineness modulus of the mixture?

13. Given the following sieve analysis data:

Sieve	Weight (oz)
No. 4	20
No. 8	20
No. 16	30
No. 30	40
No. 50	10
No. 100	10
Pan	10

What is the fineness modulus of the sample?

14. If a cubic yard of the aggregate in Problem 13 weighed 2000 lb, what percent of the material was a No. 8 grit or coarser?

15. If a cubic yard of the aggregate in Problem 13 weighed 2000 lb, what would be the weight of each grit per ton of aggregate?

16. What is the specific gravity of an aggregate weighing 2000 lb/yd^3?

17. If 200 lb of sand having a No. 16 grit was added to a ton of the aggregate having the sieve analysis in Problem 13, what would be the new fineness modulus?

18. If 10 oz of each grit were added to each tray in Problem 13, would the fineness modulus change?

19. An aggregate was made up with 200 lb of No. 4 grit material, 350 lb of No. 16 grit, and 300 lb of No. 50 (nothing else). What is the fineness modulus of this mix?

20. What would the fineness modulus of the aggregate in Problem 19 become if 100 lb of No. 100 grit fine sand were added to increase the workability of the concrete?

Besides checking for the size of aggregate particles, three other tests of the aggregate should be made. The *silt test* can be made with any quart jar. Fill the jar with about 2 in. of the aggregate. Fill the jar about three-fourths full with water, put the lid on it, and shake the mixture vigorously for about a minute. Level the aggregate and allow it to set for an hour or so. If a layer of silt more than $\frac{1}{8}$ in. thick forms on the top of the aggregate, the aggregate should be rejected or thoroughly washed before use in a concrete batch.

A colorimetric test for *organic matter* can be made using a 500-mL flask or pint bottle. Fill the bottle about one-third full with the aggregate, then fill the bottle at least two-thirds full with a 3% lye solution. Lye is sodium hydroxide (NaOH). Shake the mixture and let it set for 24 hours. The liquid above the aggregate should be colorless or slightly straw-colored at the end of that time. A darker color indicates the presence of excessive organic matter and the aggregate should be rewashed or rejected. Organic matter, especially sugars, in the aggregate can prevent the concrete from setting up.

The third test is for the amount of water in the aggregate. Aggregate, like the clay discussed in Chapter 8, can retain a monomolecular layer of water about each little rock. Further the sand and aggregate can absorb moisture in the rock itself. The rock acts like a small sponge. If the rock is thoroughly dried out, it can absorb some of the water added to the mix. This moisture cannot reach the cement and is removed from reacting with the paste. If the rock is fully saturated with water and has the additional layer of water on the outside of each rock, the outside water is available to react with the cement. If the person mixing the concrete added 6 gallons of water per sack of cement, but had 1 gallon of water brought in on the aggregate, the result would be a mix of 7 gallons per sack. This would produce weaker concrete than desired.

The ideal aggregate would be a saturated but surface dry condition. The rock in this condition would neither absorb water from the mix nor add more water to it. An aggregate is saturated and surface dry if it just starts to fall apart after being packed in the hand. If it is not saturated, the aggregate will be dusty. If it has an excess of water, it will stick together and leave your hand wet. It is necessary to conduct a water content of the aggregate test. This test can be done in several ways. In industry a moisture probe, which can be inserted into the aggregate pile to read the excess moisture content directly, is often used. The moisture probe is an *analog* meter that measures the electrical resistance between two points and gives a read out as a percent excess water. Since impure water conducts electricity, the more water, the lower the resistance.

A second method of measuring the moisture content of the aggregate is by the use of the Speedy moisture tester. A weighed amount of aggregate is placed in a stainless steel bottle. The top of the tester has a spoon in which some calcium carbide (Ca_2C) is placed. The top is then tightly screwed on and the bottle inverted and shaken. When water contacts calcium carbide, acetylene gas is given off. The more water present, the more acetylene is generated. The pressure of the excess gas is noted on a meter in the cap of the tester and is read out as a percent moisture. This is another analog type of tester in that a gas pressure is representing an amount of water present.

The old reliable method of determining excess moisture is to weigh a sample of the aggregate (about a pound), place it in an oven, and dry it to a saturated but surface-dry condition. The sample is reweighed. The difference in weights is the amount of water lost which can be calculated as a percent of the weight of the aggregate. This is a direct measurement, not an analog. The problem with this method is that it is very slow, requiring at least an hour to dry out the aggregate.

Once the percent moisture of the aggregate is known, the amount of water in a cubic yard of aggregate can be calculated and subtracted from the water that is added to mix with the cement.

MIXING WATER FOR THE CONCRETE

Since concrete will not stand an acidic environment, the water used in its mix should be checked. The water does not have to be absolutely pure. Water containing as much as 10,000 parts per million (1%) hydrochloric, sulfuric, and other inorganic acids has little affect on the concrete. Acidic pH values should be kept higher than 3.0. (A pH of 7.0 is neutral; from 0 to 7 are acids, 7 to 14 are bases.) Water containing industrial waste or sewage is to be avoided. If necessary, concrete can be mixed with seawater, but its corrosive effect on the reinforcing steel is detrimental. Algae can cause a drastic reduction of strength in concrete.

TESTING OF CONCRETE SAMPLES

After the test samples are taken from each truckload or batch of concrete delivered to the construction site, they are taken to the test laboratory, where they are cured at 100% relative humidity, either in a fog room or in curing tanks, for 28 days or for the time specified by the structural engineer, architect, or building codes. The samples are then removed from the mold and capped on both ends with a concrete capping compound (Figure 9-12). This compound is a mixture of sulfur and paraffin, which melts at around 300°F (about 150°C). The compound is poured around the ends of the test samples in a mold which makes the ends parallel and smooth. The compound hardens in seconds. The samples are placed in a compression tester having a capacity of 250,000 lb (or more) and broken (Figure 9-13). The maximum load taken by the test samples is noted. The maximum load divided by the cross-sectional area of the sample provides the strength of the concrete in pounds per square inch. This strength must equal or exceed the engineer's specifications or legally, the batch or load of concrete from which the bad samples were taken could be required to be removed and replaced. This is a result dreaded by builders and every effort is made to avoid this outcome.

FIGURE 9-12   Capping a concrete test sample.

FIGURE 9-13   Testing concrete.

PROBLEM SET 9-3

1. A standard 6-in.-diameter by 12-in.-high test cylinder of concrete took 128,000 lb to break. What was its breaking strength?

2. An architect specified a concrete that would take 3000 lb/in. What load in pounds would a 6-in.-diameter by 12-in.-high test sample have to take to meet the specification?

3. A 3-in.-diameter by 6-in.-tall test sample took 32,000 lb to break. What was its breaking strength?

4. A 3-in.-diameter by 6-in.-tall test sample was supposed to take a presure of 2800 lb/in². If it broke at 15,000 lb, did it pass the test?

5. A 6-in.-diameter by 12-in.-tall test sample was designed to take a compressive stress of 4000 lb/in². Would a test machine that could exert a load of 120,000 lb be able to break the sample?

6. A 6-in.-diameter by 12-in.-tall test sample took a load of 150,000 lb before breaking. What load would a 3- by 6-in. sample have to take to have the same compressive strength?

7. A concrete pedestal is to be designed to hold up a 42,000-lb statue on a circular area 28 in. in diameter.
   (a) What is the pressure on the base in lb/in²?
   (b) What load would a 6-in.-diameter by 12-in.-tall test sample have to take to qualify?

8. A concrete floor is designed to hold up 4000 kg/m². What load would a 15-cm-diameter by 30-cm-tall test sample have to take to meet this requirement?

9. A concrete wall was specified to have a design strength of 3600 lb/in². Three test samples, each having a 6-in. diameter and 12-in. height were tested. One broke at 110,000 lb, a second broke at 112,000 lb, and the third at 116,000 lb. Did their average strength meet the specifications?

10. A concrete has a compressive strength of 2900 lb/in². What diameter cylinder of this concrete would be required to hold up a truck that put a load of 9000 lb on it?

11. If a concrete has a density of 150 lb/ft³, how much does a cubic yard of it weigh?

12. If a concrete has a density of 145 lb/ft^3, what is its specific gravity?

13. A 15-cm-diameter by 30-cm-tall sample took a load of 1500 kg to break it. What was the strength in pascal?

14. A 10-cm-diameter by 20-cm-high test sample of concrete took 500 kg before breaking. What was the strength in pascal?

15. If a concrete required 1 MPa strength, what would be the required load on a 15-cm-diameter by 30-cm-high test sample to equal this strength?

16. A rectangular floor for a skating rink measured 60 ft by 100 ft and was to be made of concrete 6 in. thick. The specifications for the concrete are a cement factor of 5 sacks per cubic yard, a water/cement ratio of 5.5 gallons per sack, and a density of 145 lb/ft^3. How any cubic yards of concrete are needed?

17. How many sacks of cement are needed for the job in Problem 16?

18. If a 10-cm-diameter by 20-cm-high test sample of concrete took a stress of 500,000 Pa before failure, what load in kilograms would a 15-cm-diameter by 30-cm-high sample of the same concrete take?

19. A load of 5000 kg had to be supported by concrete that had a compressive strength of 1 MPa. What should be the diameter of the base in centimeters?

20. Which concrete would be stronger, one that could take 3000 lb/in^2 or one that could hold 10 MPa?

CURING OF CONCRETE

The water in the concrete does not dry out. It must remain in the concrete to react with the cement to make the concrete cure. If the water is allowed to evaporate from the concrete structure, the chemical reaction stops and the concrete ceases to get stronger. Moreover, it is difficult to restart the curing process since the water cannot get back into all of the cement once it has evaporated. Once the structure is poured, great care must be taken to prevent the water from being lost. This can be done in several ways. For highways, an old method is to cover the freshly poured concrete with tarpaper or straw. Many small contractors make a practice of purchasing old rugs from garage sales and flea markets to place over their newly cast slabs. These rugs are kept wet constantly during the required curing period. For small jobs such as driveways and slabs for small buildings, it is sufficient to keep a garden hose or sprinkler spraying water over the slab. (This method is often frowned upon by water conservation districts, however.) In larger structures and new highways, the practice now is to coat the concrete surface with a plastic that seals the surface to prevent evaporation.

REINFORCED CONCRETE

Concrete has a very low tensile strength. If any part of the concrete is placed in tension, it will fail. This can be seen by the following analogy. If a compressive force is placed on an aggregate as shown in Figure 9-14, the rocks on the sides will be forced apart, placing the cement holding those rocks in tension. Since the cement has a very low tensile strength, the concrete fails. Concrete test cylinders always fail under a compressive force with the cracks vertical or parallel to the compressive forces as shown in Figure 9-15.

Steel, on the other hand, has a very high tensile strength. It would make

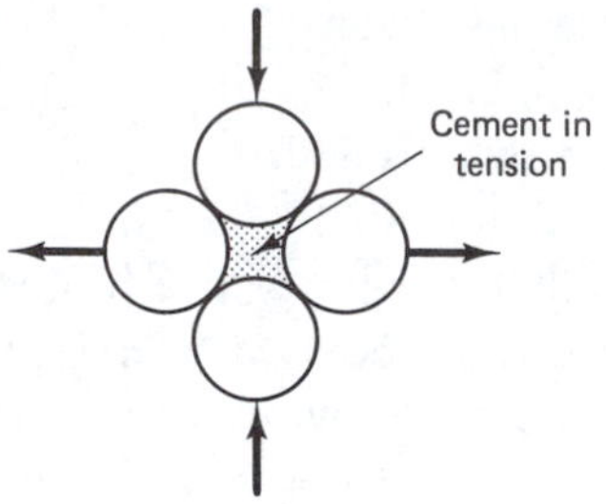

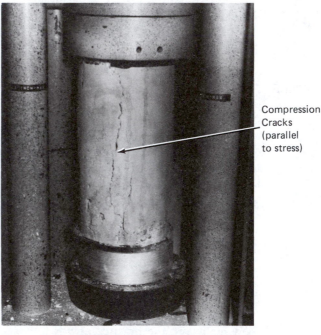

Cement in
tension

FIGURE 9-14   Aggregate in compression.

sense to put the steel in that part of the concrete which will undergo tension, and the concrete where the compressive forces lie. This is the principle of reinforced concrete, which indeed has extremely high strength. Steel could be placed entirely around a concrete cylinder or column to take the tension forces of the concrete. Steel pipes completely filled with concrete, which make extremely high strength columns, carry the trademark Lally columns.

Compression
Cracks
(parallel
to stress)

FIGURE 9-15   Concrete failure.

To be effective, steel need not encase the entire column. Almost the same effect can be obtained simply by inserting rings of steel at close intervals along the column as shown in Figure 9-16. These rings are tied to vertical steel bars prior to casting the concrete. Forms are placed around the columns and the concrete is poured in place. This is the principle by which many buildings and elevated highways are now built (Figure 9-17).

A simply loaded beam, supported on both ends and loaded in the middle, would have the bottom of the beam in tension in the center of the beam and also in tension on the top over the part of the beam that is directly over the supports (Figure 9-18).

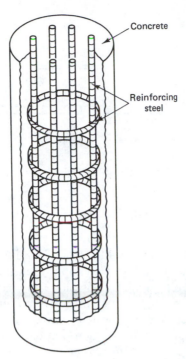

Concrete

Reinforcing
steel

FIGURE 9-16   Steel and concrete column.

The steel should be placed in the beam where there is tension. The steel should be ridged to grip the concrete firmly. Placement of the reinforcing steel or rebar in the concrete is critical. It is the job of the structural engineer to calculate exactly where it should be placed, where it should be bent, and the size and number of the reinforcing steel bars that should be placed in the concrete structure.

FIGURE 9-17   Elevated highway support.

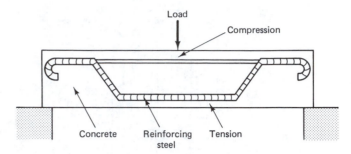

FIGURE 9-18   Reinforced concrete beam.

PRESTRESSED CONCRETE

If concrete is weak in tension but strong in compression, it is only logical that the concrete should not be allowed to go into tension at all. Concrete can be kept in compression by *prestressing*. Imagine a beam made up of a series of spools (Figure 9-19). Such a beam would have absolutely no tensile strength. It would not even hold itself together. Yet if a string or strong rubber band were inserted down the middle of the spools and drawn tight, not only would the spools stay together, but they could be used as a beam that could support considerable weight. The beam would fail only if the tensile strength of the string were exceeded or the spools crushed due to the compressive forces. This illustrates the principle of prestressed concrete.

Prestressing of concrete can be achieved in several ways. A beam could be cast around the steel bars, which have been threaded on both ends (Figure 9-20a). After the concrete has been cured, steel plates and nuts are placed over the ends of the beam and the nuts drawn tight. This places the beam in compression. The beams could then be put in place in the structure.

Another method often used is to place the steel bars under tension by the use of hydraulic jacks (Figure 9-20b). Steel plates are placed in the concrete through which the steel bars are run. After the concrete is cured the steel bars are welded to the steel plates before the tension is released. The elasticity of the steel bars

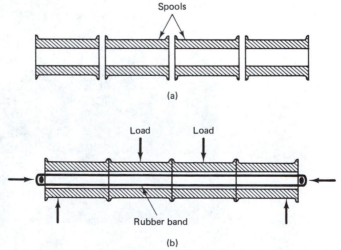

FIGURE 9-19   Principle of prestressing: (a) no-strength system; b) prestressed system.

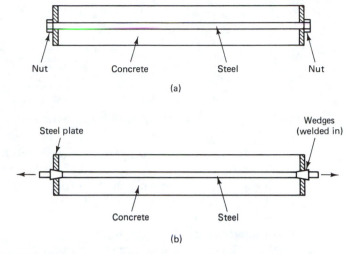

FIGURE 9-20 Prestressed concrete: (a) tension provided by threaded nuts; (b) tension provided by hydraulic jacks.

places the concrete in compression. These compressive forces must be exceeded by the loads placed on the concrete member before the concrete even goes into tension. Prestressed concrete beams have been found to have several times the strength in flexure of simply reinforced beams. Figure 9-21 shows the strength versus deflection curves for unreinforced, reinforced, and prestressed concrete test beams.

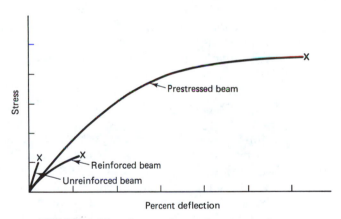

FIGURE 9-21 Comparison of concrete beams.

CONCRETE CONSTRUCTION

Since concrete is very weak in tension, even a flat slab must be well supported. If the ground under a slab is not well prepared, the slab will crack. Preparing the soil involves several steps. First the soil under the building site must be analyzed to determine if it will support the load of the building. Clay rich soils will pack down under a load and do not provide a suitable foundation. Sandy soils do not compress but will move sideways if not contained by retaining walls. It is often necessary to dig out several feet of soil and replace it with soil that will support the structure properly. When the new soil is put in place, it must be compacted with a sheepsfoot or other type of roller to get the maximum compaction.

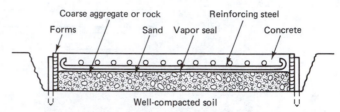

FIGURE 9-22   Properly laid concrete slab.

Once the soil is properly prepared, the foundation is dug and the necessary ground plumbing, ground electrical work, and forms to contain the concrete are put in place. To provide the best foundation, a layer of coarse aggregate should be laid down (see Figure 9-22). The aggregate should then be covered with a layer of sand that fills the voids between the aggregate rocks and provides a solid underlayment for the concrete.

Concrete is very porous. Water will seep upward through a slab from the ground under it. One only has to go into the basement of an old building and feel the dampness to realize that concrete is not a good barrier to water. If one places a rug or mat on the concrete floor of one of these old buildings, then removes it a half-hour later, the concrete under the rug will be wet, proving that the water from the ground is continually seeping through the concrete. In modern construction, a vapor seal in the form of a sheet of plastic is placed over the ground under the slab. Care must be taken to prevent holes or other tears from being put in this vapor seal during the pouring of the concrete.

After the vapor seal is put in place, the necessary reinforcing steel is laid down and tied together. The slab is now ready for pouring. Often a building inspection is required at this point. The concrete is now ordered for delivery to the site and poured within the forms. During the pouring the workers will lift the reinforcing steel slightly, allowing the concrete to work under it. The steel should still be in the lower part of the slab. Vibrators are used to make the concrete flow completely around the steel and against the form lumber. The surface of the concrete is leveled off with a screed. Often, a rotary compactor is run over the surface to pack the concrete. Troweling the concrete while it is still very wet will draw some of the cement to the surface and away from the aggregate. The resulting surface will be very smooth but will show many hairline cracks. If the concrete is allowed to flash or just begin to set before it is troweled, the cement will not be drawn to the surface as much, the surface will not be as smooth, but there will not be as many cracks. Sidewalks, driveways, and roads are usually brushed with a broom or wire brush to provide a rough surface for better traction.

Walls must be prepared in much the same manner as slabs. A proper footing is dug, reinforcing steel is put in place, and the forms are built for the walls. It must be remembered that concrete weighs roughly 150 lb/ft^3 and places a tremendous force on the forms. They must be thoroughly supported and braced. The pressure on the bottom of a column or wall 8 ft high would be over 8 lb/in^2 or about 2.5 times that of water of the same depth. Concrete ceilings and overhead casting in such structures as bridges must supported by falsework, which will be removed once the concrete has cured.

The traditional pour-in-place type of concrete construction is still widely used, but many new techniques are being used. One of these is *tilt slab* construction (Figure 9-23). A building is designed with the sections of the walls being identical. A form for these sections is made on the ground. The concrete is poured in the forms on the ground and allowed to cure. These sections are then tipped up and

FIGURE 9-23 Tilt slab construction.

anchored in place in the walls. The sections can also be lifted on top of each other by cranes for multistory buildings.

A second relatively new type of construction is *slip forming*. Here the form for a structure is made only a few feet high. Once the forms are filled with concrete and allowed to set up for a day or so, the forms are slipped up to the next level and locked in place. The next layer of concrete is poured, the forms are raised again for the next layer, and so on, until the structure is completed. Many cooling towers for electrical power plants and buildings are formed in this manner. The walls of a multistory building have been cast in a few days by this technique.

There are many modifications of the tilt-slab and the slip-forming techniques. The *lift slab method* is but one such adaptation. In this technique, columns are poured on the ground as in the tilt slab method. These columns are then erected and carefully surveyed to ensure that they are perfectly vertical. The bottom floor is cast around these columns. A sheet of plastic is placed over the slab, the forms slipped up, and a second slab cast on top of the first. Another sheet of plastic is laid, then another slab is poured. This process is continued until slabs have been poured for each floor. After curing, the slabs are lifted one at a time by means of hydraulic jacks atop each column and locked in place on the columns. Tilt slabs are then put in place for the walls, and the building needs only the finish work for completion. This type of construction requires considerable time to cast the columns and the slabs, but once this is done, the building literally leaps into its final form in just a few days. It also saves considerable money on form lumber and labor.

One method used in the construction of swimming pools and water tanks utilizes *gunite* and shotcrete. Gunite is a sprayed-on concrete; no forms are needed in this method. To build a pool, a hole is dug in the shape that is desired for the pool. The reinforcing steel is draped over the ground. Fine aggregate and dry cement are mixed at the gunite truck, which is a large blower. The dry cement–aggregate mixture is blown through the large hose and water is mixed at the nozzle. The concrete is sprayed over the reinforcing steel and against the ground. A concrete of very low slump can be made in this way. Further, since the concrete is actually blown against the surface, this is a very dense concrete. Layers of this concrete can be built up until the required thickness is reached. All that is left to do is the finish work. Entire swimming pools have been built in as little as 3 days by this technique.

Water tanks have also been made using gunite. Steel walls are erected and riveted together to form the basic shape of the tank. A 1- to 2-in. layer of gunite is applied. The entire tank is then wrapped with steel wire to prestress the entire structure. Several inches of gunite are then sprayed over the reinforcing steel to provide the mass needed to contain the water. Tanks over 300 ft in diameter and 30 ft high, containing several million gallons of water, have been built using this method.

AIR-ENTRAINED CONCRETE

Concrete can absorb water. In winter, water in the surface of the concrete can freeze and expand, cracking its outer layer. When the concrete thaws, a patch of concrete, perhaps an inch deep, can be turned to rubble. This process is called exfoliation or leafing off. In a very few cycles of freezing and thawing a road can be destroyed. One solution to this problem is *air entrainment*.

When the concrete is mixed, an additive is put in the mix which will cause millions of tiny air bubbles to form in the concrete. As many as 400 to 600 billion bubbles per cubic yard have been shown to be present in an air-entrained concrete. The diameters of the bubbles average about a hundredth of an inch. These air bubbles provide many pockets in the concrete which will allow the water room in which to expand during freezing. A test for air-entrained concrete is the 60-cycle test. Ordinary concrete, frozen and thawed 60 times, would be reduced to aggregate. Air-entrained concretes have been shown to withstand the 60 cycles of freezing and thawing without significant damage. Air-entrained concretes are used in roads in areas where weather conditions are severe.

ADDITIVES TO CONCRETE

Plain concrete is porous, basic, difficult to work with, and dull gray. Additives (called *admixtures*) can be put in the concrete to change all of these characteristics. Pozzolans are added to make the concrete less porous. There are admixtures that will give the concrete more fluidity. Additives can be put in the concrete to speed up the setting or to slow it down. Coloring agents can also be blended into the mix. The architect or engineer should select the admixtures carefully. It makes little sense to add an additive to slow down setting time and another to speed it up in the same batch. A person would be foolish to add pozzolans to cut down on the porosity of the concrete then counter it with an air-entrainment admixture.

ACCELERATED CURING TESTS OF CONCRETE

Several problems remain in the use of concrete. Unlike steel, wood, or masonry construction, concrete has no strength when it is first put in place. A steel beam can be fabricated and inspected at the factory and has its ultimate strength before it is put in a building, but much of the quality and strength of concrete depends on the workers. This places an additional burden on the builders and on the

FIGURE 9-24 Accelerated curing techniques.

Factor	Curing Technique		
	Warm-Water	Boiling-Water	Autogenous
Curing medium	Water	Water	Insulating foam
Curing temperature	95 ± 5°F	Boiling	Heat of hydration
Age accelerated curing begins	Immediately	23 hr	Immediately
Duration of curing	23½ hr ± 30 min	3½ hr ± 5 min	48 hr ± 15 min
Age at testing	24 hr ± 15 min	28½ hr ± 15 min	49 hr ± 15 min

inspectors. Further, there is no assurance that the concrete cured on the building site would have the same strength characteristics as that of the test samples cured in the lab. These problems are further compounded by the use of rapid construction techniques such as slip forming. An entire building of nine or ten floors can be poured in as little as 17 days, yet the engineers have to wait for the 28-day tests to come in to see if even the first day's pour of concrete meets specifications. Imagine the problems if the first batch of concrete, poured at ground level, is found to be defective.

Three methods of accelerating the curing of concrete test samples have been developed by the ASTM: the *warm-water method*, the *boiling-water method*, and the *autogenous method* (Figure 9-24).

In the *warm-water method*, standard 6- by 12-in. test samples are cast in metal cylinders that can be capped at both ends. These samples are immediately submerged in water held at 95°F (±5 degrees) (Figure 9-25). After a curing time of 23.5 hours (±30 minutes) the cylinders are removed from the tank and demolded. They are left standing for 30 minutes to return to room temperature. The test samples are then capped and tested at 24 hours (±15 minutes).

For the *boiling-water method*, the standard test samples are taken but allowed to cure at room temperature for a period of 23 hours (±15 minutes). The samples are then placed in boiling water for 3.5 hours (±5 minutes). The samples are removed from the boiling water and left to stand at room temperature for not less

FIGURE 9-25 Warm water accelerated curing.

than 1 hour prior to capping. The samples are capped and tested at 28.5 hours (± 15 minutes).

The *autogenous method* involves using the heat of hydration of the cement itself to accelerate the curing of the samples. Immediately after molding, the molds are covered with a metal cap and placed in a tightly fitting heavy-duty plastic bag from which as much of the entrapped air as possible is removed prior to tying the neck of the bag. The molds are then placed in a container that is well insulated with Styrofoam or other insulation material. The tank looks similar to a camp ice box that has been filled with Styrofoam. Holes are cut in the Styrofoam to allow the insulation to fit tightly around the test cylinders. The ASTM has very strict specifications as to the type of container and its heat-retention capabilities. A 6- by 12-in. cylinder of water at 180°F placed in the insulated container must be at least 100°F after 72 hours. The concrete test cylinders are left in this container for a period of 48 hours (± 15 minutes), then removed and allowed to stand for 30 minutes at room temperature. The cylinders are capped and tested at 49 hours (± 15 minutes).

In all the tests described above, the strengths obtained at the end of the accelerated curing times are approximately equal to a room temperature cure obtained from 3 to 5 days. From these data it can be determined if the concrete is on the curing line to give the required strengths at the end of 28 days of normal curing. Usually, the accelerated curing test is backed up with a standard 28-day test.

Thinker 9-2:

List several examples where accelerated curing of test samples might be specified.

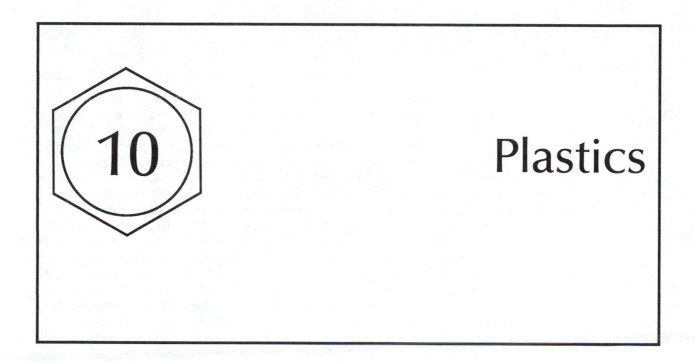

Plastics

10

INTRODUCTION

Until a century ago, people were limited to products made from naturally occurring materials. Cloth was made from cotton, flax, wool, silk, or the bark of trees. Shoes and other products were made from leather. Building materials were stone, wood, or concrete. Some easily refined metals, such as iron, copper, lead, tin, and zinc, were used in machines. If a product could not be made with these naturally occurring materials, it simply could not be made.

In July 1907, Leo H. Baekeland made one of the most important discoveries of the twentieth century. While working with the reaction of phenol and formaldehyde (which will be discussed later), he produced the first synthetic plastic, which he patented under the trade name Bakelite. The "age of plastics" was launched. Although cellulose and cellulose acetate were in common use before the beginning of the twentieth century, they were not synthetic materials but were made from cotton and wood fibers. Similarly, rayon was a regenerated cellulose fiber. By the year 1940, Bakelite, the cellulose materials (including celluloid), natural soft and hard rubber, and rayon comprised most of the plastic materials used in industry. Now, plastics number in the thousands.

The term "plastics" is confusing. Earlier, in discussing the stress–strain curve in testing metals and other materials, we used the terms "plastic deformation" and "plasticity." We expect anything that is "plastic" to be able to withstand a certain amount of permanent deformation before failing. Yet many synthetic plastics are not very "plastic." To avoid this double meaning of the word, let us refer to the synthetic materials as *polymers* or *high polymers*. The term *mer* means "body" and is the basic molecule or building block of the specific synthetic material. *Poly* is the Greek word for "many." Therefore, the word "polymer" simply means "many bodies." Plastics are made from many of the same types of molecules joined together.

BASIC ORGANIC CHEMISTRY

To understand the mechanisms by which polymers are made, as well as to comprehend the properties of a particular polymer, one must first know a little about *organic chemistry*—the chemistry of carbon. One of the confusing things about organic chemicals is the way in which they are named. Prior to 1932 and the founding of the *International Union of Chemists* (IUC), which later became the *International Union of Pure and Applied Chemists* (IUPAC), there was no set method of naming organic compounds. The discoverer of the compound named it, sometimes naming it after himself or a fellow chemist. As a result, we have such compounds as Fehling's solution, Sweitzer reagent, and many others. The IUPAC, which now sets the standards for chemistry, has developed a system of naming compounds based on the structure of the molecule. If the chemical structure of the molecule is known, the proper IUPAC chemical name can be derived from that structure. Conversely, the structure of the compound can be written from the correct IUPAC name. One problem of naming organic compounds still exists. Many organic compounds were known prior to IUPAC and are still known by their old names. Further, the same organic chemical can be known by several trade names which have nothing to do with the chemical structure. Many compounds have several names, which often adds confusion to the identification of these chemicals. Several of these multiply named compounds are used in plastics.

There is one rule in the chemistry of carbon that is almost never violated. *Every carbon atom in a compound must have four bonds satisfied*; that is, each carbon must be joined to other atoms in exactly four places. One of the simplest compound that could be formed using carbon would be to react it with four hydrogen atoms:

$$
\begin{array}{c}
\text{H} \\
| \\
\text{H—C—H} \\
| \\
\text{H}
\end{array}
$$

This is the structural formula for methane, commonly written as CH_4. Any compound with only one carbon atom in the molecule would be named as a methane derivative.

If any compound has an unsatisfied bond, it becomes a *radical*. A methane molecule minus one hydrogen becomes a methyl radical:

$$
\begin{array}{c}
\text{H} \\
| \\
\text{H—C—} \\
| \\
\text{H}
\end{array}
$$

Radicals do not exist very long in this state; they readily react with any other atom or molecule that happens to wander into their proximity. If a chlorine atom were introduced to this radical, it would immediately join with the methyl radical to form a compound called chloromethane:

$$
\begin{array}{c}
\text{H} \\
| \\
\text{H—C—Cl} \\
| \\
\text{H}
\end{array}
$$

Prior to the IUPAC standards, this compound was called methyl chloride and is still often listed under that name.

If a second hydrogen is replaced by a chlorine atom, the result would be dichloromethane:

$$
\begin{array}{c}
\text{H} \\
| \\
\text{Cl}\!-\!\text{C}\!-\!\text{Cl} \\
| \\
\text{H}
\end{array}
$$

Replacing the third hydrogen with a chlorine would produce trichloromethane:

$$
\begin{array}{c}
\text{H} \\
| \\
\text{Cl}\!-\!\text{C}\!-\!\text{Cl} \\
| \\
\text{Cl}
\end{array}
$$

The old name for this compound was methylene trichloride but it is better known as *chloroform*. It can be written as $CHCl_3$.

The proper IUPAC name for a carbon with four chlorine atoms attached is tetrachloromethane. However, most people prefer to call it carbon tetrachloride or just "carbon tet" (CCl_4). Carbon tet is an excellent solvent for grease and paints, but its use for those purposes has been outlawed by the Environmental Protection Agency (EPA) since it has been found to be a health hazard to people who use it over a long period.

Any two-carbon compound is an ethane derivative:

$$
\begin{array}{c}
\text{H} \quad \text{H} \\
| \quad\; | \\
\text{H}\!-\!\text{C}\!-\!\text{C}\!-\!\text{H} \\
| \quad\; | \\
\text{H} \quad \text{H}
\end{array}
$$

We could go through the same routine with this compound that we did with methane. Remove a hydrogen and the ethyl radical is formed:

$$
\begin{array}{c}
\text{H} \quad \text{H} \\
| \quad\; | \\
\text{H}\!-\!\text{C}\!-\!\text{C}\!-\! \\
| \quad\; | \\
\text{H} \quad \text{H}
\end{array}
$$

Replace that hydrogen with another atom (this time we use bromine) and we get bromoethane:

$$
\begin{array}{c}
\text{H} \quad \text{H} \\
| \quad\; | \\
\text{H}\!-\!\text{C}\!-\!\text{C}\!-\!\text{Br} \\
| \quad\; | \\
\text{H} \quad \text{H}
\end{array}
$$

Replace two hydrogens with bromine and the result would be dibromoethane:

$$
\begin{array}{c}
\ \ \ \ H\ \ \ H \\
\ \ \ \ | \ \ \ \ | \\
H-C-C-Br \\
\ \ \ \ | \ \ \ \ | \\
\ \ \ \ H\ \ \ Br
\end{array}
$$

Note that both bromine atoms are on the same carbon. If we were to put the bromine atoms on different carbon atoms,

$$
\begin{array}{c}
\ \ \ \ \ \ H\ \ \ H \\
\ \ \ \ \ \ | \ \ \ \ | \\
Br-C-C-Br \\
\ \ \ \ \ \ | \ \ \ \ | \\
\ \ \ \ \ \ H\ \ \ H
\end{array}
$$

we would have a different dibromoethane. These compounds are called *isomers* of each other. The prefix "iso" means "same." An isomer is therefore a compound with the "same body" (in this case the same number and type of atoms) but a different arrangement of the same atoms.

To differentiate between the various isomers, a numbering system was developed by the IUPAC. By numbering the carbon atoms the molecule would be designated as follows:

$$
\begin{array}{c}
\ \ \ \ H\ \ \ H \\
\ \ \ \ |_2 \ \ |_1 \\
H-C-C-Br \\
\ \ \ \ | \ \ \ \ | \\
\ \ \ \ H\ \ \ Br
\end{array}
\qquad
\begin{array}{c}
\ \ \ \ \ \ H\ \ \ H \\
\ \ \ \ \ \ |_2 \ \ |_1 \\
Br-C-C-Br \\
\ \ \ \ \ \ | \ \ \ \ | \\
\ \ \ \ \ \ H\ \ \ H
\end{array}
$$

1,1-Dibromoethane   1,2-Dibromoethane

The first of these isomers would become 1,1-dibromoethane, while the second would be called 1,2-dibromoethane. These iosomers have the same number and type of atoms, and the same molecular weight, but different properties as shown in Figure 10-1.

Three-carbon compounds are propane derivatives:

$$
\begin{array}{c}
\ \ \ \ H\ \ \ H\ \ \ H \\
\ \ \ \ | \ \ \ \ | \ \ \ \ | \\
H-C-C-C-H \\
\ \ \ \ | \ \ \ \ | \ \ \ \ | \\
\ \ \ \ H\ \ \ H\ \ \ H
\end{array}
$$

FIGURE 10-1 Properties of 1,1-dibromoethane and 1,2-dibromoethane.

	1,1-Dibromoethane	1,2-Dibromoethane
Melting point (°C)	−63	0.79
Boiling point (°C)	108	131.36
Density (g/cm³)	2.0555	2.792
Solubility in alcohol	Slight	Very

Source: Data from *Handbook of Chemistry and Physics*, CRC Press, Boca Raton, Fla.

which is often written by chemists as $CH_3CH_2CH_3$. Propane is a common gas used in liquefied form in torches and gas heaters, and as an internal combustion engine fuel. Propane could be substituted in much the same manner as demonstrated with methane and ethane.

Thinker 10-1:

How many different isomers of dichloropropane could there be?

Four-carbon compounds are butanes ($CH_3CH_2CH_2CH_3$). In compounds containing five or more carbon atoms, the Latin and Greek numbering prefixes are used to name the chain. Five-carbon compounds are called pentanes, six-carbon compounds are hexanes, seven-carbon compounds are heptanes, eight-carbon compounds are octanes, nine-carbon compounds are nonanes, ten-carbon compounds are decanes, eleven-carbon compounds are known as hendecanes, and so on. Some chains are often given a standard IUPAC name and the compounds are then listed as derivatives of those standardized names. Examples of these will be shown later.

It should be seen that compounds having four or more carbons can form several isomers just by the way the carbon atoms are arranged. There could be two different butanes.

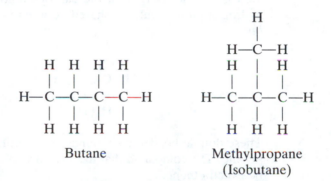

Butane   Methylpropane
                                      (Isobutane)

Prior to the UIPAC system, these were known as normal butane and isobutane, respectively. Under the new system, only the longest chain is named; the rest of the compound is listed as side chains. Isobutane would therefore be known as methylpropane since the longest chain has only three carbons.

There could be three different five-carbon compounds.

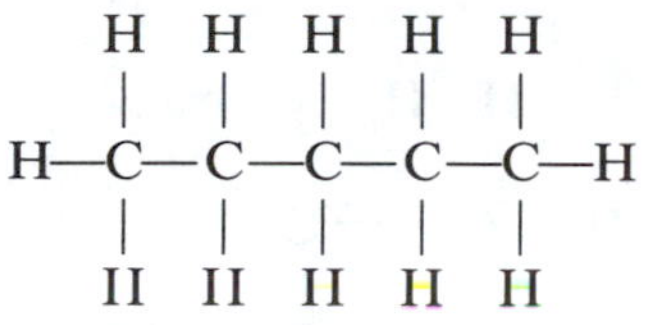

Pentane 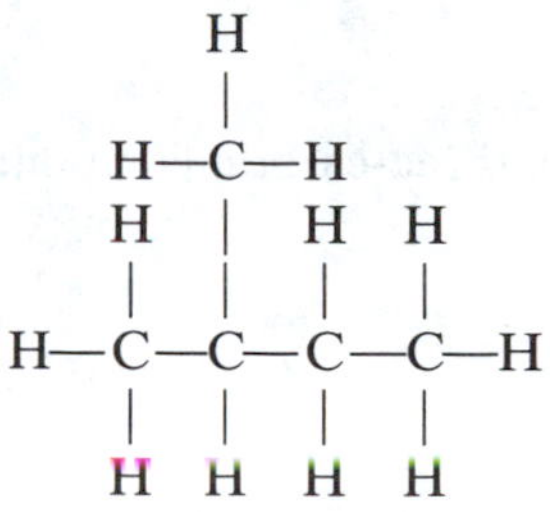   Methylbutane   Dimethylpropane
                           (Isopentane)   (Neopentane)

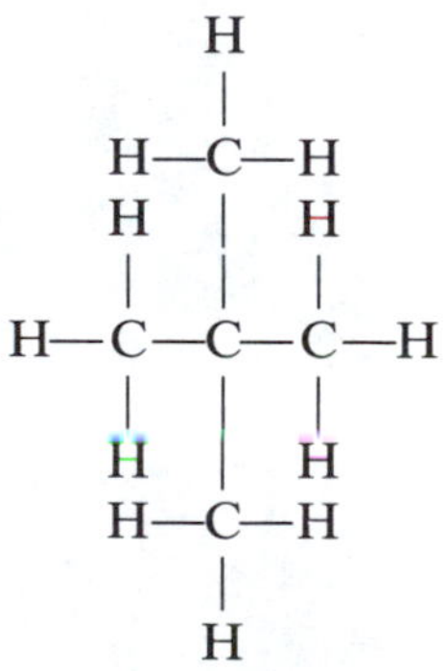

These compounds were once known as normal pentane, isopentane, and neopentane. Now they go under the names of pentane, methylbutane, and dimethylpropane. In all of these compounds each carbon atom always has four bonds around it. The bonds between the atoms are formed by sharing a pair of electrons.

$$
\begin{array}{c}
\text{H} \\
\overset{\cdot\cdot}{} \longleftarrow \text{electrons} \\
\text{H : C : H} \\
\overset{\cdot\cdot}{} \\
\text{H}
\end{array}
$$

Carbon has four electrons that it can share, and hydrogen has one. Ethane would have bonds formed by the shared electrons between the carbons and hydrogens as follows:

$$
\begin{array}{c}
\text{H} \quad \text{H} \\
\overset{\cdot\cdot}{} \quad \overset{\cdot\cdot}{} \\
\text{H : C : C : H} \\
\overset{\cdot\cdot}{} \quad \overset{\cdot\cdot}{} \longleftarrow \text{electrons} \\
\text{H} \quad \text{H}
\end{array}
$$

If there are not enough hydrogen atoms to share an electron with the carbon atoms, the electrons from the carbons that normally would be shared with the hydrogen atoms could be shared between two carbon atoms, producing a *double bond*.

$$
\begin{array}{ccc}
\text{H : C : C : H} & \longrightarrow & \text{H : C :: C : H} \\
\overset{\cdot\cdot}{} \;\; \overset{\cdot\cdot}{} & & \overset{\cdot\cdot}{} \;\; \overset{\cdot\cdot}{} \\
\text{H} \quad \text{H} & & \text{H} \quad \text{H}
\end{array}
$$

These double-bonded compounds are identified by having the chemical name end in -ene. The compound that has two carbon atoms and only four hydrogens would be called ethene.

If a hydrogen is removed from the ethene molecule, it becomes the vinyl radical:

$$
\begin{array}{c}
\text{H—C}\!=\!\text{C—H} \\
\;\;|\quad\; | \\
\text{H}
\end{array}
$$

Vinyl radical

A three-carbon compound with a double bond would become propene:

$$
\begin{array}{c}
\text{H} \quad \text{H} \\
\;| \quad\; | \\
\text{H—C—C}\!=\!\text{C—H} \\
\;| \quad\; | \quad\; | \\
\text{H} \quad \text{H} \quad \text{H}
\end{array}
$$

Propene

We could have butene, pentene, and so on. In compounds with four or more carbons, the position of the double bond must be specified. This is done by numbering the carbon atoms. The position of the double bond is noted by the number on the lower-numbered carbon next to the double bond. Thus we could have isomers of butene, 1-butene or 2-butene.

$$
\begin{array}{ccccc}
& H & H & H & \\
& | _4 & | _3 & | _2 & | _1 \\
H\!-\!C\!&-\!C\!&-\!C\!&=\!C\!&-\!H \\
& | & | & | & \\
& H & H & & H
\end{array}
\qquad
\begin{array}{ccccc}
& H & H & H & H \\
& | _1 & | _2 & | _3 & | _4 \\
H\!-\!C\!&-\!C\!&=\!C\!&-\!C\!&-\!H \\
& | & & & | \\
& H & & & H
\end{array}
$$

1-Butene   2-Butene

There could never be a 3-butene since the chain could simply be numbered from the other end and the compound would revert to 1-butene.

If two more hydrogen atoms were removed from the ethene molecule, the remaining electrons would form a triple bond. The names of triple-bonded compounds always end in -yne. The two-carbon triple-bonded compound becomes ethyne (pronounced *eth-eye-n*). This compound is still commonly known by its old name, acetylene.

$$H\!-\!C\!\equiv\!C\!-\!H$$

Ethyne
(Acetylene)

Similarly, there could be propyne, butynes, pentynes, and so on.

Compounds that have double or triple bonds are said to be unsaturated, meaning that they are not as full of hydrogen atoms as they could be. Many molecules have many double bonds and are then said to be polyunsaturated compounds. Many cooking oils and synthetic butter spreads are advertised as being polyunsaturated. Unsaturated and polyunsaturated compounds can usually be detected by the fact that they give off black smoke when burned in still air. Saturated compounds usually do not emit black smoke when burned.

RING COMPOUNDS

In 1865, a German chemist, Friedrich August Kekule Von Stradonitz, (known as Kekule) was working with a chemical found in coal tar, benzene. Benzene was determined to have only six carbons and six hydrogens in the molecule. The problem—how could six carbons with only six hydrogens be arranged so that there were four bonds associated with each carbon? Kekule suggested that the carbons were formed in a ring:

$$
\begin{array}{ccc}
& H & \\
& | & \\
& C & \\
& /\!/ \quad \backslash & \\
H\!-\!C & & C\!-\!H \\
| & & \| \\
H\!-\!C & & C\!-\!H \\
& \backslash\!\backslash \quad / & \\
& C & \\
& | & \\
& H &
\end{array}
$$

Of course, there had to be three double bonds in the ring. Since these double bonds have been found by recently developed instruments to be continually jumping between the carbon atoms, the current method of writing the structure of benzene is

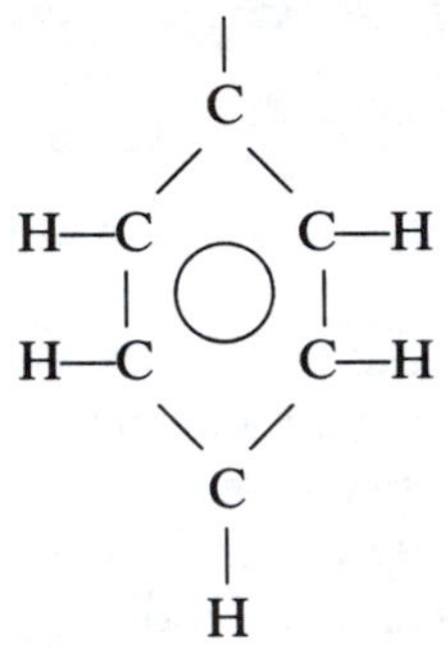

H
|
C
/   \
H—C   ◯   C—H
|
H—C   C—H
\   /
C
|
H

Further, a chemical shorthand is often used for the benzene structure—a hexagon with a circle inside.

⬡

Remember that each corner of the hexagon represents a carbon and a hydrogen atom. The compounds containing rings are called *aromatic compounds*, to distinguish them from straight-chain or *aliphatic compounds*.

    Removing a hydrogen atom from the ring also makes it a radical. If hydrogen is removed from the benzene ring, the phenyl radical is formed:

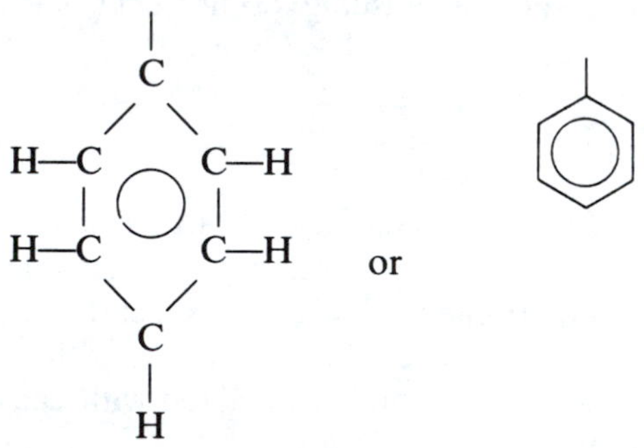

|
C
/   \
H—C   ◯   C—H
|
H—C   C—H
\   /
C
|
H

     or

Phenyl radical

If that hydrogen is replaced with a methyl group, the compound becomes toluene:

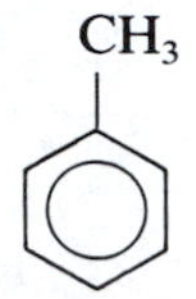

CH_3
|
⬡

Toluene

The addition of an amine (NH_2) group to the phenyl radical forms the analine molecule:

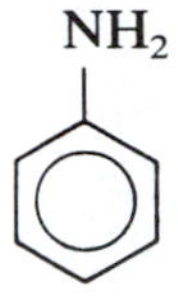

NH_2

Analine

If the replacement group is the hydroxyl (OH), the molecule becomes known as phenol. The old name for phenol was carbolic acid since it exhibits some weak acid properties.

OH

Phenol

In all the molecules above, the CH_3, NH_2, or OH are called directing groups. These molecules should be memorized since they will be used later in the discussion of polymers.

Another hydrogen from the toluene could be replaced by chlorine to form chlorotoluene. The only problem is that there are five different hydrogens on the benzene ring that could be replaced, all forming chlorotoluene. To differentiate between these sites, they are assigned prefixes. The position next to the directing group (methyl in the case of toluene) is called the ortho or "1" position. A chlorine atom in this position would make the compound become *ortho*-chlorotoluene or 1-chlorotoluene.

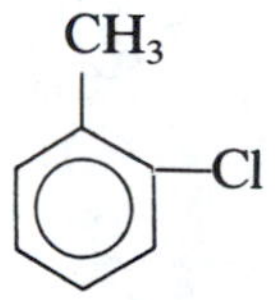

CH_3

Cl

o-Chlorotoluene

The next position down the ring is known as the meta or "2" position. Putting a chlorine in this spot would create *meta*-chlorotoluene or 2-chlorotoluene.

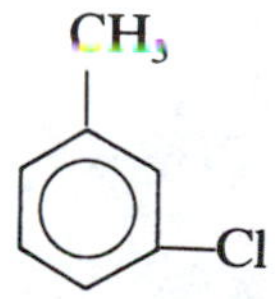

CH_3

Cl

m-Chlorotoluene

The place directly opposite the directing group is the para or "3" (sometimes called the "tere") position. A chlorine here would make the compound into *para*-chlorotoluene.

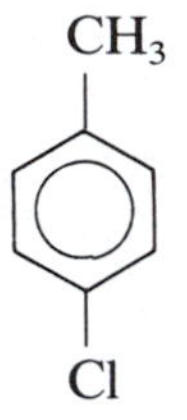

$$CH_3$$

Cl

p-Chlorotoluene

One interesting compound that combines several of these positions is *o,o',p*-trinitrotoluene:

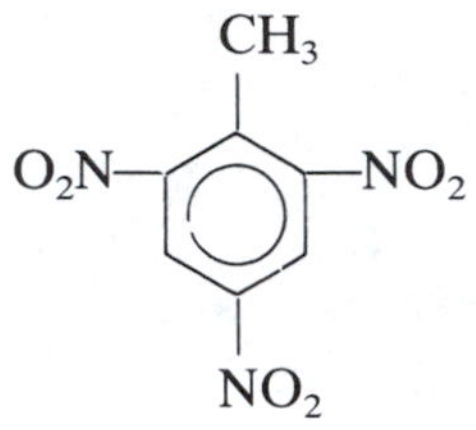

$$CH_3$$

O_2N——NO_2

$$NO_2$$

o,o',*p*-**Trinitrotoluene**

This compound is formally known as *ortho,ortho-prime,para*-trinitrotoluene. However, it is much easier to call it by its acronym, TNT.

PROBLEM SET 10-1

What is the correct IUPAC name for the compounds in Problems 1 through 7?

1. CH_3—CH_2—CH_2—CH_2—CH_2—CH_3
2. CH_3—CH_2—CH_2—CH=CH_2
3. CH_3—CH_2—CH=CH—CH_2—CH_3
4. CH_3—CH—CH_2—CH_2—CH_3
     |
     CH_3
5. 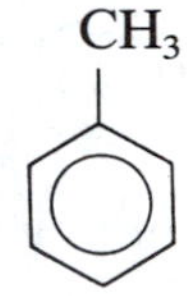 CH_3

6. 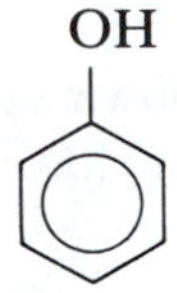   OH

                       Cl
                       |
7. CH_3—CH—CH_2—C—CH_3
     | |
     Cl   Cl

Write the formulas for the compounds in Problems 8 through 16.

8. 3,3-Dichlorohexane
9. *meta*-Bromotoluene
10. *para*-Nitrophenol
11. 1,3-Propandiol
12. 1,3-Dichlorobutane
13. 2,2,4-Trimethylpentane (This compound is the old isooctane used as a standard for fuel for internal-combustion engines.
14. 1-Butyne
15. *para*-Hydroxyphenol
16. Phenylethane
17. What could be another name for *para*-aminotoluene?
18. TCE used as a solvent in industry is 1,1,1-trichloroethane (TCE). What is the formula for this compound?
19. What is the molecular weight of TCE?
20. How many different isomers could there be of dichloropropane?

CLASSES OF ORGANIC COMPOUNDS

There are three major classes of inorganic chemicals: acids, bases, and salts. Acids are characterized by the presence of a hydrogen ion in the molecule. Some examples of inorganic acids are:

Hydrochloric acid	HCl (muriatic acid)
Nitric acid	HNO_3
Sulfuric acid	H_2SO_4 (oil of vitrol)
Phosphoric acid	H_3PO_4
Carbonic acid	H_2CO_3

Bases always have the hydroxide group (OH) in the molecule. Following are some of the better known bases:

Sodium hydroxide	$NaOH$ (lye)
Potassium hydroxide	KOH
Barium hydroxide	$Ba(OH)_2$
Calcium hydroxide	$Ca(OH)_2$ (slaked lime)
Ammonium hydroxide	NH_4OH (household ammonia)

The third class of inorganic chemicals, salts, are formed by reacting an acid with a base. This reaction always forms a salt and water. A few examples are:

$$NaOH + HCL \longrightarrow NaCl + HOH$$

$$KOH + HNO_3 \longrightarrow KNO_2 + HOH$$

$$Ba(OH)_2 + H_2SO_4 \longrightarrow BaSO_4 + 2HOH$$

The 2 before the HOH means that there are two molecules of water formed in the reaction. This balances the equation—there are just as many of each type of atom on both sides of the equation.

In organic chemistry there are many classes of compounds. The hydroxyl (OH) applied to an organic compound makes that compound an *alcohol*. The chemical names of all alcohols always end in the letters *-ol*. The alcohol formed from methane is called methanol and has the formula

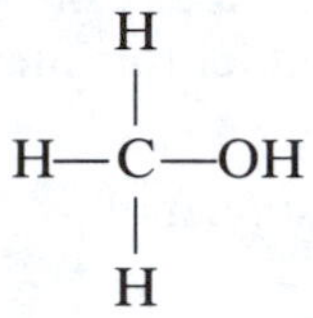

$$
\begin{array}{c}
H \\
| \\
H\!-\!C\!-\!OH \qquad \text{or} \qquad CH_3OH \\
| \\
H
\end{array}
$$

Methanol

Methanol is still called methyl alcohol. Since it can be made by distilling wood chips, it is also known as wood alcohol.

The alcohol made from ethane becomes ethanol, with the formula

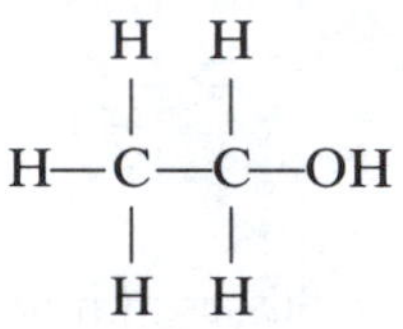

$$
\begin{array}{cc}
H & H \\
| & | \\
H\!-\!C\!-\!C\!-\!OH & \qquad \text{or} \qquad CH_3CH_2OH \\
| & | \\
H & H
\end{array}
$$

Ethanol

Ethanol, also called ethyl alcohol, is commonly made by the fermentation of starches or sugars. In the fermentation process, yeast cells (little one-celled animals) ingest or eat the sugars and starches from plants and give off ethanol as a waste product. Ethanol is therefore called grain alcohol. This alcohol is the active ingredient in whisky, beer, wine, and other alcoholic beverages.

The alcohol derived from the three-carbon chain is propanol. However, the OH group can be joined to the propane molecule in either of two places. It therefore forms two isomers, 1-propanol (sometimes called normal propyl alcohol) and 2-propanol (often named isopropyl alcohol). The latter is the rubbing alcohol used in hospitals.

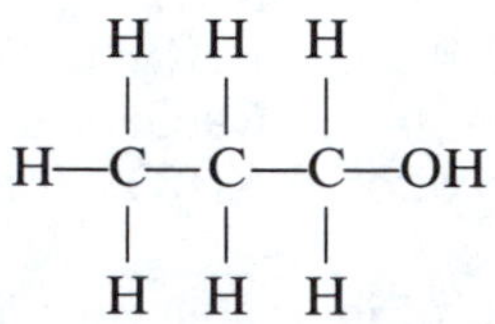

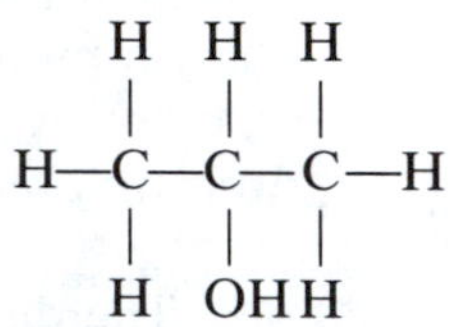

$$
\begin{array}{ccc}
H \quad H \quad H & \qquad & H \quad H \quad H \\
| \quad\; | \quad\; | & & | \quad\; | \quad\; | \\
H\!-\!C\!-\!C\!-\!C\!-\!OH & & H\!-\!C\!-\!C\!-\!C\!-\!H \\
| \quad\; | \quad\; | & & | \quad\; | \quad\; | \\
H \quad H \quad H & & H \quad OH \;H
\end{array}
$$

or

$$CH_3CH_2CH_2OH \qquad\qquad\qquad CH_3CHCH_3$$
$$|$$
$$OH$$

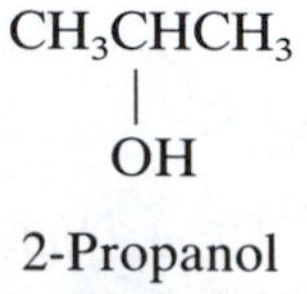

1-Propanol   2-Propanol
(*n*-Propyl alcohol)   (Isopropyl alcohol)

More than one alcohol group can be attached to a molecule. If two OH groups are attached to the ethane molecule, it becomes 1,2-ethanediol.

$$
\begin{array}{ccc}
& \text{H} \quad \text{H} & \\
& | \quad\ | & \\
\text{HO}-\text{C}-\text{C}-\text{OH} & \text{or} & \text{HOCH}_2\text{CH}_2\text{OH} \\
& | \quad\ | & \\
& \text{H} \quad \text{H} &
\end{array}
$$

1,2-Ethanediol

This compound is better known as ethylene glycol, which is used in automobile antifreeze and in the production of the synthetic fiber, Dacron.

A second class of organic compounds often found in plastics are the *aldehydes*. Aldehydes always have the functional group

$$
\begin{array}{c}
\quad\quad \text{O} \\
\quad\quad \parallel \\
-\text{C} \\
\quad\quad \backslash \\
\quad\quad\quad \text{H}
\end{array}
$$

on the molecule. The names of aldehydes always end in -*al*. Therefore, the aldehyde of a one-carbon compound would be called methanal and have the structure

$$
\begin{array}{c}
\quad\quad\quad \text{O} \\
\quad\quad\quad \parallel \\
\text{H}-\text{C} \\
\quad\quad\quad \backslash \\
\quad\quad\quad\quad \text{H}
\end{array}
$$

Methanal
(Formaldehyde)

Prior to the IUPAC naming system this compound was called either methyl aldehyde or more commonly, formaldehyde. Formaldehyde is used in making Bakelite and other plastics and as a biological specimen preservative.

The two-carbon aldehyde would be named ethanal; the three-carbon aldehyde is propanal; the four-carbon aldehyde becomes butanal; and so on.

$$
\begin{array}{ccc}
\text{H} \quad\ \text{O} & \text{H} \quad \text{H} \quad\ \text{O} & \text{H} \quad \text{H} \quad \text{H} \quad\ \text{O} \\
| \quad\ \parallel & | \quad\ | \quad\ \parallel & | \quad\ | \quad\ | \quad\ \parallel \\
\text{H}-\text{C}-\text{C} & \text{H}-\text{C}-\text{C}-\text{C} & \text{H}-\text{C}-\text{C}-\text{C}-\text{C} \\
| \quad\ \backslash & | \quad\ | \quad\ \backslash & | \quad\ | \quad\ | \quad\ \backslash \\
\text{H} \quad\ \text{H} & \text{H} \quad \text{H} \quad\ \text{H} & \text{H} \quad \text{H} \quad \text{H} \quad\ \text{H}
\end{array}
$$

or

$$
\begin{array}{ccc}
\quad\quad \text{O} & \quad\quad \text{O} & \quad\quad \text{O} \\
\quad\quad \parallel & \quad\quad \parallel & \quad\quad \parallel \\
\text{CH}_3\text{C} & \text{CH}_3\text{CH}_2\text{C} & \text{CH}_3\text{CH}_2\text{CH}_2\text{C} \\
\quad\quad \backslash & \quad\quad \backslash & \quad\quad \backslash \\
\quad\quad\quad \text{H} & \quad\quad\quad \text{H} & \quad\quad\quad \text{H} \\
\text{Ethanal} & \text{Propanal} & \text{Butanal}
\end{array}
$$

If the hydrogen of the aldehyde is replaced by any other radical, the compound becomes a *ketone*. The basic structure of ketones is

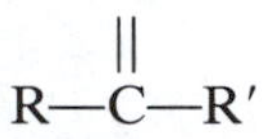

$$O$$
$$\|$$
$$R—C—R'$$

where R and R' are any radicals. Ketones are named by adding the suffix -one to the basic name. The simplest ketone would be the three-carbon structure

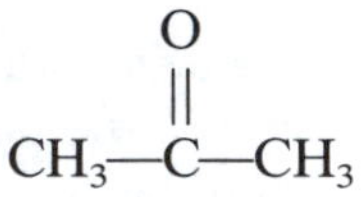

$$O$$
$$\|$$
$$CH_3—C—CH_3$$

Propanone
(Acetone)

and would be called propanone. Propanone is more commonly called by its older name, acetone or dimethyl ketone. The four-carbon ketone would be called butanone or by the name methyl ethyl ketone (commonly abbreviated MEK).

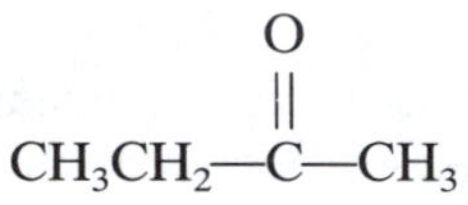

$$O$$
$$\|$$
$$CH_3CH_2—C—CH_3$$

Butanone
(Methyl ethyl ketone)

There could be two five-carbon ketones: 2-pentanone and 3-pentanone.

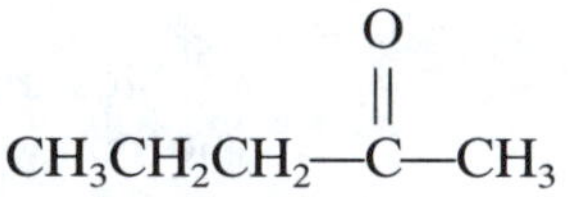

$$O \qquad\qquad\qquad O$$
$$\| \qquad\qquad\qquad \|$$
$$CH_3CH_2CH_2—C—CH_3 \qquad CH_3CH_2—C—CH_2CH_3$$

2-Pentanone 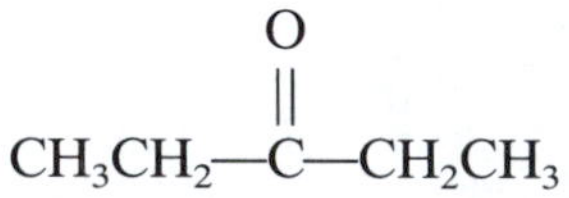   3-Pentanone
                                         (Diethyl ketone)

Unfortunately, many of these compounds are still known by their pre-IUPAC names. 2-Butanone is still purchased under the name methyl ethyl ketone, and 3-butanone still has the name diethyl ketone. Ketones are used in some polymers and are often used as paint and grease solvents in industry.

Another important class of chemicals used in polymer chemistry are the *organic acids*. The active group in these acids is the —COOH or the

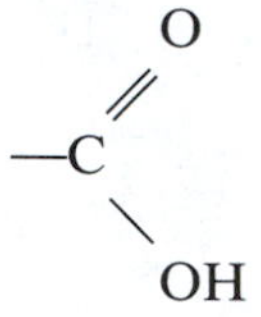

$$O$$
$$/\!/$$
$$—C$$
$$\backslash$$
$$OH$$

group. As is the case with the inorganic acids, it is the hydrogen ion that gives these compounds their acidic properties. Organic acids are named by applying the ending -oic acid to the basic structural name. The one-carbon acid becomes known

as methanoic acid. Its structure is

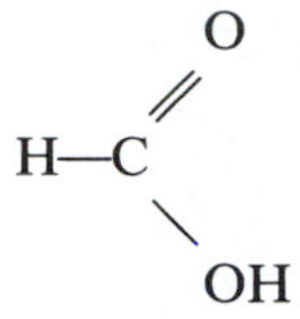

$$
\begin{array}{c}
\quad\quad\; O \\
\quad\quad\; \parallel \\
H-C \\
\quad\quad \backslash \\
\quad\quad\; OH
\end{array}
$$

Methanoic acid
(Formic acid)

The older and better known name for this acid is formic acid. The acid of the ethane molecule is ethanoic acid but is almost always called acetic acid. This is the acid found in apples and some other fruit. The weak solution of acetic acid is vinegar.

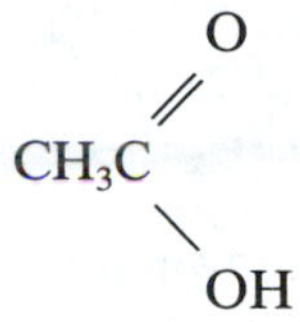

$$
\begin{array}{c}
\quad\quad\quad\; O \\
\quad\quad\quad\; \parallel \\
CH_3C \\
\quad\quad\quad \backslash \\
\quad\quad\quad\; OH
\end{array}
$$

Ethanoic acid
(Acetic acid)
(Vinegar)

It follows from this naming system that the three-carbon acids would be propanoic acid, the four-carbon acid becomes butanoic acid, and so on.

As is the case with alcohols, it is possible to have more than one acid radical on a molecule. The simplest double acid would be ethandioic acid,

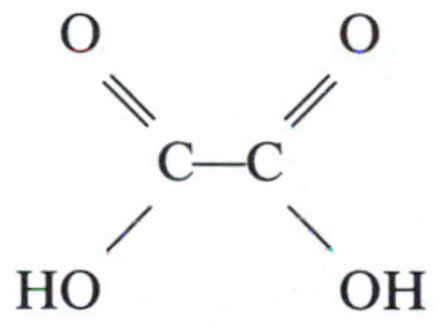

$$
\begin{array}{c}
O \quad\quad\quad O \\
\diagdown\diagdown \quad\quad \diagup\diagup \\
C-C \\
\diagup \quad\quad\quad \diagdown \\
HO \quad\quad\quad OH
\end{array}
$$

Ethandioic acid
(Oxalic acid)

better known as oxalic acid. Propandioic acid, also known as malonic acid, is

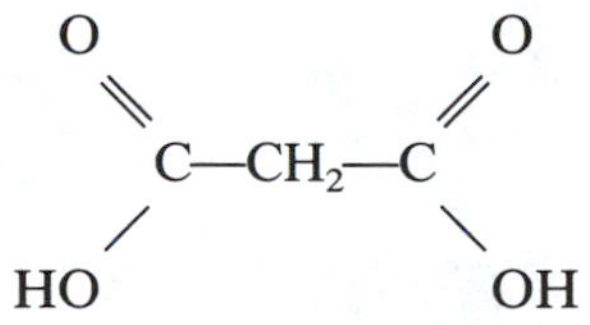

$$
\begin{array}{c}
O \quad\quad\quad\quad\quad\; O \\
\diagdown\diagdown \quad\quad\quad\quad \diagup\diagup \\
C-CH_2-C \\
\diagup \quad\quad\quad\quad\quad \diagdown \\
HO \quad\quad\quad\quad\quad OH
\end{array}
$$

Propandioic acid
(Malonic acid)

In inorganic chemistry, an acid reacted with a base produces a salt. The same reaction is possible with organic acids. The salt produced in organic chemicals is known as an *ester*. Esters have the basic structural formula

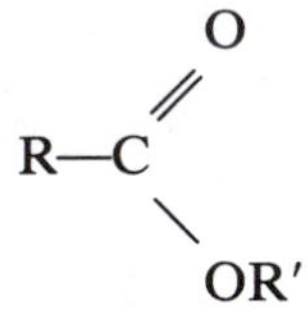

$$
\begin{array}{c}
\quad\quad O \\
\quad\quad \parallel \\
R\!-\!C \\
\quad\quad \diagdown \\
\quad\quad\quad OR'
\end{array}
$$

where R and R' are any radical. The names of esters always end in *-ate*. Therefore, the compound

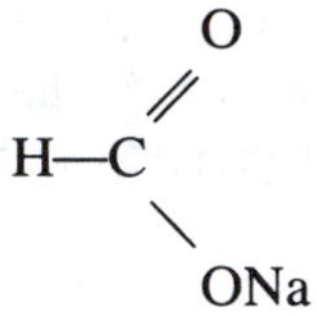

$$
\begin{array}{c}
\quad\quad O \\
\quad\quad \parallel \\
H\!-\!C \\
\quad\quad \diagdown \\
\quad\quad\quad ONa
\end{array}
$$

Sodium methanate
(Sodium formate)

would be known as sodium methanate (also known as sodium formate). If the R' is a methyl group, the compound becomes methyl methanate:

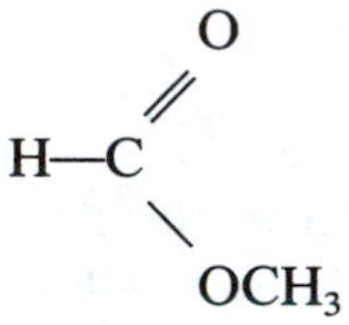

$$
\begin{array}{c}
\quad\quad O \\
\quad\quad \parallel \\
H\!-\!C \\
\quad\quad \diagdown \\
\quad\quad\quad OCH_3
\end{array}
$$

Methyl methanate
(Methyl formate)

Methyl ethanate would be

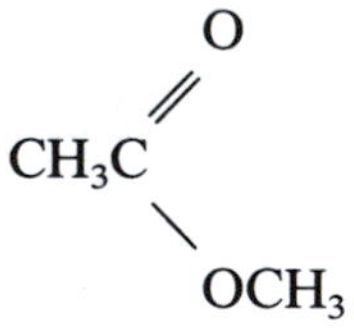

$$
\begin{array}{c}
\quad\quad O \\
\quad\quad \parallel \\
CH_3C \\
\quad\quad \diagdown \\
\quad\quad\quad OCH_3
\end{array}
$$

Methyl ethanate

Be careful not to confuse methyl ethanate with ethyl methanate, which would be

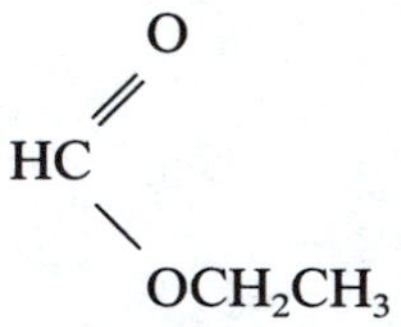

$$
\begin{array}{c}
\quad\quad O \\
\quad\quad \parallel \\
HC \\
\quad\quad \diagdown \\
\quad\quad\quad OCH_2CH_3
\end{array}
$$

Ethyl methanate

Some plastics are also made from ethers. *Ethers* are compounds that have two radicals joined together through an oxygen atom:

$$R\!-\!O\!-\!R'$$

where R and R' are any radicals. Ethers are very light volatile compounds that evaporate very rapidly. The Latin word *ether* means "nothing." Since ethers evaporate to nothing very quickly, the name is appropriate. Ethers are named by naming both radicals and tacking the word *ether* after them. The compound

$$CH_3CH_2—O—CH_3$$

would be named ethyl methyl ether. (Note that the radicals are listed in alphabetical order.) The compound

$$CH_3CH_2—O—CH_2CH_3$$

would be named diethyl ether. This is the "ether" used as an anesthetic in surgery rooms of hospitals

Two other types of organic compounds found in plastics are the *amines* and the *amides*. Both of these compounds involve the nitrogen atom. Amines have the

$$R—N{\Large\begin{matrix} \nearrow R' \\ \searrow R'' \end{matrix}}$$

group as a part of their molecule, while the amides can be identified by the

$$R—\overset{\overset{\textstyle O}{\|}}{C}—N{\Large\begin{matrix} \nearrow R' \\ \searrow R'' \end{matrix}}$$

structure. As used before, the R, R', and R" can be either hydrogen atoms or any other radical. These compounds are named as amines or amides. For example,

$$CH_3—N{\Large\begin{matrix} \nearrow H \\ \searrow H \end{matrix}}$$

would be called methanamine or methyl amine. The molecule

$$CH_3CH_2—\overset{\overset{\textstyle O}{\|}}{C}—N{\Large\begin{matrix} \nearrow H \\ \searrow H \end{matrix}}$$

would be called propanamide in the IUPAC system or propyl amide in the old system. (The carbon to which the oxygen is attached counts as the first carbon of the R group.)

Figure 10-2 summarizes the basic types of organic compounds found in plastics.

Ring compounds could also be made into alcohols, aldehydes, ketones, acids, esters, amines, and amides. We have already discussed the alcohol and the amine formed by adding the —OH or the —NH$_2$ group to the benzene ring to form phenol and analine:

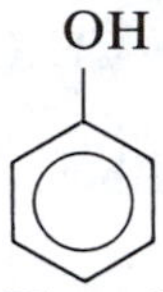

Phenol 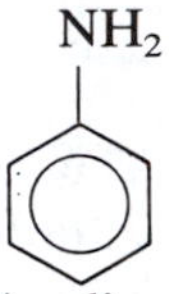   Analine

By adding —C—H to the benzene ring, benzaldehyde would be formed:

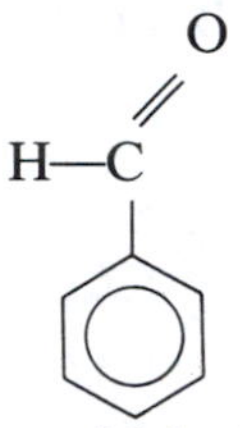

Benzaldehyde

Benzoic acid would be

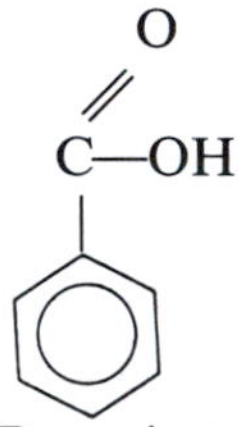

Benzoic acid

Methyl benzoate is

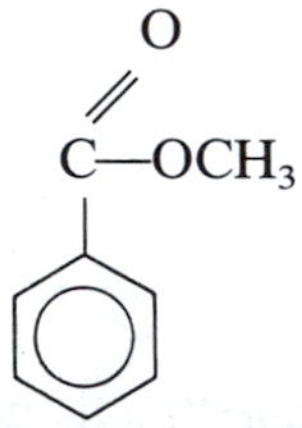

Methyl benzoate

1,2-Benzendioic acid is better known as phthalic acid and has the formula

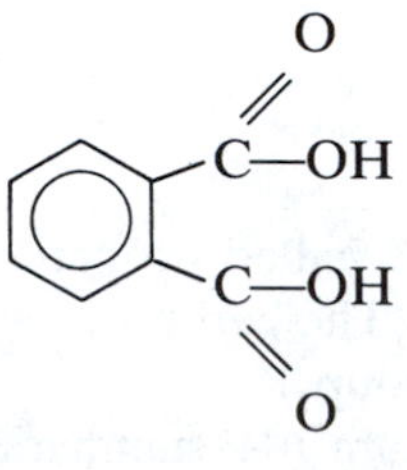

1,2-Benzendioic acid

FIGURE 10-2 Classes of organic compounds.

Class of Compound	Structure	Name Ending
Alcohols	R—OH	-ol
Aldehydes	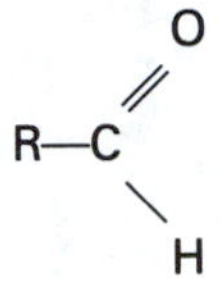 $R-C\overset{\displaystyle O}{\underset{\displaystyle H}{\Big\langle}}$	-al
Ketones	$R-\overset{\displaystyle O}{\overset{\|}{C}}-R'$	-one
Acids	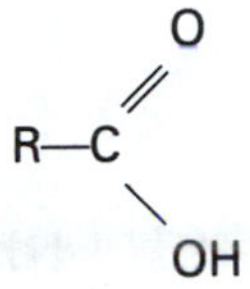 $R-C\overset{\displaystyle O}{\underset{\displaystyle OH}{\Big\langle}}$	-oic acid
Esters	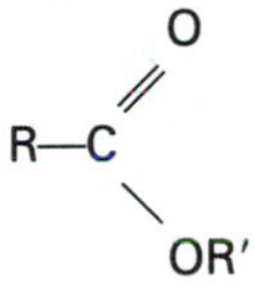 $R-C\overset{\displaystyle O}{\underset{\displaystyle OR'}{\Big\langle}}$	-ate
Ethers	R—O—R'	ether
Amines	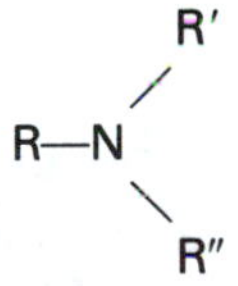 $R-N\overset{\displaystyle R'}{\underset{\displaystyle R''}{\Big\langle}}$	amine
Amides	$R-\overset{\displaystyle O}{\overset{\|}{C}}-N\overset{\displaystyle R'}{\underset{\displaystyle R''}{\Big\langle}}$	amide

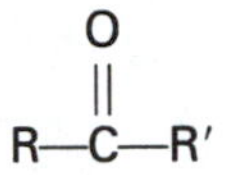

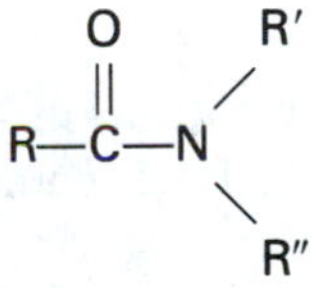

One diester used in polymers is dimethyl 1,4-benzenedicarboxylate. It is also called dimethyl terephthalate and is one of the chemicals used in making Dacron.

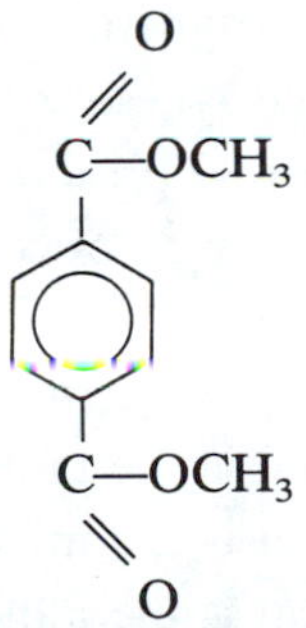

Dimethyl terephthalate

Given the fact that more and more carbon atoms can be joined together in many different configurations and geometric shapes, it can easily be seen that there is no limit to the number of organic compounds that could be made. Chemists are continually producing new compounds at a phenomenal rate. Hundreds of new compounds and plastics are being discovered weekly. Practical uses for all of these compounds may not be known for many years.

PROBLEM SET 10-2

Give the proper IUPAC name for the compounds in Problems 1 through 6.

1. $CH_3CH_2CH_2CH_2CH_2OH$

2. 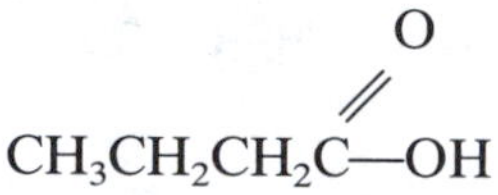 $CH_3CH_2CH_2\overset{\displaystyle O}{\overset{\displaystyle \|}{C}}-OH$

3. $CH_3CH_2CH_2CH_2\overset{\displaystyle O}{\overset{\displaystyle \|}{C}}-OCH_2CH_3$

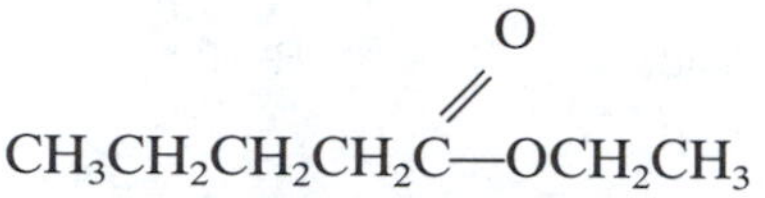

4. $CH_3CH_2CH_2-O-CH_2CH_3$

5. 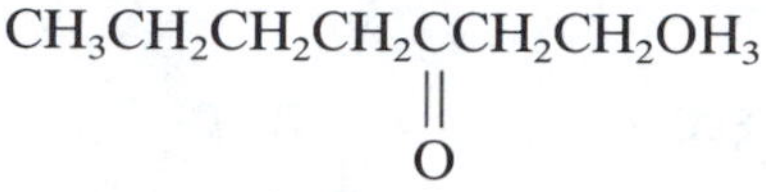 $CH_3CH_2CH_2CH_2\underset{\displaystyle O}{\underset{\displaystyle \|}{C}}CH_2CH_2OH_3$

6.

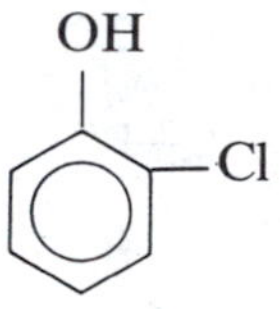

7. Write the formula for 2-pentanol.
8. Write the formula for propylethanate.
9. How many different (single) alcohols could be formed from normal pentane?
10. Look up in the *Handbook of Chemistry and Physics* or other source the IUPAC name for the following common items:
    (a) aspirin;
    (b) oil of wintergreen;
    (c) caffeine;
    (d) vinegar;
    (e) table sugar.
11. A disinfectant found in pharmacies contains hexaresorcinol. What type of chemical compound is this?
12. What type of chemical compound is methyl salicilate?
13. What type of chemical compound is catechol?
14. A wood preservative is *p*-cresol. What type of compound is it?
15. What type of compound is ethanal?
16. What type of compound is methyl methacrylate?
17. What is the molecular weight of benzene (C_6H_6)?
18. What is the molecular weight of toluene?
19. What is the molecular weight of phenol?
20. What is the molecular weight of formaldehyde (CH_2O)?

POLYMERIZATION

With a background in organic chemistry, it is now time to see how polymerization may occur and to discuss the properties of plastics. Polymers are, by their physical properties, divided into three classes: *thermoplastic* high polymers, *thermosetting* high polymers, and *elastomers*. Thermoplastic high polymers get soft or melt when heated. Thermosetting high polymers will not melt when heated. They may burn, char, crack, or get harder, but they will not melt. Elastomers are polymers that stretch—the synthetic rubber compounds.

THERMOPLASTIC HIGH POLYMERS

There are two chemical conditions necessary for polymerization to occur:

1. *There must be a molecule that has at least two reaction sites to start.* That is, there must be at least two points in the molecule which can have unsatisfied bonds and which will easily join with other molecules. All that is needed to get these two reaction sites is to start with a molecule that has a double bond. The simplest of these molecules would be ethene:

$$
\begin{array}{cc}
H & H \\
| & | \\
C & :: & C \\
| & | \\
H & H
\end{array}
$$

Ethene

Since each bond is a pair of shared electrons, one of the bonds in the ethene can open up, leaving a single bond between the carbon atoms, and the other two electrons will become unsatisfied radicals.

$$
\begin{array}{cc}
H & H \\
| & | \\
-C-C- \\
| & | \\
H & H
\end{array}
$$

If another ethene molecule that has opened up its double bond comes by, the two ethene radicals can join together:

$$
\begin{array}{cccc}
H & H & & H & H \\
| & | & & | & | \\
-C-C- & & -C-C- \\
| & | & & | & | \\
H & H & & H & H
\end{array}
$$

Note that this reacted product of the two ethenes still has two reaction sites or unreacted bonds on the ends. This produces the second condition necessary for polymerization to occur.

2. *After each reaction at least two open reaction sites must remain.* Now another opened-up ethene, or even another reacted group, can join with this four-carbon

chain to lengthen it even more. After this reaction the two reaction sites remain and further lengthening of the chain can occur.

$$CH_2$$
$$\backslash$$
$$CH_2$$
$$\backslash$$
$$CH_2\!-\!CH_2\!-\!CH_2\!-\!CH_2$$
$$\backslash$$
$$CH_2 \quad CH_2$$
$$\backslash \quad /$$
$$CH_2\!-\!CH_2$$

$$CH_2\!-\!CH_2\!-\!CH_2$$
$$/ \qquad \backslash$$
$$CH_2 \qquad CH_2$$
$$/ \qquad \backslash$$
$$CH_2$$
$$/$$
$$CH_2$$

This reaction can continue to occur many times. As the chain gets longer, the overall material gets thicker or more viscous. It is customary to write the formula for this reaction simply as

$$-\!(CH_2\!-\!CH_2)_n$$

Polyethylene

where n indicates the number of those molecules joined together. It can be in the thousands.

In the reaction above, we started with ethene. Ethene is the mer or basic structural building block for this particular plastic. Since we have polymerized ethene, the polymer produced would be called polyethene or more popularly, polyethylene. Polyethylene is formed in long chains of ethene molecules and can be envisioned as a matting of many threads or strings.

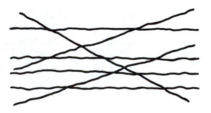

This material can be extruded or forced through a die having small holes to form threads or fibers, or it can be rolled into sheets. It can further be molded into many forms and shapes. Garment bags, ropes, threads, boxes, and tubes are but a few of the uses of polyethylene. One can imagine these long strings being pulled apart by pulling on them from the ends.

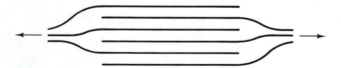

Since the only forces holding the long strings or chains together is their entwining about each other, there would not be much tensile strength to the mat or bundle. The long chains in polyethylene are similar to these strings. For this reason polyethylene does not have a very high tensile strength and is a very pliable material. Further, heating this polymer would simply allow the fibers to fall apart more easily. It would melt upon heating and is therefore a thermoplastic high polymer.

Returning to the example of the strings, if the strings were knotted, more force would be required to pull them apart.

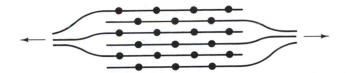

The same effect as knotting the string could be put in the polymer by replacing one or more of the hydrogen atoms with atoms of larger size. The diameter of the chlorine atoms is about 2.3 times that of hydrogen. If one started the reaction with chloroethene instead of straight ethene, there would be an atom of larger size on the chain, which would make it have a higher density and a higher tensile strength. Chloroethene is also known as vinyl chloride. The polymerization of vinyl chloride would produce polyvinyl chloride, commonly abbreviated as PVC.

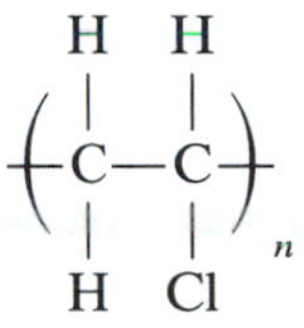

$$\left(\begin{array}{c} \overset{\displaystyle H}{\underset{\displaystyle H}{\vert}} \\ C \\ \end{array} - \begin{array}{c} \overset{\displaystyle H}{\vert} \\ C \\ \overset{\vert}{\displaystyle Cl} \end{array}\right)_n$$

Polyvinyl chloride
(PVC)

Polyvinyl chloride is used in making plastic pipe and other items which require more rigidity and strength than can be provided by polyethylene.

Carrying the logic above a step further, if two chlorines are used in the mer, even more resistance to mechanical deformation would be provided. Polymerizing dichloroethene, also known by the name vinylidene, produces polyvinylidene.

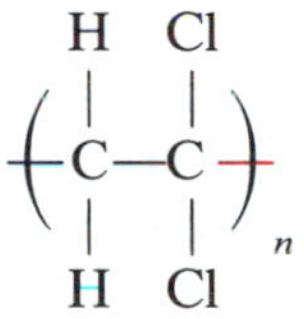

$$\left(\begin{array}{c} \overset{\displaystyle H}{\underset{\displaystyle H}{\vert}} \\ C \\ \end{array} - \begin{array}{c} \overset{\displaystyle Cl}{\vert} \\ C \\ \overset{\vert}{\displaystyle Cl} \end{array}\right)_n$$

Polyvinylidene
(Saran, Kynar)

Polyvinylidine is produced under several trade names, including Kynar and Saran.

The hydrogens on the ethene mer could also be replaced by the phenyl radical, a block of material even larger than the single chlorine atom. This phenyl ethene becomes known as the *styrene* molecule. Polymerizing this molecule would produce polystyrene:

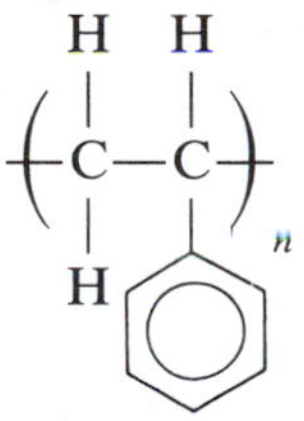

$$\left(\begin{array}{c} \overset{\displaystyle H}{\underset{\displaystyle H}{\vert}} \\ C \\ \end{array} - \begin{array}{c} \overset{\displaystyle H}{\vert} \\ C \\ \vert \end{array}\right)_n$$

Polystyrene
(Styrofoam)

To get the unsaturated mers to start to react and join together, an initiator must be used. The initiator is usually a peroxide (there are many commercially available) which causes heat to be emitted during the reaction. The more heat, the faster the material will polymerize. It has been found that too much heat (caused by using too much peroxide) can actually cause the material to boil and foam as it reacts. Styrene in a material that is often foamed on purpose to produce a rigid lightweight material marketed under the name *Styrofoam*. The foam is now made using a blowing agent that liberates a gas as the material polymerizes.

If one of the hydrogen atoms of the ethene is replaced with a cyanide group (—C≡N), the *acrylonitrile* mer is formed. The polymer of this molecule is marketed under the trade name Orlon.

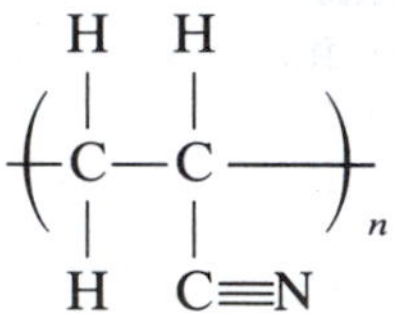

$$\left(\!\!\begin{array}{c} H \;\; H \\ | \quad | \\ C-C \\ | \quad | \\ H \;\; C\equiv N \end{array}\!\!\right)_{n}$$

Acrylonitrile
(Orlon)

In addition to the ethene derivative, other compounds can be used for the mer for plastics. One could start with just about any chain that has a double bond— propene, for example. The polymer of this molecule would be *polypropylene*:

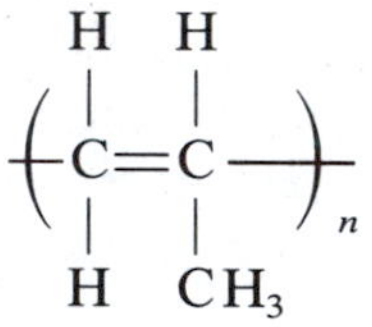

$$\left(\!\!\begin{array}{c} H \;\; H \\ | \quad | \\ C=C \\ | \quad | \\ H \;\; CH_3 \end{array}\!\!\right)_{n}$$

Polypropylene

Another molecule that would polymerize was discovered in the early 1940s. It was methyl-2-methylpropenate, known also by the name methyl methacrylate.

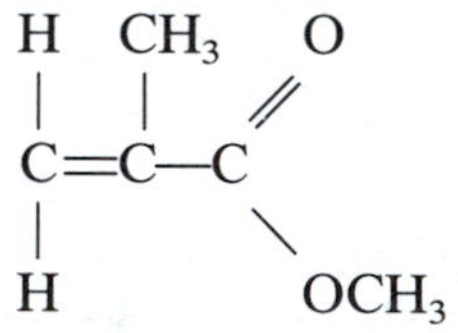

$$\begin{array}{c} H \quad CH_3 \quad O \\ | \quad\quad | \quad\quad \| \\ C=C-C \\ | \quad\quad\quad\quad \backslash \\ H \quad\quad\quad\quad OCH_3 \end{array}$$

Methyl 2-methylpropenate
(Methyl methacrylate)

The double bond on this compound would open up to polymerize into *polymethyl methacrylate*:

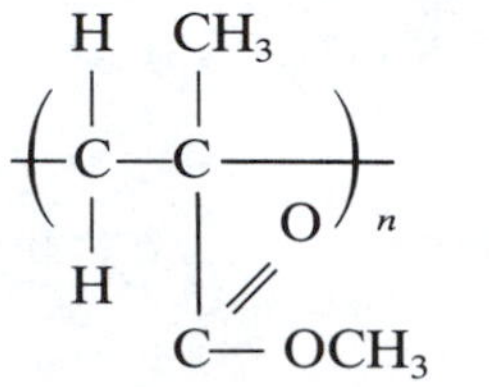

$$\left(\!\!\begin{array}{c} H \;\; CH_3 \\ | \quad\; | \\ C-C \\ | \quad\; | \quad O \\ H \quad | \;\; \| \\ \quad\;\; C-OCH_3 \end{array}\!\!\right)_{n}$$

Polymethyl methacrylate

Many companies now produce this polymer under such trade names as

PMMA
Lucite
Plexiglas
Acrylic
Acrylite
Acrylan
Perspex
Methacral
Zerlon

The uses of this polymer include windshields for aircraft, camera lenses, casing for underwater instruments, hose fittings, and many other applications.

The 2-butene molecule will also polymerize to form a plastic now known as *polybutylene*:

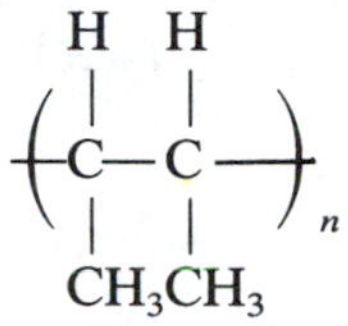

$$
\left(\begin{array}{c} \overset{\text{H}}{\underset{\text{CH}_3}{\overset{|}{\underset{|}{C}}}} - \overset{\text{H}}{\underset{\text{CH}_3}{\overset{|}{\underset{|}{C}}}} \end{array}\right)_n
$$

Polybutylene

One drawback to the use of polyethylene-based polymers is that they will react chemically with elements such as chlorine and bromine and other chemicals. Chemists reasoned that by replacing all the hydrogen atoms on the ethene molecule with the most negative element on the periodic chart, fluorine, this problem could be solved. The molecule formed by replacing the hydrogens with fluorine would be tetrafluoroethene.

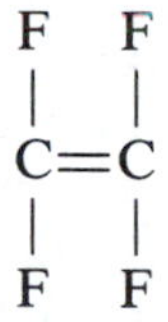

$$
\overset{\text{F}}{\underset{\text{F}}{\overset{|}{\underset{|}{C}}}} = \overset{\text{F}}{\underset{\text{F}}{\overset{|}{\underset{|}{C}}}}
$$

Tetrafluoroethene

Tetrafluoroethene will open its double bond and can be polymerized to form *polytetrafluoroethene*, which is sold under the trade name *Teflon*:

$$
\left(\begin{array}{c} \overset{\text{F}}{\underset{\text{F}}{\overset{|}{\underset{|}{C}}}} - \overset{\text{F}}{\underset{\text{F}}{\overset{|}{\underset{|}{C}}}} \end{array}\right)_n
$$

Polytetrafluoroethene
(Teflon)

This compound is also sold as a polyfluorocarbon and used as a chemically resistant coating which has a very low coefficient of friction. Some applications are in cooking ware that needs no grease or cooking oil, chemically resistant pipes and gaskets, coatings on roofing tiles, and anodized aluminum surfaces.

Not all polymers are derived from a single compound. In many cases two molecules react to form the mer. One of the earliest of these was the formation of nylon. To form the mer of nylon, 1,6-diaminohexane, often called hexamethylene diamine, was reacted with 1,6-hexandioic acid, better known as adipic acid:

$$
\begin{array}{c}
\text{H—N—H} \\
| \\
\text{CH}_2 \\
| \\
\text{CH}_2 \\
| \\
\text{CH}_2 \\
| \quad\quad + \\
\text{CH}_2 \\
| \\
\text{CH}_2 \\
| \\
\text{CH}_2 \\
| \\
\text{H—N—H}
\end{array}
\quad
\begin{array}{c}
\text{HO—C=O} \\
| \\
\text{CH}_2 \\
| \\
\text{CH}_2 \\
| \\
\text{CH}_2 \\
| \\
\text{CH}_2 \\
| \\
\text{HO—C=O}
\end{array}
\longrightarrow
\left(\!\!\begin{array}{c}\text{H}\\|\\\text{N}\end{array}\!\!-(\text{CH}_2)_6-\begin{array}{c}\text{H O}\\|\ ||\\\text{N—C}\end{array}\!\!-(\text{CH}_2)_4-\begin{array}{c}\text{O}\\||\\\text{C}\end{array}\!\!\right)_n + \text{H}_2\text{O}
$$

Hexamethylenediamine +   Adipic $\longrightarrow$ Nylon   + Water
                                                        acid

In this reaction one of the hydrogen atoms from the hexamethylenediamine reacts with the —OH from the adipic acid to form water. The two radicals then join to form the mer of nylon. The acid end of the mer would then react with the amine end of another mer and polymerization could occur. Since water is given off in this reaction, it is called a *condensation reaction*.

Soon chemists found that any diamine would react with any diacid to form a mer. The rest of the molecules were just carried along in the reaction. Decamethylenediamine could react with adipic acid just as well as the hexamethylenediamine but the mer would be heavier and have different properties. The reaction of any diamine with any diacid is a nylon. The reaction of the six-carbon diamine with the six-carbon diacid is known as nylon 6,6. The six-carbon acid would react with a ten-carbon amine and be known as nylon 6,10. Similarly, there could be nylon 6,12 and others. Each of the nylons would have different characteristics and uses. A light nylon might be used for stockings, while a heavier molecule would be used in carpets and tires. All nylons are *polyamides* since they all possess the

$$
\begin{array}{c}
\text{O H} \\
||\ \ | \\
\text{—C—N—}
\end{array}
$$

group.

Another type of polyamide that has found great use in composites and as a reinforcement for tires are the aromatic polymides. Instead the straight chains (as in nylon 6,6) a benzene ring is inserted in both the diamine and the diacid. This produces the mer

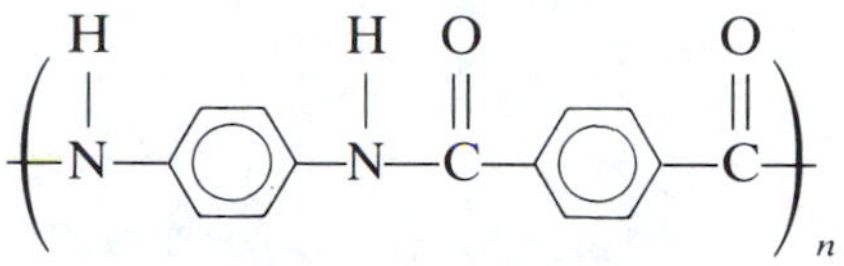

Polyaramid
(Kevlar)

This type of polymer is classed as a *polyaramid*. It is marketed under the name Kevlar and has extremely high tensile strength and light weight. This compound is discussed further in Chapter 11.

Thinker 10-2:

Nylon and Kevlar are reactions involving a diamine with a diacid. Write an equation for another possible set of compounds that would react similarly to nylon and Kevlar.

A slightly different reaction from the diamine–diacid mer could be formed between dimethyl terephthalate and ethylene glycol. The dimethyl terephthalate is a double ester and the ethylene glycol is a double alcohol. In this reaction the methyl radical of the ester reacts with the —OH of the alcohol to form methanol. The two parent compounds then join to form the mer for Dacron.

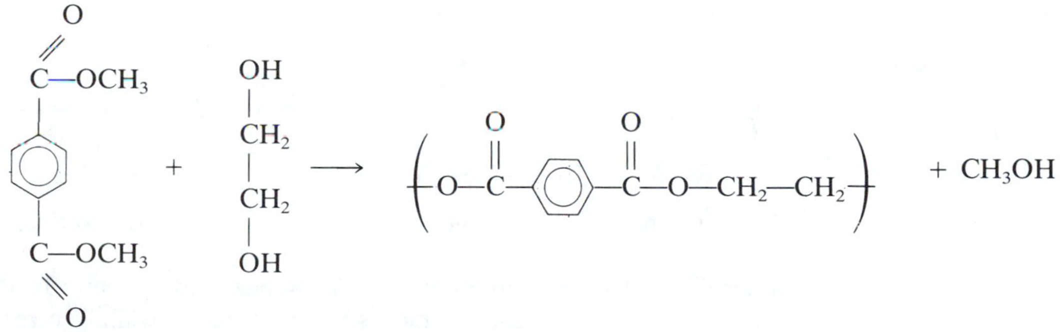

| Dimethyl terephthalate | + Ethylene glycol ⟶ | Dacron | + Methanol |

Dacron is one example of a polyester.

Thinker 10-3:

Dacron is formed by a reaction of a dialcohol with a diester. Write another equation for compounds that would produce a Dacron-like reaction.

Polycarbonates are a class of polymer which have many practical uses. The mer of polycarbonates is

$$\left(O-\overset{\displaystyle \overset{O}{\|}}{C}-O-\bigcirc-\overset{\displaystyle \overset{CH_3}{|}}{\underset{\displaystyle \underset{CH_3}{|}}{C}}-\bigcirc\right)_n$$

Polycarbonate
(Lexan)

Polycarbonates have a relatively low melting temperature and are rather soft. Although they can be cut or scratched fairly easily, they will not break under impact. This property, combined with good optical transparency, makes them ideal for shatter-proof windows. One trade name by which polycarbonates are marketed is Lexan. Lexan has even been advertised as a bullet-resistant shield to be mounted over automobile window glass.

Polyurethanes, especially in the foam form, are good thermal insulators and have reasonably high strength. Urethanes are formed by reaction of a diisocyanate with a dialcohol. Diisocyanates have the structure.

$$\begin{array}{ccc} \overset{O}{\|} & & \overset{O}{\|} \\ C & & C \\ \| & & \| \\ N & \!\!-C-\!\! & N \end{array}$$

Diisocyanate

The reaction of urethanes would be

$$\begin{array}{cc} \overset{O}{\|} & \overset{O}{\|} \\ C & C \\ \| & \| \\ N-R-N \end{array} + HO-R'-OH \longrightarrow \left(\begin{array}{cc} \overset{O}{\|} & \overset{O}{\|} \\ C & C-O-R'-O \\ | & | \\ -N-R-N-H \end{array}\right)_n$$

Diisocyanate +   Dialcohol $\longrightarrow$ Polyurethane

where R and R' are any radical. R is often a benzene ring. In this reaction the hydrogen from the alcohol attaches to the nitrogen while the rest of the double alcohol attaches to the carbon atom.

One polymer that is becoming popular in composites is polyphenyl sulfide. The mer of this polymer is

$$\left(\bigcirc-S\right)_n$$

Polyphenyl sulfide is a thermoplastic polymer that forms long chains. The sulfur reacts with the para position of the phenyl radical to extend the chain. Polyphenyl sulfide is used as the matrix with glass and carbon fibers to produce a very strong composite. Composites are discussed in Chapter 11.

All of the polymers discussed so far are thermoplastic hipolymers. It should be understood that all of these polymers form long-chained molecules with little or no chemical bonding between the chains. It is the lack of bonding between the chains that makes them melt when heated.

POLYMER CRYSTALS

In Chapter 2 a crystal was defined as a long range orderly array of particles. Crystals can be detected by x-ray diffraction and have definite melting points. Until recently it was thought that polymers were noncrystalline in form, similar to glass. Now there is x-ray evidence that some polymers do show traces of crystallinity. It is believed that certain polymers can have places in the structure where the randomly oriented fibers can align themselves into a short range orderly arrangement giving a semblance of a crystal structure.

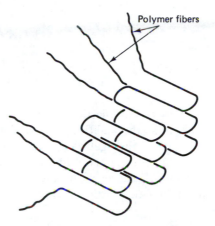

Polymers that do exhibit this crystallinity are more brittle and have higher strength at elevated temperatures than would be expected from their molecular structure.

THERMOSETTING HIGH POLYMERS

In Baekeland's experiment, he was reacting phenol with formaldehyde. The result was a white solid material. The chemical equation for this reaction is

Phenol + Formaldehyde + Phenol $\longrightarrow$ Bakelite + Water

In this condensation reaction the hydrogen from the ortho position on the two phenol molecules reacts with the oxygen of the formaldehyde, producing water. This allows the phenols and the methyl radical left when the oxygen is removed

from the formaldehyde to react to form the Bakelite mer. However, the same
reaction could occur with the hydrogen in the para position of the phenol. The
phenyl radicals could also rotate about the methyl radical in the molecule. This
gives rise to a three-dimensional molecule rather than the straight chains found in
all the previous polymers. The three-dimensional effect is called *cross-linking*:

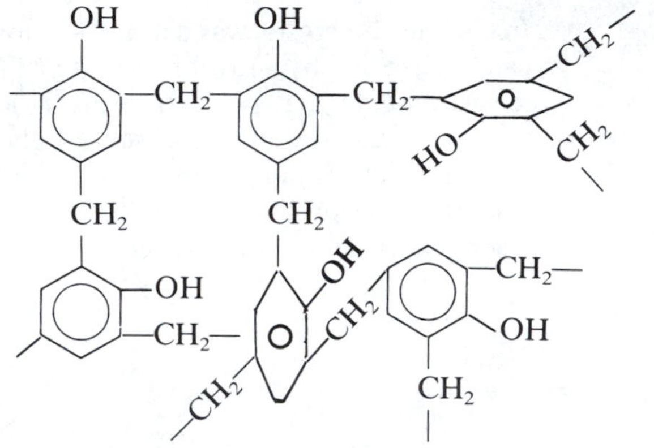

In thermoplastic high polymers there is no cross-linking. In polymers that
cross-link, the addition of heat simply makes them cross-link even more. Instead
of melting, they get harder.

Once the phenol and formaldehyde were found to polymerize, other modifications of the reaction were tried. Aniline replaced the phenol to give the aniline–
formaldelyde plastics.

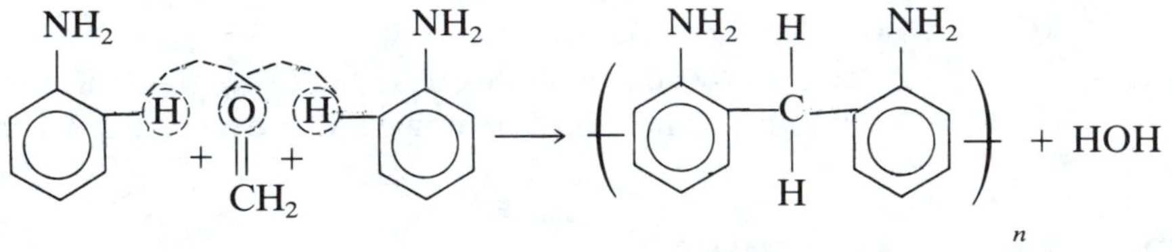

Another variation uses melamine with the formaldehyde. Melamine has three
nitrogen atoms in the ring instead of all six being carbon atoms. This, too, forms
a thermosetting high polymer.

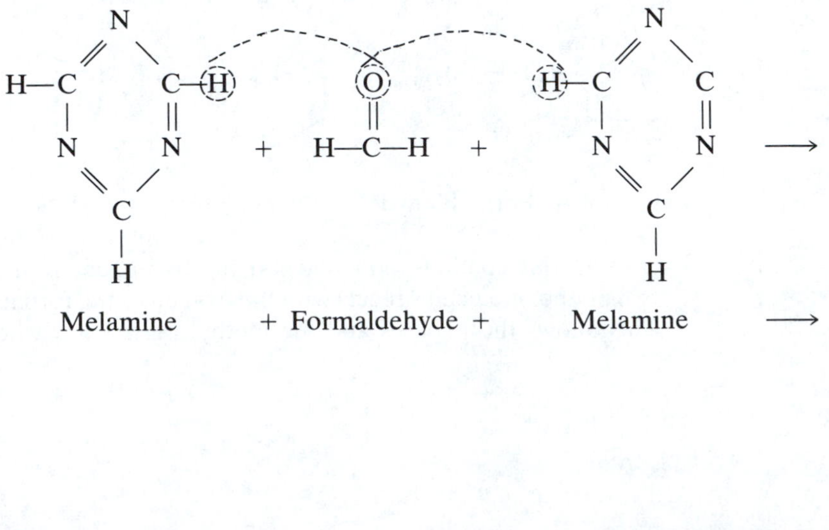

Melamine   + Formaldehyde +   Melamine $\longrightarrow$

$$\left(\begin{array}{c} \text{melamine-formaldehyde structure} \end{array}\right)_n + HOH$$

Melmac   + Water

Dishes are often made from this product and sold under the name Melmac.

Thermosetting high polymers are rarely used in their pure form. They are usually mixed with a filler or extender. Extenders not only take up space, making the product cheaper, but they add desired properties to the plastic. Bakelite can be mixed with asbestos, glass fibers, sawdust, or other material to provide good thermal or electrical insulation properties. Skillet handles, electrical outlets, and plugs have been made in this fashion. The melamine–formaldehyde polymer is mixed with vermiculite (a form of the rock, mica) and applied to particleboard or other backing materials. The formaldehyde–melamine–mica product is known as *Formica*.

Urea can also react with formaldehyde to form a polymer:

$$\text{Urea} \quad \text{Formaldehyde} \quad \text{Urea} \longrightarrow \left(\begin{array}{c} \text{urea-formaldehyde structure} \end{array}\right)_n + HOH$$

All of these are thermosetting highpolymers and are all cross-linked molecules and therefore thermosetting.

One type of thermosetting highpolymer that has found use both as an adhesive and as a matrix for other materials is epoxy. The epoxy molecule always involves the ethylene oxide structure:

$$\begin{array}{c} O \\ / \setminus \\ -C-C- \end{array}$$

The general formula for epoxies is

$$\left(\begin{array}{c} \text{epoxy structure with R} \end{array}\right)_n$$

where R is any radical. In industry many different radicals are used to give the properties needed for the specific applications.

Two or more mers are often combined to make a specific plastic. One example is the *acrylonitrile–butadiene–styrene* structure produced under the name *ABS*. In this molecule the butadiene serves as the chain with the acrylonitrile and the styrene added to it as side chains. Both the acrylonitrile and the styrene can then be polymerized and cross-link the chain. ABS is an example of a *copolymer*. In copolymers more than one molecule is joined together to form the basic mer.

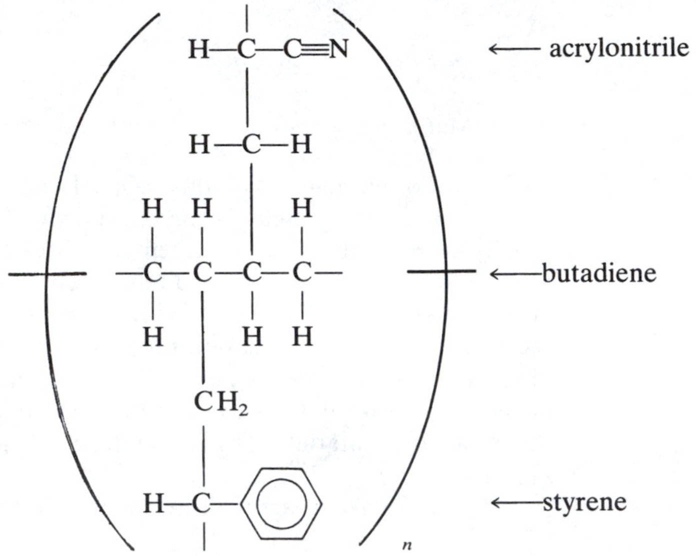

$$
\begin{array}{c}
\text{H—C—C} \equiv \text{N} \quad \longleftarrow \text{acrylonitrile} \\
\text{H—C—H} \\
\text{—C—C—C—C—} \quad \longleftarrow \text{butadiene} \\
\text{CH}_2 \\
\text{H—C—} \bigcirc \quad \longleftarrow \text{styrene}
\end{array}
$$

ABS

ELASTOMERS

Shortly after the discovery of the New World, Spanish explorers discovered a native toy ball made from the sap of the caoutchouc (weeping) tree. A century and a half later, Joseph Priestly found that this gummy material would erase pencil marks and called it rubber. By impregnating cotton with rubber, a water-repellant material was made. By 1819 Charles MacIntosh had produced a raincoat made from this coated fabric. But the raincoat would get hard and brittle in the winter and soft and tacky in the heat of the summer. The gummy material would get brittle when left out in the weather and would swell and decompose when brought in contact with any organic material such as oil or kerosene. Working independently, Charles Goodyear and Thomas Hancock found that mixing sulfur with the rubber would make it impervious to heat and cold as well as preventing the rubber from cracking due to exposure to air. The process of *vulcanization* had been discovered.

Chemists set to work to determine the structure of the rubber molecule. By the year 1860 it was determined that the formula was that of isoprene:

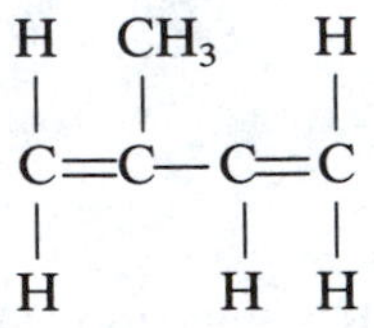

Isoprene

Although several chemists had succeeded in partially polymerizing isoprene in the laboratory, it was not until World War I (1914), that the first synthetic rubber was made. This rubber was of poor quality and still would not withstand attack by organic materials. It was therefore useless as hoses for fuels or oils. The year 1930 saw the development of thiokol rubber, which was able to resist attack by gasoline and other organic solvents.

Since natural rubber was so cheap, there was little need seen for a synthetic product. But when the Japanese overran the Malaysian peninsula in 1941, the United States lost the vast majority of its rubber supplies. Spurred by this need, the rush for synthetic rubber was accelerated. Companies that had been competing in the development of synthetic rubber were now ordered to pool their information. The federal government financed the construction of synthetic rubber plants and by the middle of World War II these plants were in full production. Although the first synthetic rubber products were of poor quality, the new polymers were soon found to be better than natural rubber. We have never returned to the wide use of natural rubber seen before the 1940s.

To understand the mechanism of rubber, let us return to the isoprene molecule found in nature. Note that there are two double bonds in the molecule. As in the case of ethene, the electrons of the double bonds can open up, shifting the double bond to the center (the 2-carbon), making a reaction site at both ends of the molecule. This molecule now has the two reaction sites required for polymerization.

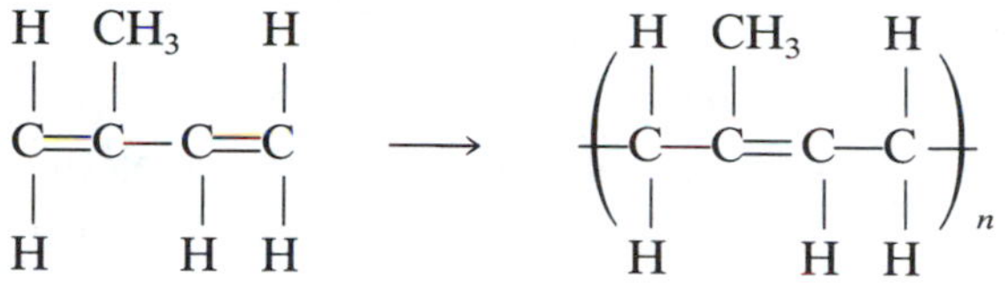

Isoprene → Polyisoprene

One synthetic rubber was made from a slightly simpler molecule, butadiene. The polymerization process similar to the isoprene reaction can occur with butadiene:

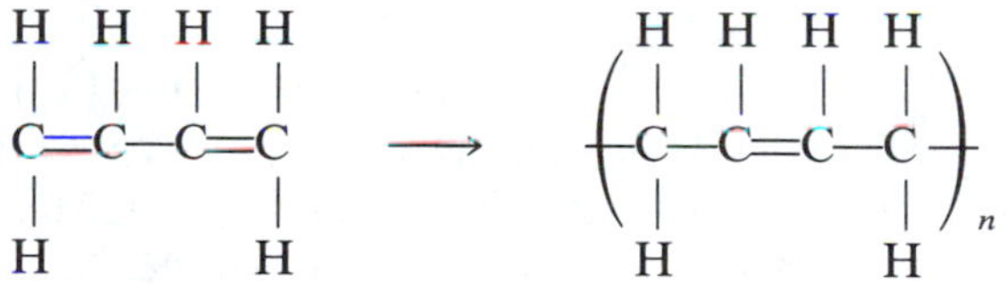

Butadiene → Buna rubber

Butadiene was the basis for Buna (trademark) rubber. It was found that butadiene mixed with other unsaturated compounds produced even better properties for use in tires, water and fuel hoses, gaskets, and similar products. Mixing butadiene with styrene gave rise to Buna-S rubber.

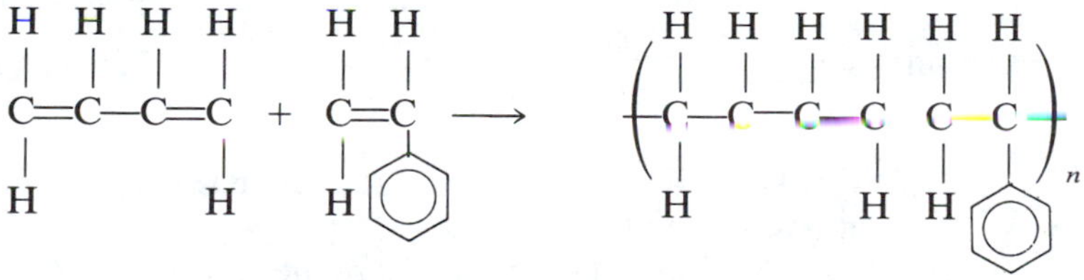

Butadiene + Styrene ⟶ Buna-S rubber

Buna-N rubber was a mixture of butadiene with acrylonitrile:

$$
\underset{\text{Butadiene}}{
\begin{array}{c}
\text{H} \;\; \text{H} \;\; \text{H} \;\; \text{H} \\
| \quad | \quad | \quad | \\
\text{C}=\text{C}-\text{C}=\text{C} \\
| \qquad\qquad | \\
\text{H} \qquad\quad \text{H}
\end{array}}
\; + \;
\underset{\text{Acrylonitrile} \longrightarrow}{
\begin{array}{c}
\text{H} \;\; \text{H} \\
| \quad | \\
\text{C}=\text{C} \\
| \quad | \\
\text{H} \;\; \text{C}\equiv\text{N}
\end{array}}
\; \longrightarrow \;
\underset{\text{Buna-N}}{
\left(
\begin{array}{c}
\text{H} \;\; \text{H} \;\; \text{H} \;\; \text{H} \;\; \text{H} \;\; \text{H} \\
| \quad | \quad | \quad | \quad | \quad | \\
\text{C}-\text{C}=\text{C}-\text{C}-\text{C}-\text{C} \\
| \qquad\qquad\qquad | \quad | \quad | \\
\text{H} \qquad\qquad\;\; \text{H} \;\; \text{H} \;\; \text{C}\equiv\text{N}
\end{array}
\right)_n}
$$

One synthetic rubber that was found to hold air better and be more resistant to the action of oxygen was made from 1,1-dimethyl ethene (isobutylene). This product was produced under such names as GR-1, Butyl, and Polybutylene.

$$
\underset{\text{Isobutylene}}{
\begin{array}{c}
\text{CH}_3\text{H} \\
| \quad | \\
\text{C}=\text{C} \\
| \quad | \\
\text{CH}_3\text{H}
\end{array}}
\; \longrightarrow \;
\underset{\text{Polybutylene}}{
\left(
\begin{array}{c}
\text{CH}_3\text{H} \\
| \quad | \\
\text{C}-\text{C} \\
| \quad | \\
\text{CH}_3\text{H}
\end{array}
\right)_n}
$$

The synthetic rubber that has been called the nearest rival to natural rubber is made from the *chloroprene* molecule and is trademarked as *Neoprene*. It was the first commercially available synthetic rubber.

$$
\underset{\text{Chloroprene}}{
\begin{array}{c}
\text{H} \qquad\quad \text{H} \\
| \qquad\quad\; | \\
\text{C}=\text{C}-\text{C}=\text{C} \\
| \quad | \quad | \quad | \\
\text{H} \;\; \text{Cl} \;\; \text{H} \;\; \text{H}
\end{array}}
\; \longrightarrow \;
\underset{\text{Neoprene}}{
\left(
\begin{array}{c}
\text{H} \qquad\quad \text{H} \\
| \qquad\quad\; | \\
\text{C}-\text{C}=\text{C}-\text{C} \\
| \quad | \quad | \quad | \\
\text{H} \;\; \text{Cl} \;\; \text{H} \;\; \text{H}
\end{array}
\right)_n}
$$

Neoprene is inferior to natural rubber in physical properties such as tensile strength and elasticity, but it does resist attack from oil, grease, and other organic materials better than natural rubber. Many hoses, tubes, and household and medical supplies are made from Neoprene.

Thiokol rubber is made by polymerizing 1,2-dichloroethane (ethylene di-chloride) with sodium polysulfide:

$$
\underset{\substack{\text{1,2-Dichloroethane} +}}{
\begin{array}{c}
\text{H} \;\; \text{H} \\
| \quad | \\
\text{Cl}-\text{C}-\text{C}-\text{Cl} \\
| \quad | \\
\text{H} \;\; \text{H}
\end{array}}
\; + \;
\underset{\substack{\text{Sodium} \\ \text{polysulfide}} \longrightarrow}{\text{NaS}_4}
\; \longrightarrow \;
\underset{\text{Thiokol}}{
\left(
\begin{array}{c}
\text{S}-\text{CH}_2-\text{CH}_2-\text{S} \\
| \qquad\qquad\qquad | \\
\text{S} \qquad\qquad\quad\; \text{S} \\
| \qquad\qquad\qquad |
\end{array}
\right)_n}
\; + \; \text{NaCl}
$$

Thiokol is a self-vulcanizing rubber that leaves no room for organic molecules to attack the polymer.

Not all butadiene or isoprene molecules form an elastomer or pliable substance. It should be observed that the carbon chains are not straight-line chains.

The carbon-to-carbon bonds are always at an angle to each other. The chain of polyethylene would actually look as follows:

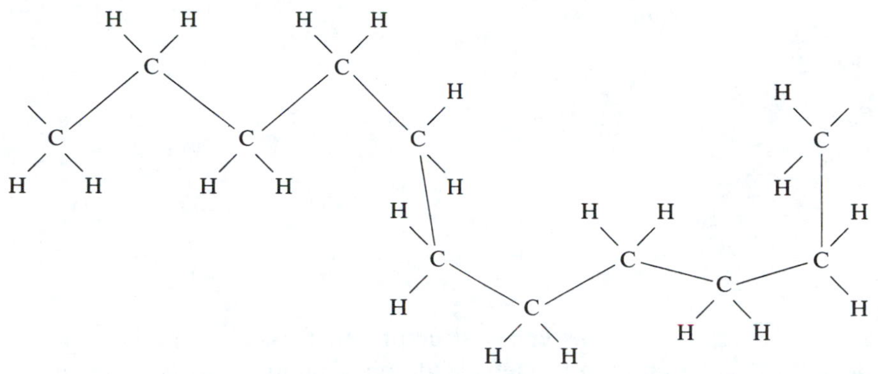

These chains can rotate and spin about the single carbon bonds. However, when a double bond is formed, the carbons are not free to rotate. Therefore, the butadiene, isoprene, and other polyunsaturated molecules can take on two configurations. If the end methyl groups are on the same side of the double bond, the molecule is said to be in the cis configuration. Putting the methyl radicals on opposite sides of the double bond makes it into the trans configuration:

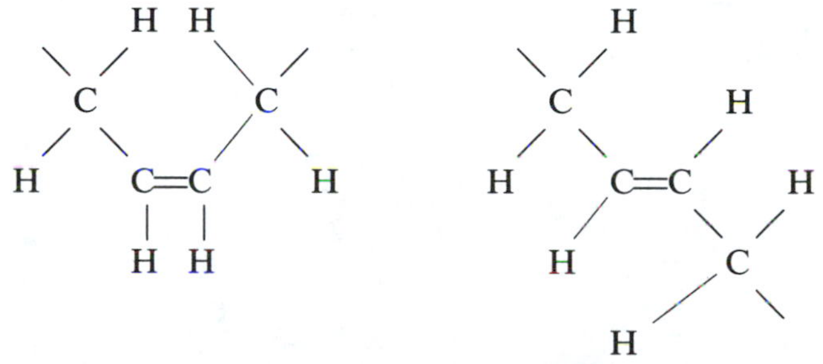

cis-Butadiene

trans-Butadiene
(Gutta Percha)

The trans form of isoprene can produce only relatively straight molecules and forms the hard rubber product known as gutta percha. Gutta percha is used to make bowling balls, billiard balls, hard rubber canes, and similar products.

The cis form can form either a straight-line molecule or can form into coils. It is these coils in the molecule that give rubber its elasticity.

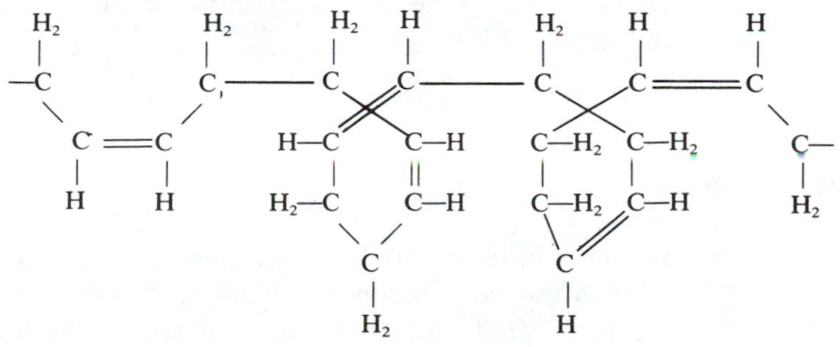

Since the polymerized butadiene, isoprene, or similar mers still have a double bond in the molecule, they are susceptible to attack by other organic mers or oxygen. Oxygen can cause this double bond to open up and form an epoxy configuration.

$$\text{epoxy structure}$$

This epoxy structure embrittles the rubber and makes it crack and check. Vulcanization, the addition of sulfur to the molecule, partially cross-links the rubber molecules, giving them more mechanical rigidity and removing some of the double bonds which are sites for attack by oxygen.

$$\text{vulcanized rubber structure}$$

After the butadiene or other mer is polymerized, it is removed from the reaction vessel, coagulated, and shipped to the manufacturing plants. Rubber in this stage is called latex. It looks something like rubber ceoment that has been spread on paper, allowed to dry partially, then rolled into balls. It is a gummy and very tacky substance. To make tires from latex, about 30% by weight of carbon black is added. The carbon black gives the rubber its body and desired abrasive characteristics. The more carbon black that is added, the harder and more lasting the tires will be.

SILICON POLYMERS

Silicon is immediately below carbon in the periodic table. Chemically, silicon can react in the same manner as carbon. As with carbon, silicon must also have four bonds satisfied with each atom. Silicon is a very abundant element on the earth.

It is found in sand as silica (SiO_2). The reaction of silicon with four hydrogen atoms produces the silane molecule:

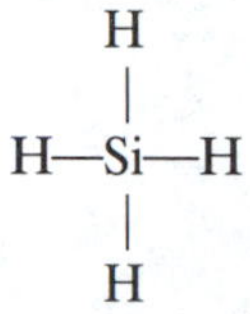

$$
\begin{array}{c}
H \\
| \\
H-Si-H \\
| \\
H
\end{array}
$$

Silane

Two silicon atom molecules are the disilanes:

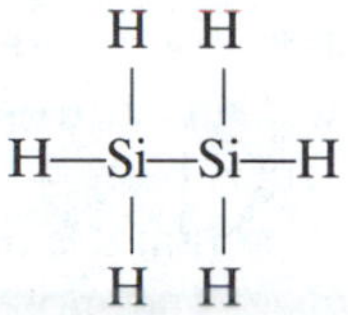

$$
\begin{array}{c}
H \quad H \\
| \quad | \\
H-Si-Si-H \\
| \quad | \\
H \quad H
\end{array}
$$

Disilane

Carbon and silicon can be combined in a molecule. Methylsilane is one example.

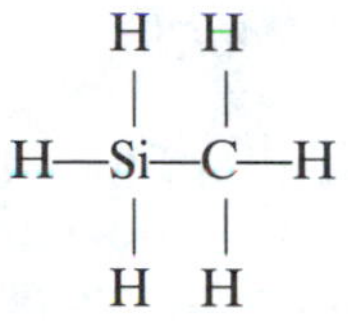

$$
\begin{array}{c}
H \quad H \\
| \quad | \\
H-Si-C-H \\
| \quad | \\
H \quad H
\end{array}
$$

Methylsilane

The silicone polymer can be made by starting with dimethylsilane and reacting it with cold water. This produces dimethyldihydroxysilane, which quickly polymerizes by giving off water.

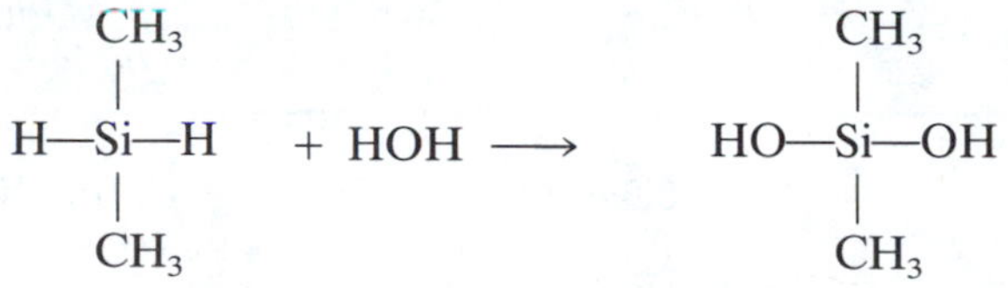

$$
\begin{array}{c}
CH_3 \\
| \\
H-Si-H \\
| \\
CH_3
\end{array}
\; + \; HOH \longrightarrow
\begin{array}{c}
CH_3 \\
| \\
HO-Si-OH \\
| \\
CH_3
\end{array}
$$

Dimethylsilane + Water $\longrightarrow$ Dimethyldihydroxysilane

$$
\begin{array}{c}
CH_3 \\
| \\
HO-Si-OH \\
| \\
CH_3
\end{array}
\; + \;
\begin{array}{c}
CH_3 \\
| \\
HO-Si-OH \\
| \\
CH_3
\end{array}
\longrightarrow
\left(\begin{array}{c}
CH_3 \quad CH_3 \\
| \qquad | \\
O-Si-O-Si-O \\
| \qquad | \\
CH_3 \quad CH_3
\end{array}\right)_n
\; + \; H_2O
$$

Silicone polymer

If the polymerization is stopped before the molecule can get very large, high-boiling-point oils which can withstand temperatures to 800°C are produced. If the polymerization is allowed to continue, thermosetting high polymers and elastomers can result.

Two silicon polymers which are used in composites are trichlorosilane and trihydroxysilane. These would have the formulas

$$
\begin{array}{cc}
\quad\;\; Cl & \quad\;\; OH \\
\quad\;\; | & \quad\;\; | \\
Cl-Si-H & HO-Si-H \\
\quad\;\; | & \quad\;\; | \\
\quad\;\; Cl & \quad\;\; OH
\end{array}
$$

Trichlorosilane   Trihydroxysilane

These compounds are very similar to their carbon counterparts, trichloromethane (chloroform) and methantriol.

The polymers listed above are but a very small fraction of the known plastics. The principles of reaction are the same in all of them. Often, several of these polymers are mixed to give a plastic with specifically desired characteristics. We now have the ability to produce a material with the properties needed for just about any function the design may require. Many of the accomplishments in aeronautics, space, and more common earth-bound endeavors were made possible by the creation of polymers. Many new uses of old polymers are being discovered and thousands of new polymers are being developed for new needs.

PHYSICAL PROPERTIES OF POLYMERS

Besides being thermoplastic, thermosetting, or elastomers, high polymers have different tensile strengths, moduli of elasticity, and other physical properties. Many of these properties depend on the degree of polymerization. The tensile strength of polyethylene can vary from 1000 to about 6000 lb/in^2. Figure 10-3 gives some comparative values for a few of the high polymers available.

FIGURE 10-3   Summary of selected hipolymers.

Type	Density (g/cm³)	Tensile Strength (lb/in²)	Youngs Modulus (lb/in² × 10³)	Ultimate Percent Elongation	Trade Names
Thermoplastics Polyethylene	0.91–0.965	1000–5500	14–160	15–700	Alathon Ameripol Brea Dulan Durex Ger-Pac Microthene Poly-eth
Polyvinyl chloride	1.1–1.4	1500–9000	200–600	2–400	Blacar Dacovin Geon GP Exel Koroseal Plioflex Pliovic PVC Ryertex-omicron Ultron Upalon Kynar Saran
Polyvinylidine					

FIGURE 10-3   (cont.)

Type	Density (g/cm³)	Tensile Strength (lb/in²)	Youngs Modulus (lb/in² × 10³)	Ultimate Percent Elongation	Trade Names
Polypropylene	0.9	4300–5500	1400–1700	220	Escon Maplen
Polystyrene	1.04–1.08	5000–10,000	400–600,000	1–2.5	Amphenol Dylene Grace
Polymethyl methacrylate	1.18–1.20	7000–11,000	350–500	2–10	Acrilan Acrylic Lucite Methacrol Perspex Plexiglas PMMA Zerlon
Polytetrafluoroethylene	2.1–2.3	2000–4500	33–65	200–400	Teflon Tetran
Polyamide	1.13–1.15	9000–9500	—	60–300	Nylon
Polyaramid fibers	1.45	410,000	20,000	—	Kevlar
Polycarbonate	1.2	8000–9500	290–325	20–100	CR-34 Lexan Merlon
Polyurethane		5000		600	Estone Multrathane Vibrathane
Polyester					Dacron Orlon
Thermosetting					
Phenol–formaldehyde	1.36–1.43	6000–9000	900–1300	0.5–1.0	Bakelite Durez Dyphene Plyophen
Urea–formaldehyde	1.47–1.52	5500–13,000	1300–1400	0.6	Avisco Gabrite Rholite
Epoxy	1.114	4000	300	2–6	C-8 Epoxy DER Devran Epikote Maraglas Red Thread Unex
Melamine– formaldehyde	1.47–1.78	5500–13,000	1300–1950	0.6–0.9	Formica Melantine Melmac Permelite Ultra Pas
Elastomers					
Butadiene	0.94	3500	—	3000	Ameripol CB Budene Buna Cis-4
Butadiene–styrene	1.00	600–3000	—	600–2000	Ameripol SBR Buna-S Carbomix
Isoprene	0.93	3000	—	800	Natural rubber
Chloroprene	1.24	3500	—	800	Neoprene
ABS	1.06	7000	—	1–2	Abson Lustron
Silicone	1.5–2.8	3000–4000	1200	0–700	

PROBLEM SET 10-3

Use the following list of high polymers to answer Problems 1 through 10:

Polyethylene
Polyvinyl Chloride
Polystyrene
Polycarbonate
Teflon
Kevlar
Bakelite
Dacron
Buna-S
Neoprene
Thiokol

1. Which of the above are thermoplastic?
2. Which are elastomers?
3. Which contain fluorine atoms?
4. Which one would be the best for making a handle of a skillet?
5. Which one would be the best for making a hose to pump gasoline?
6. Which one will not break on impact?
7. Which one would liberate chlorine gas if burned?
8. Which elastomer was designed to hold air the best?
9. Which thermoplastic high polymer is the most pliable?
10. Which fiber has the highest tensile strength?
11. What is the molecular weight of the ethene mer?
12. If the n of polyethylene was 70,000, what would be the molecular weight of the polyethylene?
13. Which has the higher molecular weight, isoprene or neoprene?
14. What percent of Teflon is fluorine?
15. What is the molecular weight of the nylon 6,6 mer?
16. What is the molecular weight of the Dacron mer?
17. What is the molecular weight of the styrene mer?
18. Would nylon be a good choice of material for use as a handle for a skillet? Defend your answer.
19. If the chlorine from the vinyl chloride were replaced with a bromine atom, what would be the percent change in molecular weight?
20. Would it be possible to polymerize vinyl bromide? Why?

TESTS OF PLASTICS

The question habitually arises: What is it made of? There are several simple tests by which one can determine the type of polymer.

Density tests are not always valid with polymers. Polyethylene may be in sheet form or polystyrene may be in a foam, which would make accurate density determinations difficult. However, rough estimates of the density may be made by determining if the plastic will sink in water, chloroform, or ethylene chloride. These liquids have specific gravities of 1.00, 1.49, and 1.26, respectively.

Plastics can be tested against several solvents as a means of identification.

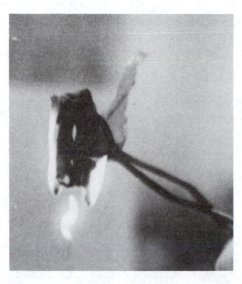

FIGURE 10-4 Flame test of polyethylene. (Note the dripping flame).

Place a small piece of an unknown plastic in a test tube or beaker and cover it with one of the following organic compounds to see if the polymer will soften or dissolve.

> Butyl alcohol
> Acetone
> Carbon tetrachloride
> Toluene
> Diethyl ether

FIGURE 10-5 Flame tests of hipolymers.

Type of Polymer	Ease of Ignition	Self-Extinguishing	Flame	Behavior	Odor
Polyethylene	Fair	No	Yellow, no smoke	Melts, drippings burn	Paraffin candle
Vinyl chloride	Slow	Yes	Yellow, white smoke	Softens, no char or drippings	Acrid, chlorine
Polyvinylidine chloride	Slow	Yes	Yellow, green spurts	Softens, no char or drippings	Pungent, chlorine
Polytetrafluoroethene	Slow	Yes	None	Inert, slight char	None
Polystyrene	Fair	No	Orange, black smoke	Softens, chars, drips	Flowers (geraniums?)
PMMA	Fair	No	Yellow-blue	Softens, no char, froths	Sweetish
Nylon	Very slow	Yes	Blue, yellow top	Softens, drips, froths	Burning hair
Bakelite	Slow	Yes	Yellow	Swells, cracks	Phenol
Urea–formaldehyde	Fair–slow	Yes	Pale yellow, some blue	Swells, cracks	Urea
Melamine	Fair–slow	Yes	Pale yellow	Swells, cracks	Urea
Analine–formaldehyde	Fair	Yes	Yellow, some smoke	Swells, cracks	Charcoal
Polyester	Easy	No	Yellow, blue edge, black smoke	Cracks, sparks	Sweetish
Neoprene	Easy	No	Orange, black smoke	Softens, chars	Acrid

The first test that should be made is to determine if the plastic is thermosetting or thermoplastic. Place the plastic sample in the flame of a bunsen burner, match, or cigarette lighter and see if it melts or gets harder. While doing this test, note the following:

Ease of burning
Color of the flame if the plastic burns
Odor
Color of smoke, if any
Whether the plastic self-extinguishes any flame
Any behavior of the polymer in the flame

This is a destructive test, so caution must be used. The flame test of polyethylene is shown in Figure 10-4.

Figure 10-5 outlines some of the results of the flame tests for several polymers. You might try some of these simple tests on a Styrofoam cup or polyethylene clothes bag to become familiar with the tests.

Composites and Wood

COMPOSITES

Composites are defined as two or more judiciously combined materials that will achieve better properties than they would have separately. This property of the whole being greater than the sum of its parts is called *synergy*. Certain plants achieve this synergistic property and can be called natural composites. Celery and cornstalks are but two plants that reinforce pulpy material with long fibers.

Many composite systems are now available. All of these involve a matrix and a reinforcing fiber. If the fibers are continuous and directionally oriented, the system is called an *advanced composite*. An early example of a composite is reinforced concrete. The concrete becomes the matrix and the steel is the fiber. One of the better known composites is fiberglass.

Fiberglass can be used to illustrate the principle by which the synergistic properties are obtained. As discussed in Chapter 8, glass has a theoretical tensile strength of about 1 million pounds per square inch (over 70,000 kg/cm^2). Yet if a tensile test is run on a glass rod, the tensile strength could be found to be in the low hundreds of pounds per square inch. Further, the tensile strengths found by testing many specimens of glass would not be consistent. Larger-diameter pieces might show a higher tensile strength than that of thinner samples. Longer rods might show a lower tensile strength than would shorter pieces. The reason for this unpredictable behavior is that glass has many very minute cracks caused during the cooling process. When stress is applied, the glass will break at its largest crack. Drawing the glass into a fiber tilts the cracks to a position more along the axis of the fiber, diminishing its weakening effect and making the fiber stronger (see Figure 11-1).

If glass is embedded in a matrix that has a lower tensile strength but is more elastic than the glass, the matrix will transfer the load from a weak point on one fiber to its surrounding fibers. The glass fibers will not break at its weakest point

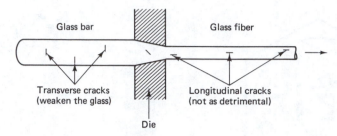

Glass bar   Glass fiber

Transverse cracks   Longitudinal cracks
(weaken the glass)   (not as detrimental)

Die

FIGURE 11-1   Effect of a crack on a glass fiber.

but at the average strength of the fibers, which is a much higher and more pre-
dictable stress. This is illustrated in Figure 11-2.

Composites can be made from any system of two or more components. Some
of these systems are:

Fiber–resin composites
Fiber–ceramic composites
Carbon–metal composites
Metal–concrete composites
Metal–resin composites
Wood–plastic composites

There are many materials used as fibers in composites: boron, graphite,
aramid polymers, glass, and various metals. Most of today's advanced composites
use either glass, Kevlar (an aramid), or various types of graphite fibers. (Although
listed as graphite, these fibers are actually carbon since they do not have the
hexagonal close-packed structure of graphite.) These fibers combine high strength
with light weight. Kevlar (see Chapter 10 for the chemical structure) has a tensile
strength of 400,000 lb/in², but its density is only 1.45 g/cm³. The density of steel is
7.86 g/cm³ and its maximum tensile strength is usually less than 300,000 lb/in².
Figure 11-3 summarizes the properties of these fibers.

In the early days of composites, carbon fibers were made by heating asphalt
threads as they were pulled through a die. Now carbon fibers are made by heating
a rayon or other organic precursor fiber in an inert atmosphere to temperatures
above 1800°F. The carbon fibers are always black, Kevlar is yellow, and the glass
fibers and cloth are white.

The resins can be just about any polymer. They are selected for the specific
properties that they will lend to the composite. They can be thermoplastic, ther-

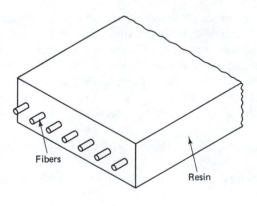

Fibers

Resin

FIGURE 11-2   Composite matrix.

FIGURE 11-3 Properties of high-strength fibers.

Fiber Type	Density (g/cm³)	Tensile Strength (lb/in²)	Modulus of Elasticity (lb/in² × 10⁶)
Graphite			
Low density	1.76	410,000	30–37
High density	1.94	300,000	50–75
Kevlar	1.45	400,000	20
Glass	2.54	500,000	10.5
Steel (reference)	7.86	60,000–320,000	29

mosetting, high polymers, or even elastomers. Figure 11-4 lists a few of the resins used with fibers.

Every industry has its own language and the composites industry is no exception. New words have been derived to deal with the materials and manufacturing processes of composites. Many old words have been given new or different meanings as applied to composites. *T-set* and *T-plastic* are used for thermosetting and thermoplastic high polymers used as a matrix. A resin is said to be in the *A stage* while it is still of low molecular weight and while the resin is readily soluble and fusible. In the *B-stage* the resin is more viscous, with a higher molecular weight, but is insoluble, yet plastic, and fusible. A *C-stage* resin is solid, with high molecular weight, and is insoluble and infusible. To *lay up* a composite is to place the fibers (usually woven into a cloth) in a mold or around a mandrel, or core, in the shape of the final product. The lay-up may be done by hand (known as hand lay-up) or by computer-controlled, laser-guided machines. Robots are entering the composite industry very rapidly. Foams are called *open-celled* if the cells are interconnected. A *closed-cell* foam refers to foams in which the cells are completely individual and are not interconnected (Figure 11-5).

Thinker 11-1:

Take a look at a Styrofoam cup. Is it a closed-cell or open-cell foam?

The design of a composite demands careful selection of the fibers and the resin. Each fiber has advantages and disadvantages. Glass may provide a strong fiber, but it is heavier than Kevlar or carbon. Graphite and Kevlar are more expensive than glass. The best solution to the selection of fibers is to hybridize

FIGURE 11-4 Resin types used in composites.

Thermoplastics	Thermosetting
ABS	Epoxy
PMMA	Bakelite
Fluorocarbon (e.g., Teflon)	Melamine
Nylon	Polyester (e.g., Dacron)
Polycarbonate	Urea
Polyethylene	Urethane
Polyphenylene sulfide	Silicone
Polypropylene	
Styrene	
Vinyl	

 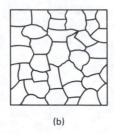

(a)   (b)

FIGURE 11-5 (a) Open-cell and (b) closed-cell foams.

them. A *hybrid* is a combination of fibers used in a single matrix. Hybrids may take various forms. *Interply* (Figure 11-6a) hybrids have two or more different fibers or cloths made from those fibers stacked in alternate layers. *Intraply* (Figure 11-6b) hybrids have the different fibers woven or intermixed in the same ply or layer.

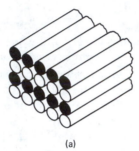 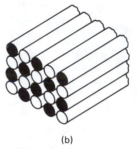

(a)   (b)

FIGURE 11-6 Hybrid composities: (a) interply hybrid; (b) intraply hybrid.

Fibers can be placed selectively in a hybrid, as shown in Figure 11-7. Here the more expensive carbon fibers can be placed in positions where greater stiffness is required while the cheaper glass fibers can be used in the middle. Under deflection, the graphite carries the principal load. Blending of two fibers results in an overall product that is cheaper than one that has an entirely carbon fiber part.

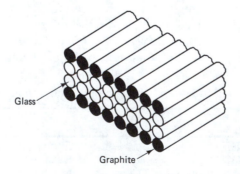

Glass

Graphite

FIGURE 11-7 Selective placement hybrid composites.

The way a fabric is woven (its fiber orientation) can have a decided effect on its properties. If strong fibers are used in one direction and weak fibers perpendicular to them, the fabric will have its greatest strength in the direction of the stronger fibers. The direction in which the stress is applied is called the *warp* (Figure 11-8). The direction perpendicular to the warp is the *weft*. Between these two directions are the bias directions.

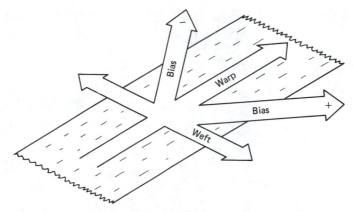

FIGURE 11-8   Direction of fibers in composites.

Another method in which the hybrid reinforcing fabric is made is by *interply knitting* (Figure 11-9). Longitudinal (warp) strands of Kevlar are stitched to latitudinal (weft) strands of graphite by very fine filaments of Kevlar. Two to five plies can be sewn together in this manner, reducing the time for lay-up while providing resistance to shear between the plies.

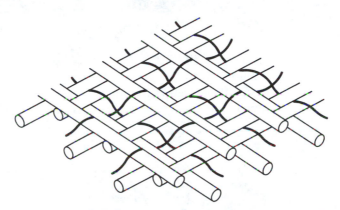

FIGURE 11-9   Interply knitting.

FABRICATION

The oldest method of composite manufacture is hand lay-up. The resin is poured into a mold and the fabric is rolled or brushed into the resin. This technique is very popular with boat builders, auto-body repairers, and home project enthusiasts. It requires very little equipment and the cure is at room temperature.

Pre-preg refers to woven fibers that have been impregnated with a resin prior to the manufacture of a product. Pre-preg tapes are kept refrigerated to prevent them from curing before they are laid into their desired shapes. The pre-preg fabric is heated while it is being wrapped (Figure 11-10) or laid up, then cooled by a blast of cold air or carbon dioxide to set the mold. The formed product is then put under pressure and heated to cure the T-set polymers.

One popular method of applying pressure to a laid-up mold is *vacuum bag molding*, the so-called "poor man's press" (Figure 11-11). The completed layup is placed in a vinyl, polyethylene, or rubber bag and a vacuum is pulled on the bag. Theoretically, this could place a pressure of 14.7 lb/in² on the mold. Practically, a

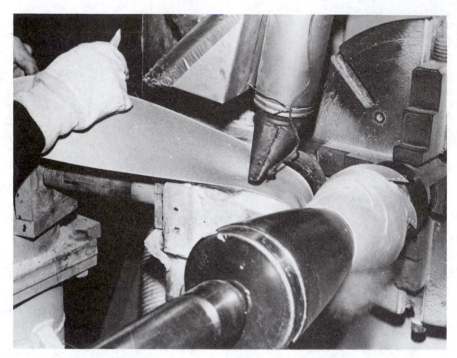

FIGURE 11-10   Wrapping a pre-preg winding.

pressure of about 12 lb/in² is used. If higher pressures are needed, the entire unit is placed in an autoclave, where pressures as high as 20,000 lb/in² and temperatures up to 1000°F can be obtained. Curing times may require from a few minutes to 24 hours. The longer-curing-time requirement is a hindrance to mass production.

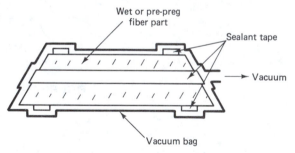

FIGURE 11-11   Vacuum bag molding.

The composites industry also has its fair share of acronyms or initials. In *resin transfer molding* (RTM) a two-piece match-cavity mold is used, as shown in Figure 11-12. The reinforcing material, often chopped-up mat, is draped in the mold cavity.

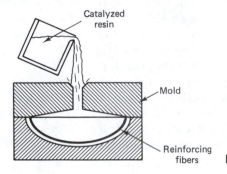

FIGURE 11-12   Resin transfer molding.

The mold halves are then clamped together and the resin is pumped through the injection port. This system uses low pressure, so the tooling and molds need not be expensive to manufacture. Resin transfer molding offers faster curing times and less labor than some of the other techniques.

A manufacturing method used in the production of aircraft wings, missile fins, and other aircraft-related structures is the *ultimately reinforced thermoset reaction injection* (URTRI) process (Figure 11-13). In this process a core is cast from a high-temperature epoxy foam with hollow glass microspheres around a fitting. The core is then wrapped with several layers of graphite fabric and placed in a mold. Epoxy resin is then injected into the heated mold, where it is cured for about 5 minutes. The foam made by using the hollow microspheres and resin is called *syntactic foam*.

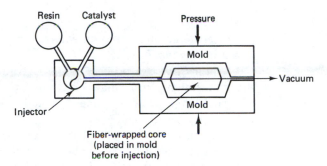

FIGURE 11-13 Ultrareinforced thermoset reaction injection molding.

Reaction injection molding (RIM) is a process applied to polyurethane, epoxy, or other T-plastic systems (Figure 11-14). Mixing of the mer and initiator is done in an impingement mixing head outside the mold. An impingement mixing head is a device that mixes streams of the resin and the initiator under high velocity. (The two streams impinge on each other to get the proper mixing.) Once the material is mixed it is delivered into the mold at low pressures. Some automobile parts on the newer composite cars are made this way.

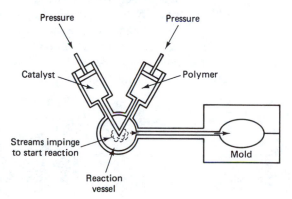

FIGURE 11-14 Reaction injection molding.

A process used to produce nose cones for missiles and aircraft as well as cylindrical pressure bottles and industrial storage tanks is *filament winding* (Figure 11-15). In filament winding, fibers are drawn through a resin bath and wound onto a rotating mandrel which has the shape and size of the desired product. This rather slow process does have the advantage of controlling the direction of the fibers to

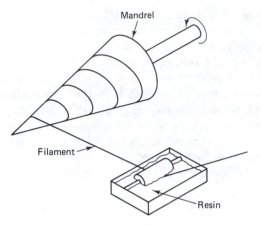

FIGURE 11-15   Filament winding.

produce the greatest strength where it is needed. Once wound, the object can be cured in an *autoclave*.

Pultrusion is a process whereby continuous fibers or bundles of fibers are drawn through a resin matrix bath and then pulled through a heated die (Figure 11-16). Flange beams, tubing, channels, flat bars, and box beams are made by pultrusion.

Composites are often used in a sandwich configuration. A honeycombed lightweight composite can be bonded between two layers of woven high-strength composites (Figure 11-17). The strength-to-weight ratio provided by this geometry is far greater than that of any commercial product on the market today. The core serves to separate and stabilize the outside layers against buckling and bending. It provides a very rigid structure. The honeycomb can be made of polymer-impregnated paper, expanded aluminum, fiberglass, or other composite material. Besides being extremely light weight and strong, the trapped air in the honeycomb gives the sandwich good thermal insulation, corrosion resistance, and vibration resistance properties. Many truck bodies, decks of commercial aircraft, and helicopter blades are made of these honeycombed, sandwich composites. Other sandwich structures involve overlaying fiberglass or other composite over solid cores of Styrofoam or balsa wood. Ice chests and sailboats are examples of this technique.

Composites seem to be the material of the future, but they do have some problems. Quality control is very essential in the manufacture of composite products. Air bubbles can get between the laminates of composites if care is not taken. Further, the layers can "delaminate" or come apart through failure of the matrix or of the adhesive between layers. Metals rust or corrode and composites can degrade if not maintained properly. The polymers can be attacked by organic

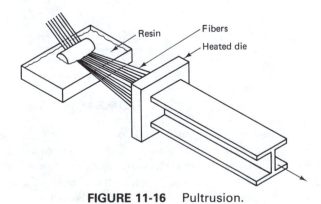

FIGURE 11-16   Pultrusion.

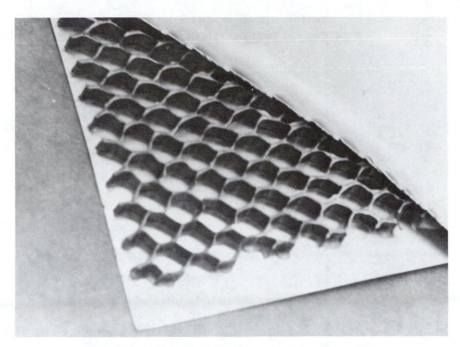

FIGURE 11-17   Sandwich honeycomb composites.

solvents. They can weather by reacting with oxygen and other gases of the atmosphere, and can even absorb moisture over a period of time. Gases can be trapped in composites during curing and form a blister on the surface.

Presently, the major drawback to the use of composites in industry is the lengthy curing time required in their manufacture. In the automotive industry metal parts are stamped out in a matter of seconds. Mass production depends on this speed. Currently, parts that could be made of metal in a few seconds to a few minutes might take several hours to make from composites. This excessive curing time makes their price prohibitive to the average buyer. However, modern technology is reducing to a few minutes the curing time required for many of these composites. Many automobile parts are already being made of plastics and composites. By the year 2000, it is predicted that a competitively priced totally composite car will be on the market.

PROBLEM SET 11-1

Use Figure 11-3 for the data to answer these problems.
1. What is the tensile strength of Kevlar in gigapascal?
2. What is the tensile strength of low-density graphite in gigapascal?
3. What is the tensile strength of glass fibers in gigapascal?
4. What is the tensile strength-to-density ratio of Kevlar?
5. What is the tensile strength-to-density ratio of high-density graphite?
6. If a 50,000-lb block of concrete were to be suspended on a steel rod, what diameter of rod would be necessary if the steel had a tensile strength of 90,000 lb/in^2?
7. If a 50,000-lb block of concrete were to be suspended on a Kevlar rope, what diameter of rope would be required?
8. If a 10,000-kg load were to be suspended by a steel bar having a tensile strength of 100,000 lb/in^2 what diameter of rod (in centimeters) would be required?

9. If a 10,000-kg load were to be suspended by a low-density graphite rope, what diameter of rope (in centimeters) would be required?

10. What is the modulus of elasticity of Kevlar in gigapascal?

11. What is the modulus of elasticity of high-density graphite in gigapascal? (Use the highest value of E.)

12. What is the modulus of elasticity of steel in gigapascal?

13. If a bar of Kevlar were 0.5 in. in diameter and 2.000 in. long, how much would it elongate under a load of 5000 lb?

14. If a bar of low-density graphite were 0.5 in. in diameter and 2.000 in. long, how much would it elongate under a load of 7000 lb? (Use the highest value of E.)

15. If a bar of high-density graphite was 1.25 cm in diameter and 10 cm long, how much would it elongate under a load of 3000 kg? (Use the highest value of E.)

16. If a bar of steel had a diameter of 1.25 cm and was 10 cm long, how much would it elongate under a load of 3000 kg?

17. It is desired to suspend a load of 6000 kg with a piece of Kevlar 3 m long and not have it elongate over 1 mm. What diameter must it be?

18. It is desired to suspend a load of 6000 kg with a piece of steel 3 m long and not have it elongate over 1 mm. What diameter of bar would be needed? Compare this diameter with the Kevlar of Problem 17.

19. A single strand of high-density graphite measures 0.002 in. in diameter. What is the maximum load it would take before breaking?

20. A single strand of fiberglass measures 0.05 mm. What is the maximum load in grams it could take before failure?

WOOD

Wood is nature's composite. The trees from which commercial grades of wood are taken are divided into two types. The *hardwoods* are from *deciduous* trees called *angiosperms* (seed in a jar). Oak, ash, elm, walnut, and maple are examples of hardwoods. Deciduous trees lose their leaves in the winter. The *softwoods* come from *gymnosperms* (naked seed) or *conifers*, which are evergreens. Softwoods are represented by pine, hemlock, redwood, spruce, and cedar. The terms "hardwood" and "softwood" are often confusing: for example, the balsa wood used for model airplanes and ships is a hardwood that comes from a deciduous tree in South America, yet it has a much softer surface than those of many species of pines and other softwoods.

A growing tree has three different parts in the trunk and limbs (Figure 11-18). The outer protective layer, commonly called the bark, is known scientifically as the *phloem* (pronounced "flo-m"). The part of the tree that generates new cells and causes the never-ending growth of a tree is the *cambium* layer, which lies just under the bark. The wood inside the cambium layer is called the *xylem*. The cambium layer generates new cells by dividing both radially and tangentially. The outer cells become new bark or phloem, while the inner cells become new wood or xylem cells. The xylem is composed of several types of cells. The long pointed cells that run vertically in the tree are called tracheids. In softwoods the tracheids average about 3.5 mm (about ⅛ in.) in length, and from 0.03 to 0.05 mm in diameter. In hardwoods these vertical cells are somewhat shorter and thinner than in the softwoods.

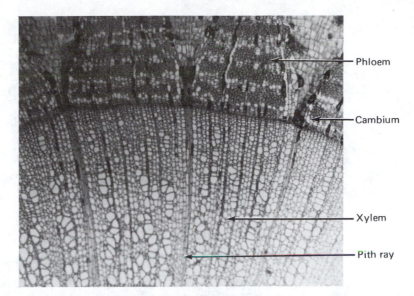

Phloem

Cambium

Xylem

Pith ray

FIGURE 11-18 Structure of a tree.

The layer of phloem just outside the cambium consists of living cells that carry water and nutrients from the roots to the leaves. The leaves convert the carbon dioxide from the air and the water from the ground into sugars through the process of photosynthesis. The sugar (glucose) is carried down the limbs and trunk by the sapwood, which is the part of the xylem lying just beneath the cambium. Wood rays or pith rays are radial cells in the tree which carry the sugars, transported from the leaves by the sapwood to the cells farther inside the tree. When they are first formed, the phloem and xylem cells are hollow. As they age, the cells convert the sugar to starch, then to cellulose, which is added to the cell walls and thickens them. Gradually, the cell walls thicken until the entire cell is cellulose. At this point the nucleus of the cell dies and the cell is dead. These filled xylem cells, known as the heartwood, are used for lumber. The filled phloem cells on the outside of the tree become the bark, which protects the tree from the environment. In some trees removal of the bark will also result in removal of the cambium layer. To take the bark from these trees kills them. In other trees, such as the birch and cork oaks, removal of the bark does not remove the cambium, so removing the bark of these trees does not kill the tree.

The greatest growth of the tree occurs in the spring when there is plenty of moisture and sunlight. New cells form and grow quite large. The cells continue to divide, but later in the summer and on into the winter the growth of the cells taper off, leaving rows of small cells. The layer of large, softer cells is known as the spring wood, while the smaller, more dense cells form the summer wood. One layer of spring wood and a layer of summer wood are grown each year to form an annual ring (Figure 11-19). It is the annual rings that give the wood its grain. These rings not only let a researcher determine the age of the tree, but permit a trained person to ascertain the type of weather in the area of the tree for any given year during the tree's life.

The cells of the xylem are joined together by a gummy substance called *lignin* (Figure 11-20). The cellulose cells in a lignin matrix form a composite. The lignin makes up about 15 to 35% of the wood, while the cellulose comprises 40 to 50%. A substance known as hemicellulose, from which both the lignin and the cellulose is made, makes up most of the remaining part of the wood. There are small amounts of mineral matter and other organics in wood.

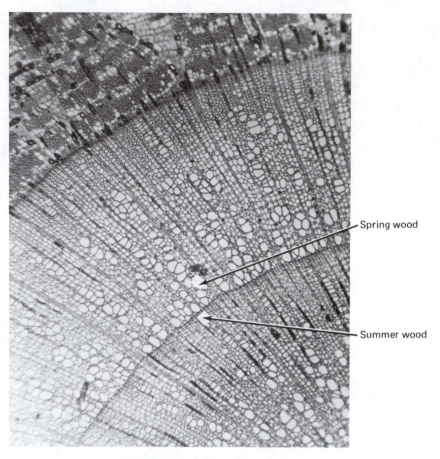

FIGURE 11-19   Annual rings.

After a tree is cut, the trunk must be cut into lumber. Wet wood will dry out over a period of time. In doing so it will shrink, and it will shrink more radially than in circumference. Boards cut along the radius of a tree will shrink along the width of the plank and will not warp significantly. Planks cut tangentially will have more shrinkage in the middle than on the ends and will warp considerably. Figure 11-21 illustrates this principle.

It would be ideal if every board could be cut along the radius of the log, but this would result in a great waste of wood. Several techniques of sawing logs have been developed which are compromises to give the most economical use of the wood with the least warpage. Figure 11-22 shows some of these techniques.

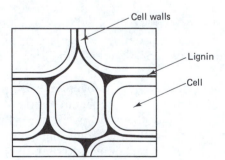

FIGURE 11-20   Cells and lignin.

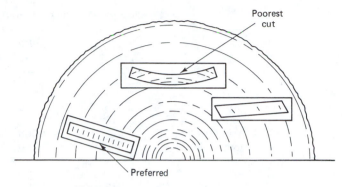

FIGURE 11-21 Shrinkage of wood.

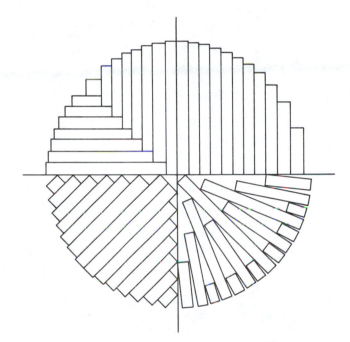

FIGURE 11-22 Sawing patterns for lumber.

PROPERTIES OF WOOD

The properties of lumber depend on several factors. The density of wood is usually less than that of water, but ironwood and a few others can have very high densities, approaching 1.45 g/cm³ or 90 lb/ft³. This wood will not even float on water. Wood is a hygroscopic material, that is it absorbs water easily. It should be dried prior to use to prevent warpage and splitting. Green wood will have from 40 to 100% of its oven dry weight in water. Air-dried wood still contains 12 to 15% moisture. Lumber dried in an oven at 100 to 105°C (212 to 221°F) until it reaches a constant weight is considered to have zero percent moisture. The tensile strength, density, modulus of elasticity, compressive strength, and other properties depend on the type of wood, its water content, the quality of the board (whether it contains knots, etc.), and the direction of the test.

Wood is an *anisotropic* material, which means that it has different properties

FIGURE 11-23 Properties of selected woods.

Wood	Specific Gravity	Modulus of Elasticity (lb/in² × 10³)[a]	Modulus of Rupture (lb/in²)[a]	Compressive Strength (lb/in²)[a]
Hardwoods				
Maple, sugar	0.676	1,830	15,500	7,800
Oak, white	0.710	1,770	15,100	7,440
Poplar, yellow	0.427	1,500	9,200	5,540
Walnut, black	0.562	1,680	14,000	7,580
Softwoods				
Cedar, red	0.34	1,100	7,200	5,020
Douglas fir	0.512	1,900	11,900	7,430
Hemlock	0.44	1,100	9,800	6,210
Pine, white	0.373	1,270	8,900	4,800
Redwood	0.436	1,340	10,700	6,150
Spruce, white	0.431	1,420	9,000	5,470

[a]All figures are for parallel to the grain.

in different directions. The modulus of elasticity of most woods tested parallel to the grain is around 1.1 to 1.9×10^6 lb/in². The densities of wood vary from 0.25 g/cm³ (15 lb/ft³) to 1.45 g/cm³ (90 lb/ft³). Tensile strengths of wood vary considerably depending on the type of wood and the direction of the test. The tensile strength measured parallel to the grain is far different from that measured cross-grain. Usually, the strength of wood is quoted as the *modulus of rupture*, which is the tensile strength as determined by a flexure test of the wood. Wood usually has knots and other imperfections that require large safety factors. The allowable design tensile strengths of the wood may be as low as one-tenth of the values shown in Figure 11-23.

PROBLEM SET 11-2

Refer to Figure 11-23 for data in working these problems.

1. How many pounds would a piece of black walnut measuring 36 in. by 4 in. by 8 in. weigh?

2. It is desired to support 10,000 lb on a post of redwood. What should be the size of a square post to support this load in compression?

3. Would a 2- by 4-in. post of Douglas fir support a load of 8000 lb in compression?

4. A railroad tie measuring 8 in. by 8 in. by 8 ft and made of white pine would weigh how much?

5. How much more would a block of white oak measuring 12 in. by 18 in. by 6 in. weigh than a block of white spruce of the same dimensions?

6. Which is stronger in compressive strength per density, Douglas fir or yellow poplar?

7. Which is stronger in compressive strength per density, sugar maple or black walnut?

8. What would be the compressive strength of white oak in kilograms/m²?

9. What would be the modulus of elasticity of hemlock in megapascal?

10. What would be the modulus of rupture of cedar in megapascal?

11. How many kilograms would a piece of Douglas fir 20 cm by 10 cm by 4 m weigh?

12. What would be the weight in kilograms of a board of hemlock 20 cm by 20 cm by 2.5 m?

13. How many kilograms would a board of redwood 2 in. by 4 in. by 10 ft weigh?

14. Would a 4- by 4-in. redwood post hold a compressive load of 30 tons?

15. What would be the safety factor if a 20- by 20-cm yellow poplar post had a compressive load of 5000 kg?

16. What size block of black walnut, square in cross section, would be needed to support a compressive load of 70,000 lb if a safety factor of 2 were desired?

17. What size block of white oak, square in cross section, would be required to hold up a compressive load of 100,000 lb if a safety factor of 3.0 were needed?

18. Which is stronger in compression, a 2- by 4-in. block of white pine or a 5- by 10-cm block of Douglas fir? How much stronger?

19. What is the modulus of elasticity of black walnut in gigapascal?

20. What is the modulus of elasticity of sugar maple in gigapascal?

Hardwoods are used for furniture and decorative purposes. Hardwoods take longer to grow and the supply is more limited, which results in a higher price. Housing and other structures are framed with softwoods. Redwood or cedar is required in many states as mudsills which come in contact with concrete or the ground. Redwoods and cedars resist dry rot and insect (termite) attack better than do other softwoods.

MEASUREMENTS OF WOOD

Wood is sold by the *board foot* in the United States. A board foot of lumber is defined as "a board 1 foot by 1 foot by 1 inch thick." It amounts to 144 in.³ of wood. For example, a 2- by 4-in. board (a 2 × 4) 8 ft long would contain 5.33 board feet of lumber.

$$\frac{(8 \text{ ft}) (12 \text{ in./ft}) (2 \text{ in.}) (4 \text{ in.})}{144 \text{ in}^3/\text{bd ft}} = 5.33 \text{ bd ft}$$

The number of board feet can be figured a second way. Each foot of a 2 × 4 carries $\frac{2}{3}$ board foot. A foot length of 2 × 6 is a board foot of lumber. A 1 × 4 is $\frac{1}{3}$ board foot per linear foot and a 1 × 6 has $\frac{1}{2}$ board foot per linear foot. Just multiply the length of the lumber by the number of board feet per foot of length to get the total board feet. For example, 10 boards each 2 by 4 by 16 ft long would contain 53.3 board feet of lumber.

$$(10) (16 \text{ ft}) (\tfrac{1}{3} \text{ board feet/foot}) = 53.3 \text{ board feet}$$

The board feet of lumber are measured in the rough lumber. A 2 × 4 cut rough is dressed to $1\frac{1}{2}$ by $3\frac{1}{2}$ in. Construction lumber is sold in standard board lengths of 2-ft increments. It would be necessary to buy 12-ft-long lumber if 11-ft lengths were needed. In planning houses and other structures, good design would be kept in these 2-ft increments to minimize waste.

Prices are often quoted in dollars per thousand, which is shortened to $/M

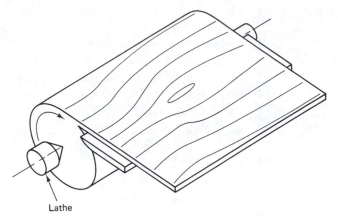

Lathe

FIGURE 11-24   Plywood "peeling."

in lumberyards. Redwood quoted as $500/M means that the price is $500 per 1000 board feet, which is $0.50 per board foot.

To calculate the price of a piece of lumber, the number of board feet in the lumber must be calculated, then multiplied by the price per board foot. For example; if a construction grade of Douglas fir were quoted at $600/M, how much would a beam of Douglas fir 4 in. by 10 in. by 18 ft long cost? The board feet of lumber in the wood beam is

$$BF = (4 \text{ in.} \times 10 \text{ in.} \times 18 \text{ ft} \times 12 \text{ in./ft})/144 \text{ in}^3/BF = 60 \text{ BF}$$

The price per board foot of lumber is

$$\text{price} = (\$600/1000 \text{ BF}) = \$0.60/BF$$

The cost of the beam is then

$$\text{Cost} = (60 \text{ BF})\ (\$0.60/BF) = \$36$$

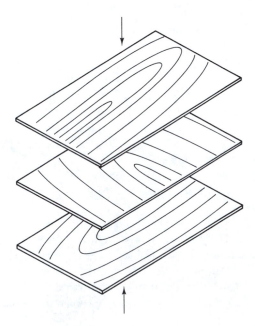

FIGURE 11-25   Plywood.

LAMINATES OF WOOD

Since wood is an anisotropic material having different tensile strengths, moduli of elasticity, and other properties in each direction, a more uniform material can be achieved by gluing panels of wood together with the fibers in different directions. This is the principle of *plywood*. Thin sheets of wood are "peeled" from a log on a large lathe (Figure 11-24).

The sheets are cut to approximate size, inspected, knots removed and replaced if necessary, and glued with the grains or fibers perpendicular to each other (see Figure 11-25). Plywood may have as few as three layers or "plys" or up to 13 or more on special order. One-eighth to three-eighth-inch-thick plywood is generally three-ply, while $\frac{1}{2}$- to $\frac{3}{4}$-in. plywood is usually five-ply. Thicker sizes may have more plys.

A water-soluble glue that will deteriorate if allowed to get wet is used in interior plywood. A waterproof glue is used in exterior plywood. The interior plys of plywood may have the knots removed but not refilled, leaving holes in the inside part of the panel. Not only is a waterproof glue used in marine plywood, but all plies are solid.

Besides the exterior and interior ratings, each side of the plywood is graded. A grade A surface has no knots and is sanded. A grade D side will be unsanded and have open spots where the knots were cut out. Plywood marked "A-A" has both sides filled and sanded. It is the type of plywood used for cabinets and other work where both sides of the wood will be finished and open to view. An A-D plywood might be used for paneling where only one side is to be finished and exposed. A C-D plyscore would be used for structural work where the wood will not be seen.

When large beams of wood are needed, they are made up of several boards glued or laminated together (Figure 11-26). These are known in the trades as *glu-lams*. Besides being able to make larger beams than could be cut from a single log, glu-lams can be made from "clear" lumber with no knots or weak spots in the wood. The superior strength, combined with the pleasing appearance of the wood, make it ideal for construction of large homes, churches, office buildings, and other structures having exposed beams.

FIGURE 11-26   Glu-lam.

PROBLEM SET 11-3

1. How many board feet are there in a 2 × 4 that is 12 ft long?

2. How many board feet are there in a 4 × 4 fence post 8 ft long?

3. Flooring is often 2 × 6 tongue-and-groove lumber. How many board feet is in a 10-ft-long piece of 2 × 6?

4. A floor measuring 16 ft by 20 ft is to be laid. How many board feet of 2 × 6 lumber should be ordered to do the job?

5. A ceiling joist is a 16-ft-long 4 × 12. How many board feet of lumber is in each joist?

6. A patio deck measuring 12 ft by 14 ft is to be made of redwood having a price of $450/M. If it is made of 2 × 6 lumber, what would be the cost of the deck?

7. A load of 2 × 4 lumber for a house had 190 boards each 8 ft long. How many board feet of lumber was in the load?

8. If the price of the lumber in Problem 7 was $600/M, how much did it cost?

9. To build a small garage a contractor ordered thirty 2 × 6's each 14 ft long, seventy-two 2 by 4's each 8 ft long, and forty-two 1 by 4's each 16 ft long. The price of the construction grade lumber was $425/M. What was the cost of the lumber?

10. A board foot of walnut costs $2.25. What would be the price of a 2 × 8 board of walnut 6 ft long?

11. A log is 15 ft long by 18 in. in diameter after the bark is removed. Estimate the board feet of lumber in the log.

12. A fence 80 ft long and 6 ft high required 160 ft of 2- by 4-in. redwood, and nine 4- by 4-in. by 8-ft-long redwood posts. The rest of the lumber was 1- by 8-in. by 6-ft-long boards. How many board feet of lumber is required for the fence?

13. At a cost of $400/M, what would be the cost of the fence in Problem 12?

14. A hardwood floor of white oak measured 16 ft by 20 ft. The wood was 0.5 in. thick. How many board feet of white oak flooring is needed?

15. What would be the cost of the white oak flooring in Problem 14 if the wood cost $1200/M?

16. To construct the forms for a concrete foundation, 180 linear feet of 2 × 4 form lumber is required. If the lumber cost $450/M, what would be the cost of the lumber?

17. A new home builder has the choice of laying carpet over subflooring at a cost of $35.00 per square yard or installing $\frac{1}{2}$-in. hardwood flooring at a cost of $1100/M. Which would be cheaper to do for a 14- by 18-ft room?

18. If the price of redwood is $600/M, how much would it cost to make a rectangular pool deck 6 ft wide around a rectangular pool that measures 12 ft by 20 ft? The lumber of the deck is 2- by 6-in. boards. (Figure just the deck surface, not the lumber required for the supports.)

19. A roof can be sheeted by either $\frac{3}{4}$-in. plywood or 1 × 4 lumber. The area of the roof is 18 ft by 40 ft. A 4- by 8-ft. sheet of $\frac{3}{4}$-in. C-D plyscore costs $7.00, while the 1 × 4 lumber is $500/M. For which method would the lumber be cheaper?

20. What would a load of 1000 board feet of Douglas fir weigh?

Glossary

Acm line: The line on the iron–carbon phase diagram that lies between the austenite and the austenite–cementite two-phase region.

Ac₁ line: The horizontal line on the iron–carbon phase diagram that is at the eutectoid temperature.

Ac₃ line: The line on the iron–carbon phase diagram which lies between the austenite region and the austenite–ferrite two-phase region.

ABS: A polymer made from acrylonitrile, butadiene, and styrene.

Abscissa: The x axis or horizontal axis on a graph.

Absolute zero: The lowest possible theoretical temperature. At this temperature there is absolutely no heat. It is equal to $-273°C$.

Accelerated curing of concrete: Involves methods of speeding up the curing of test samples of concrete for early testing.

Accuracy: The degree to which test data agree with the true value for that test. (Note that the true values may never be known. The values generally agreed upon should be called the "accepted values.")

Acids (inorganic): Compounds that contain hydrogen ions.

Acids (organic): Compounds that contain the carboxyl or —COOH group.

Acrylic: Polymers that contain methyl methacrylate.

Acrylonitrile: Compounds that contain the vinyl cyanide structure.

Actinide series: The elements with atomic numbers between 89 and 103 (inclusive): actinium, thorium, protactinium, uranium, neptunium, plutonium, americium, curium, berkelium, californium, einsteinium, fermium, mendelevium, nobelium, and lawrencium.

Activity series: A list of chemical elements ranked in the order by which they will react chemically. An element that is higher in the activity series will replace a lower element in a chemical compound.

Advanced composite: A composite that has continuous fiber reinforcements.

Age hardening: The precipitation of one phase of a two-phase system into microscopic particles evenly distributed in the other phase. It is also known as precipitation hardening and can be accelerated by heat treatments.

Air quench: The heat treatment by which steel or other metals are rapidly cooled by a blast of cold air.

AISI: American Iron and Steel Institute.

Alcohols: A class of organic chemicals that have an —OH group somewhere in the molecule.

Aldehydes: A class of organic chemicals that have the —CH:O group in the molecule.

Aliphatic compounds: Organic chemicals that have only long-chain carbons. Aliphatic compounds do not have ring structures.

Alkali metals: A group of elements consisting of hydrogen, lithium, sodium, potassium, rubidium, cesium, and francium.

Alkaline earth metals: A group of elements consisting of calcium, strontium, barium, and radium.

Alloys: Metals formed by blending two or more metals together.

Alpha iron: Ferrite or the single-phase composition containing dissolved carbon in solid solution with iron. For all practical purposes it is pure iron.

Alumina: Aluminum oxide (Al_2O_3).

Aluminosilicates: High-temperature, chemically resistant glasses made primarily from alumina and silica.

Amalgam: An alloy of an element and mercury.

Ames hardness: A measure of surface plasticity using a portable Rockwell-type tester. It gives readings in Rockwell units.

Amides: A class of organic chemicals that have the —C:ONRR′ group in the molecule.

Amine: A class of organic chemicals that have the —NH_2 group in the molecule.

Amorphous: The word *amorphous* means "without body." Amorphous materials lack a definite crystal structure.

Ampere: The electrical flow rate of 1 coulomb per second.

Amphoteric elements: Elements that can react chemically either as a metal or nonmetal. These include the elements aluminum, silicon, carbon, phosphorous, titanium, vanadium, germanium, arsenic, niobium, molybdenum, tin, antimony, tellurium, tantalum, tungsten, lead, and bismuth.

Analog: A measurable substitute for another value. A watch uses a mechanical escapement or an electrical oscillation as an analog for time.

Angiosperms: The class of plants (and trees) that have seeds surrounded by nutrient bodies. Literally the word means "seed in a jar."

Angle of incidence: The angle measured from the perpendicular to a surface at the point which a light ray hits the surface.

Angstrom unit: 1×10^{-8} centimeter or 1×10^{-10} meter.

Anion: The element or radical that is negatively charged.

Anisotropic: Materials that have different properties in different directions.

Annealing: The heat treatment of a metal that renders it in its softest condition.

Annealing twins: The mirror-image crystal structures that form while material is forming its original crystals. They can form while the material is cooling from the melt, as a result of heat treatments or electroplating.

Anode: The pole in either a plating cell or electrolytic cell that gives up electrons. It is the negative pole of a battery or the positive pole of a plating cell.

Anodizing: A plating process whereby the anode material is combined with ions of the electrolyte to form hard, chemically resistant, attractive protective surfaces on metal.

ANSI: American National Standards Institute.

Aqua regia: A mixture of one-fourth nitric acid and three-fourths hydrochloric acid. It means "king of waters" because it will dissolve gold (hot).

Aromatic compounds: Compounds that have a ring structure in the molecule.

ASM: American Society for Metals.

A-stage: The stage in the curing of composites in which the resin is readily soluble, fusible, and has a low molecular weight.

ASTM: American Society for Testing and Materials.

Atom: The smallest particle to which an element can be divided.

Atomic number: The number of protons in the nucleus of an atom.

Atomic weight: The sum of the protons and neutrons in the nucleus of the atom.

Austempering: A heat treatment of steel in which the metal is quenched at a rate which will miss the knee of the TTT curve to a temperature just above the Ms line, then held at that temperature until all transformation is complete. The structure will be mostly bainite.

Austenite: The face-centered cubic structure of steel.

Austenitic stainless steel: Stainless steels whose composition keeps them in a face-centered cubic structure at room temperatures. They are nonmagnetic stainless steels.

Autoclave: A heating oven that can be sealed and have a vacuum or pressure applied in it.

Autogeneous method: Uses the heat of hydration of cement to speed up curing a test sample of concrete.

Avogadro's number: The number of atoms in 1 gram-atom of an element or 1 gram-molecule of a compound. It is equal to 6.023×10^{23} atoms per gram-atom or molecules per gram-molecule.

Bain S curve: *See* TTT curve.

Bessemer converter: A machine that changes iron into steel by blowing air through the melt from the bottom.

Bias: In the reinforcing plies of a composite, the angle the fibers make with the longitudinal direction of the material.

Billet: A section of an ingot used for rolling or other manufacturing process in steel or other metal mills.

Blister copper: The form of the metal immediately after it comes out of a converter, when it is 99% + pure copper. It is further refined by electrolytic processes.

Bloom: A slab of metal whose width generally equals its thickness. It is used for hot rolling into further shapes.

Blowing agent: An additive to polymers that causes them to form a foam.

Board foot: A piece of lumber 12 inches by 12 inches by 1 inch thick. It is equal to 144 cubic inches of wood.

Body-centered cubic: Crystal structure of equal-length axes with all angles equal to 90° and a complete atom in the center of the cube.

Boiling point: The temperature at which the vapor pressure of a liquid is equal to the pressure of the atmosphere surrounding it.

Boiling-water method: A method of accelerated curing of concrete test samples that lets the cast samples stand for 23 hours, then cures them for $3\frac{1}{2}$ hours in boiling water before capping and testing.

Borosilicate glass: Glass that has boron atoms instead of sodium atoms in a silicate matrix. It has a very low coefficient of thermal expansion, so will take considerable thermal shock before breaking. It is marketed under the names Pyrex, Kymex, and others.

Brass: An alloy of copper and zinc.

Breaking strength: The stress a material has at the point of failure; also known as the rupture strength.

Bright annealing: An annealing process done in an inert-gas atmosphere in a muffle furnace.

Brine: A saltwater solution used as a quenching medium in the heat treatment of steel.

Brinell hardness: The measure of resistance to surface plastic deformation as measured by a static load being applied to a 10-millimeter steel ball placed on the surface.

British thermal unit: The amount of heat required to heat 1 pound of water 1 degree Fahrenheit.

Brittle material: A material that breaks with little or no permanent deformation. Its breaking strength is its tensile strength.

Bronze: An alloy of copper and just about anything else. Traditional bronzes are copper–tin alloys.

B-stage: The state of a resin in which it is insoluble but still pliable and fusible. A resin is more viscous and has a higher molecular weight in the B-stage than in the A-stage.

Btu: *See* British thermal unit.

C curve: *See* TTT curve.

Calorie: The amount of heat required to raise 1 gram of water 1 degree Celsius. Specifically, it is the heat required to raise 1 gram of water from 15°C to 16°C.

Cambium: The cell layer in a tree which contains the cells that divide to form the xylem and the phloem.

Carbonitriding: A heat treatment of steels by which they are case hardened by being heated in a gaseous atmosphere that contains both carbon and nitrogen. The carbon and nitrogen are absorbed simultaneously. The carbonitriding can be followed by suitable quenching.

Carburizing: A case-hardening treatment of steel whereby the metal is buried in a high-carbon atmosphere to allow the carbon to diffuse into the outer layers of the metal. Carburizing is generally followed by quenching to produce a hardened surface on the steel.

Case hardening: A process of hardening the surface of steel so that the outer layer will be significantly harder than the interior of the metal.

Cast iron: An iron–carbon alloy that has more than 2% carbon.

Cathode: The pole in an electrolytic cell which accepts electrons. It is the negative pole of a plating cell or the positive pole of a battery.

Cathodic protection: The process of protecting a metal by connecting it to a more anodic material or metal which has higher chemical activity than the one being protected. Cathodic protection can also be provided by placing a negative charge on the material.

Cation: An element or radical that is positively charged.

Cement factor: The number of sacks of cement per cubic yard of concrete.

Cementite: Iron carbide (Fe_3C).

Cermets: Ceramics that have metals mixed in with them.

Charpy test: An impact test that uses a swinging pendulum to hit a notched specimen held at both ends and hit in the middle.

Chemical properties: Those properties of a material that pertain to the way the material reacts with other materials. A function of the electronic structure of the elements involved.

Chemtempered glass: Soda-lime glass that has been soaked in potassium salts to replace the outer layer of sodium atoms with larger potassium atoms. This makes the outer surface denser, greatly strengthening the glass.

Chloroprene: 2-Chlorobutadiene that is polymerized to make neoprene rubber.

Closed cell: In a polymer foam, that form which has the individual cells completely sealed off from its adjacent cells.

Coefficient of thermal expansion: The change in length a material will undergo per unit length for each degree of temperature change.

Coke: Anthracite or hard coal that has had all hydrogen and other atoms removed by heating it in the absence of oxygen in the coke ovens.

Complex cubic system: An overlapped face-centered cubic system. It is sometimes called the diamond structure or the face-centered close-packed system.

Component: Any element, ion, or molecule of a system that has an affect on the behavior of the system.

Composites: Two or more judiciously combined materials that give better properties to the system than are available separately.

Compression: The application of forces directly toward each other.

Compressive strength: The maximum stress a material can take in compression before failure.

Concrete: Any aggregate bonded together by a cementing agent.

Condensation reaction: The reaction of two compounds which results in a by-product of a small molecule, usually water or alcohol.

Conifers: Cone-bearing trees; also called gymnosperms.

Continuous-pour process: The pouring of a single slab, billet, or bloom from an entire melt of metal.

Converters: Used in metallurgy to remove the excess carbon in a metal by blowing air over or through the melt, or by electrical means. They are used to convert pig iron into steel and are also used in the copper and other metals industries.

Cooling curve: A temperature versus time plot of the rate of cooling of a material.

Copolymer: A polymer in which the mer is made from two or more different molecules.

Corrosion: The oxidation of a metal; the reaction of oxygen or other nonmetallic element with a metal.

Coulomb: That amount of electricity which will plate exactly 0.0011180 gram of silver from a solution of silver nitrate; equals 6.24×10^{18} electrons.

Creep: The change in length per unit length under a static load over a period of time.

Critical strain: In fatigue tests, the strain above which the number of cycles to failure is greatly reduced.

Cross-linking: In polymers, the formation of a chemical bond between the linear chains which produces a thermosetting high polymer.

Cryogenic temperatures: The temperatures of freezing; usually referred to as the temperatures produced by liquid oxygen, nitrogen, or helium.

Crystal: That state of matter which has a definite long-range arrangement of atoms, radicals, or molecules.

Crystal glass: A lead–alkali glass decoratively used as cut glass.

C-stage: The state of a high polymer in which the resin is insoluble and infusible and has an extremely high molecular weight.

Cubic system: The crystal structure in which all the axis lengths are of equal units and all three axes are mutually perpendicular.

Cupola malleable iron: *See* Pearlitic malleable iron.

Cyaniding: The process of case hardening a steel by burying it in sodium cyanide (NaCN). In this process both the carbon and the nitrogen from the cyanide are absorbed in the outer layers of the steel. Further quenching will provide the metal with a very hard case.

Dacron: The trade name for the polyester formed by the condensation reaction of dimethyl terephthalate and ethylene glycol.

Deciduous trees: Trees that lose their leaves in the winter.

Delta iron: The solid solution of carbon in pure iron formed at temperatures above 1400°C (2522°F). It is the high-temperature form of alpha iron.

Density: The mass per unit volume of a material.

Destructive test: A test in which a part is damaged to the degree that it can no longer be used for the purpose for which it was designed.

Deuterium: The isotope of hydrogen that has one proton and one neutron in the nucleus.

DHP copper: *See* Phosphate deoxidized copper.

Diamagnetic: A material that has less magnetic permeability than does a perfect vacuum. It would be weakly repelled by a magnet in air. Diamagnetic materials conduct lines of magnetic flux with only slight resistance.

Die casting: The manufacturing process whereby liquid metal is poured into a steel mold or die, allowed to cool, then punched out of the mold.

Diffusion: The migration of one type of particle spreading evenly throughout a second material.

Doping: The addition of trace amounts of materials with more or fewer electrons that those of the parent material in semiconductors.

Double bond: A bond in an organic compound in which two elements share two pair of electrons.

Drawing: *See* Tempering.

Ductile cast iron: Cast iron alloyed with a small amount of magnesium or cesium, which causes the graphite to form in spheres, thus reducing the brittleness caused by the excess carbon.

Ductile material: A material in which the breaking stress is lower than the tensile strength and the breaking strain is greater than the tensile strain.

Ductility: A metal's ability to be drawn through a die into a wire. It is also defined as the property of a metal that allows it to undergo permanent deformation prior to fracture.

Durometer: A hardness tester used mainly for plastics which measures the elastic deformation of the surface of the material as an indication of hardness. There are two types of durometers, type A and type D.

Dye penetrant: A test to find surface flaws. A dye is allowed to seep into the flaws, which will show up after a developer is applied.

Dynamic equilibrium: The state of a system of two or more phases in which the overall amount of each phase stays the same but the individual particles may change.

Eddy current: A current induced in a conductor by varying a magnetic field about it.

Eddy current tests: Tests that detect the disturbance to the flow of an induced electric current by internal cracks or flaws in a material.

Elastic limit: The point in the stress–strain curve where stress ceases to be linearly proportional to strain.

Elastic region: The area of a stress–strain curve in which stress is linearly proportional to strain.

Elasticity: The ability of a material to return to its original size and shape once a load is removed.

Elastomers: Polymers that stretch and return to their original size and shape; the natural and synthetic rubbers.

Electric arc converter: Apparatus that uses large carbon electrodes to carry an arc to molten pig iron to burn out the excess carbon and convert it to steel.

Electrical conductance: The ability of a material to carry an electrical current. It is the reciprocal of electrical resistance.

Electrical properties: The properties that involve the behavior of the electrons of a material. They include electrical resistance and conductance of the material.

Electrical resistance: Opposition to the flow of electrons in a material; a function of the thickness of the material.

Electrical resistivity: The ability to oppose the flow of electrons; not a function of the thickness of the material. It can be affected by temperature.

Electricity: The flow of electrons.

Electrodes: The positive and negative poles of a plating cell or battery.

Electrolysis: The process of passing a current through a solution and the plating and deplating of the metals in that solution.

Electrolytic tough pitch copper: Copper that has been refined by electroplating and poled to remove oxygen.

Electromotive series: A list of metals in order of their ability to react with acids or nonmetals.

Electronegativity: A measure of the ease with which an element gives up its electrons.

Electron (imperfection):   An extra electron at a lattice point in a crystal. Electrons make N-type semiconductors.

Electron (particle):   An atomic particle outside the nucleus which carries a unit negative electrical charge and is approximately 1/1840 the mass of a neutron.

Electronic structure: The number, arrangement, and energy levels of the electrons about the nucleus.

Electroplating: The deposition of metal on the cathode of a plating cell from the electrolyte by the flow of an electric current.

Element: One of the fundamental particles or materials from which all other materials is composed. There are 92 naturally occurring elements and 12 artificially created elements at the present time, with the possibility of more artificial ones being created.

Elongation: The extension of a piece of material due to an external load being applied.

Energy: The ability to do work.

Equilibrium: The state when everything is in balance and at its minimum energy level.

Error: Inaccuracies in measurements. They can be minimized but never eliminated.

Esters: The salts of organic acids with the structure RC:OOR'.

Ether: Organic compounds that have two radicals joined through an oxygen atom, with the structure R—O—R'.

Eutectic: The composition of a system of two or more components which has the lowest melting temperature. It can also be defined as that composition of a system of two or more components which will go directly from a liquid to a two-phase solid upon cooling without passing through a two-phase region.

Eutectic matrix: A two-phase solid structure that forms at the eutectic composition.

Eutectic temperature: The lowest melting point of a system of two or more components.

Eutectoid: The temperature and composition for which a single-phase solid will, on cooling, pass directly to a two-phase solid without passing through a two-phase region.

Eutectoid matrix: The two-phase solid structure that forms at the eutectoid composition.

Extensometer: An instrument that measures the elongation of a member under a tensile load.

Face-centered cubic: The crystal that has axes of equal unit lengths, mutually perpendicular axes, and an extra particle in the center of each side or face of the cube.

Fatigue: The failure of materials due to cyclic strains.

Ferrite: Alpha (α) iron, the solid solution of less than 0.025% carbon in iron. For all practical purposes it is pure iron.

Ferritic malleable iron: Iron made from white cast iron heated to 1700°F, then quenched to 1350°F and cooled very slowly to room temperature. The graphite in the iron is spheroidized.

Ferritic stainless steel: Low-carbon steels containing 20 to 37% chromium. They are designated the 400 series of stainless steels.

Ferromagnetic: Materials that can be magnetized in the same manner as iron.

Filament: A fine wire or thread often used as a heating element.

Fineness modulus: A measure of the gradation of an aggregate. A well-graded aggregate will usually have a fineness modulus that is not an integer or a whole number. Well-sorted aggregates will have integer or whole-number fineness moduli.

Flame hardening: The hardening of the surface of a metal by heating it with a gas flame, then quenching it. Only the surface is heated into the austenitic range, so only the surface will go to martensite and be hardened.

Flame point: The temperature at which oil will ignite and stay lit.

Flash point: The temperature at which oil will first ignite, but will not maintain a flame.

Flexure: The bending of a material.

Flint glass: The name given to optical-quality lead glass.

Flotation: The process of separating a mineral from extraneous material in the ore by mixing it in a solution that will make a froth or foam. Air is then bubbled through the solution, which causes the foam to attach to the mineral and carry it to the surface. The froth is skimmed off and the mineral recovered from the foam.

Flux: A material that breaks down surface tensions between two other materials, causing them to flow together.

Force: Mass times acceleration.

Formica: A trade name for a formaldehyde–melamine polymer with a mica filler.

Frank–Read source: The mechanism of fatigue which is initiated by the generation of new dislocations which pile up to cause a crack in the material. The crack then propagates to failure.

Free machining brass: Brass which has a lead content so that it can be turned, milled, or cut easily and leave a smooth surface after cutting.

Full annealing: Heat treatment of steel which puts it in its softest condition. It is done by heating the steel about 50° into the austenitic region then very slowly cooling it.

Galvanizing: The process of dipping steel into a molten bath of zinc. The zinc provides both a coating and cathodic protection to the steel.

Gamma iron: Austenite.

Gas: A state of matter that has no structure at all. There is no bonding between the molecules.

German silver: A bronze having approximately 50% copper, 30% nickel, and 20% zinc and other metals.

Glass: The amorphous structure of silicates and other oxides. It is a super viscous liquid generally made from silicates having one or more metallic oxides in the structure.

Glu-lams: Wooden beams made from many boards being glued together.

Grain: An individual crystal in a metal; in wood, the direction in which the tree had grown.

Grain size number: The log of the number of grains found in 0.0645 square millimeter or surface area divided by the log of 2, all increased by one.

Gram-atom: The weight in grams of an element, numerically equal to the atomic weight of that element.

Gray cast iron: Iron with about 4% carbon which appear as flakes of graphite in the structure.

Grog (ceramics):   A mixture of coarse metallic oxides that are bonded by finer particles to make refractory materials.

Gunite: A fine aggregate concrete that is applied by spraying over the reinforcing steel.

Gymnosperms: The class of plants or trees that have "naked seeds"; the cone-bearing trees.

Halogens: The chemical family that include the elements fluorine, chlorine, bromine, iodine, and astatine.

Hardenability: The depth to which a steel can be hardened.

Hardness: The ability of metals to resist surface deformation. Different hardness tests measure hardness in terms of resistance to plastic deformation, elastic deformation, resistance to surface abrasion.

Hardwoods: Woods derived from deciduous trees.

Heat: The energy of atoms or molecules in motion.

Heat capacity: The amount of heat, measured in calories or Btu required to heat a unit weight 1 degree of temperature. In the metric system the units of heat capacity are calories per gram per degree Celsius. In the English system the units are Btu per pound per degree Fahrenheit.

Heat of combustion: The amount of heat released by a compound upon reaction with oxygen.

Heat of fusion: The amount of heat required to change a unit weight of solid to a liquid at its melting temperature.

Heat of vaporization: The amount of heat required to change 1 unit weight of liquid to gas at its boiling temperature.

Hematite: An iron ore consisting of ferric oxide (Fe_2O_3).

Hexagonal close-packed: A crystal system in which the individual unit cells are overlapped.

Hexagonal structure: A system that has at least two of the axes of equal unit length, with the angles between two of the crystal planes being 90° while the third angle is 120°.

High polymer: A material that has combined many of the same molecules to a point where the structure has an extremely high molecular weight.

Hole: The absence of an electron at a lattice point in a semiconductor.

Holography: The art and science of using lasers to produce interference patterns or true three-dimensional photographs.

Hot short: A property of some steels which causes them to lose strength at high temperatures. It is caused by sulfur reacting with the iron at the grain boundaries.

Hybrid: Composite that has two or more different reinforcing fibers.

Hypereutectoid steel: Steels that have more than the eutectoid composition (0.8%) of carbon.

Hypoeutectoid steel: Steels that have less than the eutectoid composition (0.9%) of carbon.

Impact: The sudden application of force to a material.

Imperfection: Anything that upsets the orderly array of a crystal lattice.

Inclusion: Macroscopic imperfection in a crystal lattice. They cover many lattice points and are held in place mechanically in the structure.

Index of refraction: The ratio of the speed of light in a vacuum to the speed of light in a substance. It can be calculated as the sine of the angle of incidence divided by the sine of the angle of refraction.

Inert gas: Helium, neon, argon, krypton, xenon, and radon. They are called inert gases because they do not react readily with anything.

Intermetallic compounds: Chemically combined metals of two or more types.

International union of pure and applied chemists: A body that sets standards for chemical research, headquartered in Geneva, Switzerland.

Interply hybrid: A composite with different fibers in different layers or plys.

Interply knitting: The sewing together of the various layers or plys of reinforcing fibers.

Interstitial: An atom or particle not at a lattice point in a crystal. It is a crystal imperfection.

Intraply hybrid: A composite with different types of fibers within each layer.

Ion: A charged atomic particle. It is an atom or group of atoms that is not electrically neutral. Ions are either positive or negative depending on whether they have more or fewer electrons than the number of protons in the nucleus of the atoms.

Ionic bond: The result of metals and nonmetals giving up and accepting electrons, respectively, to form positive and negative ions. These ions then attract each other to form a chemical compound.

Isomer: Molecules that have the same number and type of atoms but have different structures or arrangements of the atoms. Different isomers have the same molecular weight, but other properties may be different.

Isothermal tranformation curve: *See* TTT curve.

Isotope: Elements that have the same number of protons but a different number of neutrons on the nucleus of the atom. Different isotopes have the same atomic number but a different number of neutrons.

Isotropic: A material that has the same properties in different directions.

IUC: International Union of Chemists.

IUPAC: International Union of Pure and Applied Chemists.

Izod test: An impact test that uses a swinging pendulum to strike a notched bar which is held at one end.

Jominy depth: The depth in a steel to which 50% martensite can be formed by quenching from the austenitic range.

Jominy test: An end quench test of steel designed to determine the Jominy hardenability depth.

Ketones: Organic compounds that have the —C—:O group in the molecule.

Kevlar: The trade name of a polyaramid which is used as a reinforcement fiber in composites. It has a very high strength-to-density ratio.

Kilocalorie: One thousand calories; sometimes called the large calorie.

Lanthanide series: The chemical elements that have atomic numbers between 58 and 71 inclusive: cerium, praseodymium, neodymium, promethium, samarium, europium, gadolinium, terbium, dysprosium, holmium, erbium, thulium, ytterbium, and lutetium.

Lattice constant: The unit lengths of the axes in the three major directions.

Lattice point: The intersections of the major crystal planes.

Lay-up: In composites, the position and direction of the reinforcing material.

Lead glass: Glass made from lead oxide, silica, and potassium oxide. It has a very high index of refraction and is used in lenses and as protective glass for radiation shielding.

Ledeburite: The eutectic composition of the austenite–cementite in the iron–iron carbide phase diagram.

Lift slab construction: The type of concrete building in which all the floors are cast by slip forming on the ground, then individually raised into place on columns.

Lignin: The gummy substance that holds the xylem cells together in wood.

Line defect: An imperfection in a crystal which involves a plane or partial plane of atoms or particles.

Line dislocation: An extra partial plane of atoms in a crystal structure.

Liquid: A state of matter that has a random structure.

Liquidus: A line that lies between a liquid and a liquid–solid two-phase region on a phase diagram.

Local strain: (1) The concentration of the forces at a point in a tensile test specimen which results in the narrowing (necking) at one place on the test specimen. Local strain occurs after the specimen has taken loads beyond its tensile strength. (2) Local strains are formed by hard spots or other nonuniformities in a metal which prevent good machinability.

Lodestone: An iron ore consisting of magnetite which has been made into a permanent magnet.

Lucite: A trade name for a polymer of methyl methacrylate.

Magnaflux: A nondestructive test that uses a magnetic field and iron filings to detect cracks in a material.

Magnesia: Magnesium oxide.

Magnetic properties: Material properties that refer to how it will conduct or retain lines of magnetic flux.

Magnetite: An iron ore consisting of a mixture of ferric and ferrous oxide (Fe_2O_3 and FeO).

Malleability: The property of a material that allows it to be plastically deformed by rolling or hammering.

Malleable iron: High-tensile-strength white cast irons which are produced by a lengthy heat treatment of castings. There are two types: ferritic malleable iron and pearlitic malleable iron.

Maraged steel: A high-nickel-content low-carbon steel that has been heated to 1500°F, air cooled, then aged at 900°F. It is considered a superalloy.

Marquenching: *See* Martempering.

Martempering: A heat treatment in which a steel is from the austenitic temperature to about 500°F, held at that temperature until the metal is a uniform temperature throughout, then quenched to room temperature.

Martensite: The body-centered tetragonal structure of steel. It is the hardest form of steel that is formed by quenching a steel from the austenitic range very rapidly to room temperature.

Martensitic stainless steels: Stainless steels that can be hardened easily by oil quenching. They are very resistant to weak acids and bases and are made from the 400 series of stainless steels.

Matte: The form of copper obtained from the smelting furnace. It has about 30% copper at this stage.

Mechanical properties: Material properties that refer to how it will behave under applications of force. Tensile strength, compressive strength, modulus of elasticity, impact strength, and others are examples of mechanical properties.

Mechanical twins: The imperfections of twins formed by placing stresses on the material. Mechanical twins can be formed by bending, rolling, or other methods of deforming a material.

Mechanisms of slip: The ways in which a crystal can fail. The total number of mechanisms of slip possible in a crystal is determined by multiplying the number of possible slip planes by the number of directions of slip in each plane.

Melting point: That temperature at which the long-range crystal structure of a solid is destroyed.

Mer: The basic molecule or building block of plastics.

Miller index: A method of numbering a specific plane or family of planes in a crystal structure. It is determined by taking the reciprocal of the x, y, and z intercepts of the plane.

Mistake: A "goof-up," "wrong move," or miscalculation. Mistakes do not need to happen. Mistakes are not to be confused with errors.

Mixture: A combination of two or more materials in which no chemical reaction occurs. The components of mixtures can be separated by physical means.

Modulus of elasticity: The stress divided by the strain in the elastic region of a material. It can also be defined as the slope of the elastic portion of the stress–

strain curve for a material. It is also called Young's modulus and the letter E is used as its symbol.

Modulus of resilience: The maximum recoverable energy per unit volume of a material. It can be determined by finding the area under the linear portion of the stress–strain curve.

Modulus of rupture: The tensile strength of a material as determined through a torsion or flexure test.

Modulus of shear: Defined as the shear strength divided by the shear stress in the elastic region. It is usually denoted by the letter G.

Modulus of toughness: The energy needed per unit volume to fail a material. It can be calculated by finding the area under the total stress–strain curve.

Mohs hardness: A measure of the resistance to surface abrasion of a material. It uses 10 different minerals, from No. 1, talc, to No. 10, diamond, as the standards for the test.

Molecular weight: The sum of all the atomic weights in the molecule.

Molecule: The smallest part to which a compound can be divided and still retain the characteristics of that compound. Molecules are composed of atoms.

Mond process: A method of refining nickel. It uses carbon monoxide to form nickel carbonyl, which is then decomposed into nickel.

Monel: An alloy of nickel and copper.

Monoclinic system: A crystal structure that has two of the crystal angles at 90° but no axes lengths are equal.

Monotectic: The point on a system of two or more components in which one of the pure components is the lowest melting point on the phase diagram.

Muffle furnace: An oven which can be sealed so that the air can be replaced with other gases.

NDE: Nondestructive evaluation

NDT: Nondestructive testing.

Neoprene: A synthetic rubber made from chloroprene. It is inferior to natural rubber in some properties but is better in its resistance to oil and other organic materials. It is also good in its ability to hold air and other gases.

Neutron: A nuclear particle in the atom which has no electric charge but carries one atomic mass unit.

Newton: A unit of force in the metric system equal to 1 kilogram-meter per second squared.

Nickel silver: A white bronze containing up to 70% copper, up to 30% nickel, and the rest zinc.

Nitinol: An intermetallic compound of nickel and titanium. It has the ability to return to a preformed shape when heated. It is advertised as the "metal with a memory."

Nitriding: A case-hardening technique in which a steel is placed in a high-nitrogen atmosphere (usually ammonia), allowing the nitrogen to diffuse into the outer layers of steel to pin the slip planes and harden it significantly.

Noble gases: *See* Inert gases

Noble metals: Metals that do not react readily. They include gold, platinum, iridium, rhodium, osmium, and ruthenium.

Nodular cast iron: Also known as ductile cast iron. *See also* Ductile cast iron.

Nondestructive evaluation: The qualitative and quantitative checking of a material by methods which do not keep the product from being used for its intended purpose.

Nondestructive testing: The process of determining the properties of a material or product by methods which do not prevent the material from being used for its intended purpose.

Nonferrous metals: Metals that do not contain iron in their composition.

Normalizing: A heat treatment of steels in which the steel is heated about 100°F into the austenite range, then cooled in still air to room temperature. It gives the steel an even grain size, resulting in good machinability. It is also used prior to other heat treatment.

N-semiconductor: Semiconductors that have electron imperfections. They become the negative part of a transistor or integrated circuit.

Nucleation: The grouping together of the first few atoms or molecules to form a liquid from a gas or a solid from a liquid.

Nucleus: The central part of the atom, containing neutrons and protons.

Numerics: Numbers that have no dimension or units. A few numerics are $\frac{1}{2}$, π and $\sqrt{2}$.

Nylon: A polyamide made from a diamine and a diacid.

Objective test: A test in which the person conducting the test has no influence on the results of the test.

Oil quench: A heat treatment of steel in which the steel is heated into the austenitic region, then cooled very rapidly in oil.

Open cell: In a polymer foam, that forms in which the individual grains are connected.

Open hearth converter: A method of changing pig iron into steel by blowing air across the top of a pool of the molten metal.

Optical absorption: The material property of preventing light from passing through.

Optical properties: Properties of a material which are affected by or affect light. Color, index of refraction, and optical absorption and transparency are but a few of them.

Optical transparency: The ability of a material to pass light without being absorbed or diffused.

Ordinate: The *y* axis of a graph.

Organic chemistry: The chemistry of carbon.

Orthorhombic system: A crystal structure system in which all three crystal axes are perpendicular to each other but no axes have the same unit length.

Oxidizing agent: A compound that will take electrons away from another element, radical, or compound.

Oxygen lance converter: Apparatus that changes pig iron into steel by using a pipe to blow pure oxygen through the melt.

Paramagnetic: Materials that can be slightly attracted to a magnet in air. The magnetic permeability of paramagnetic materials is slightly greater than that of a perfect vacuum.

Pascal: A unit of pressure or force per unit area in the metric system. One pascal is a newton per square meter.

Patenting: A heat treatment of steel in which the steel is quenched from the austenitic region in a bath of molten lead. It is used to give the metal good ductility prior to drawing it out into wire.

PCA: Portland Cement Association.

Pearlite: The eutectoid composition of steel. It is formed as alternate layers of ferrite and cementite in the grain structure.

Pearlitic malleable iron: Malleable iron made by a controlled heat treatment. It has 0.3 to 0.9% carbon, a tensile strength of up about 40,000 lb/in^2, and averages about 8% less in density than a forged steel.

Peel: The application of a force to a material in such a manner that one layer is pulled away from another layer, breaking one line of bonds at a time.

Percent elongation: The length of a tensile test specimen at any load minus the original length divided by the original length multiplied by 100. The maximum percent elongation is the percent elongation at the breaking point.

Periodic table: A chart of the elements listed in order of increasing atomic numbers, with elements of similar chemical characteristics placed in the same vertical line.

Peritectic: The temperature at which a single-phase solid will decompose into a liquid–solid two-phase region upon heating.

Phase: A macroscopically, physically distinct portion of matter.

Phloem: The part of the tree that becomes the bark.

Phosphate deoxidized copper: Copper that has about a 0.02% phosphorous content to control the amount of oxygen in the metal. It is also known as DHP copper.

Phosphate glass: Glass made from phosphorus pentoxide (P_2O_5), alumina, and zinc oxide. It is transparent to infrared wavelengths.

Photochromic glass: Glass that has a small amount of silver in the glass which will darken in bright sunlight and become clear again when in the darker areas.

Photosensitive glass: Glass that has a gold, cerium, or copper sensitizing agent included in the formula, which makes it light sensitive. These glasses can be used as photographic films and as bases for chemical machining.

Physical properties: Properties of a material which are affected by heat, light, magnetism, electricity, or other forces. They include the mechanical properties, optical properties, electrical properties, and others of this nature.

Pig iron: The ingots that are the product of the first stage of refining of iron ore. It contains about 4% carbon and many other impurities.

Pinning: The anchoring or tying together of slip planes by interstitials or other imperfections which prevents slip from occurring between the two planes.

Plasma: An ionized gas.

Plastic deformation: The permanent deformation that occurs after a material has been strained beyond the elastic limit of a material.

Plastic region: The part of a stress–strain curve that is beyond the elastic region or the elastic limit.

Plasticity: The ability of a material to undergo permanent deformation prior to breaking.

Plexiglas: A trade name for polymethyl methacrylate.

Plywood: Thin sheets of wood that have been glued together in layers, with the wood going in different directions in each layer or ply.

PMMA: Polymethyl methacrylate.

Point defect: Imperfections that affect only one or at most a few atoms or molecules in a material.

Poisson's ratio: The negative of the lateral strain to the longitudinal strain taken in the elastic region of a tensile test.

Polymer: A compound made by joining together many of the same types of molecules.

Portland cement: A glue for aggregate made from limestone. When reacted with water and mixed with aggregate it will produce a monolithic rocklike material not unlike stone.

Precipitation hardening: *See* Age hardening.

Precision: The degree to which repeated readings taken in an experiment agree with each other.

Pre-preg: A reinforcement fiber or cloth that has been impregnated with a liquid thermosetting resin and cured to the B-stage. It will later be used as a lay-up, formed and cured.

Pressure: A force per unit area applied to a material.

Prestressed concrete: Concrete that has been placed in compression by the steel or other tension-bearing materials.

Prestressed glass: Glass that has been heated to just below its softening point, then chilled with a blast of air or oil quench. This sets the outer layers of the glass, allowing the inside to cool more slowly, putting the outside of the glass in compression. Prestressed glass has greater strength than unstressed glass, but will break into small grains when scratched or punctured.

Primitive cell: *See* Unit cell.

Process annealing: The heating of steel to a temperature just below the Ac_1 line and slowly cooling it. It is used in the sheet and wire industry to prepare the steel for the next step in rolling or drawing.

Proportional limit: An arbitrary point on a stress–strain curve that is just past the elastic limit of the material. It is found by constructing a line parallel to the elastic portion of the curve but offset by a factor of 0.02%.

Proton: A nuclear particle having a mass of one atomic mass unit and a single positive charge.

P-semiconductors: Semiconductors that have holes (imperfections) built into the crystal structure. They become the positive-type semiconductor in a transistor or integrated circuit.

Pultrusion: The process of forming a composite by pulling continuous fibers through a resin, then through a heated die to form them into shapes.

Pyrites: Iron sulfide. It has a golden color and is sometimes called "fool's gold."

Quality assurance: The process of checking incoming products to ensure that they meet specifications.

Quality control: The process of checking products to ensure that they meet specifications.

Quenching: The rapid cooling of a metal or other product from high temperatures in a very short period of time.

Radical: A molecule with an unsatisfied (or open) bond.

Rare-earth glass: Glasses that contain boron oxides and oxides of tungsten, lanthanum, tantalum, or thorium, but no silicates. They have the highest index of refraction of all glasses, which makes them suitable for lenses.

Rare earths: The oxides of metals with atomic numbers from 57 through 71. *See also* Lanthanide series.

Reaction injection molding: The mixing of two to four components in the proper chemical ratio in a high-pressure chamber, then forcing the mixed material into a mold. It is used in polyurethanes, epoxys, and other liquid chemical systems.

Reaction rate: The speed with which a chemical reaction occurs.

Refraction: Bending light by passing it from one medium to another.

Refractory: A material that will resist heat. It is used to line furnace walls and provide heat insulation for other high-temperature operations.

Reliability: The repeatability of a test. A reliable test will give the same results each time it is done.

Resilience: The tendency for a material to return to its original size and shape after the release of a load applied in its elastic region.

Resin transfer molding: The transfer of a catalyzed resin into an enclosed mold in which the reinforcement has been placed.

Resistance (electrical):   The opposition to the flow of electrons in a conductor. The resistance is dependent on the gemoetry and thickness of the conductor. Electrical resistance is measured in ohms.

Resistivity: The ability of a material to oppose the flow of electricity; not dependent on the size or shape of the material.

Rhombohedral: Crystal structure that has all three axes of equal unit length; the three angles between the axes are equal but not 90°.

RIM: Reaction injection molding.

Roasting: With ores (copper or nickel), a process used primarily to convert sulfides to oxides so that they can be reduced to metals by smelting. Also used to eliminate volatile components of orcs.

Rockwell hardness: A measure of surface plasticity as determined by forcing an indenter into the surface of a metal under a specified load. The loads vary from 60 to 150 kilograms and the indenters from a 120° black diamond cone to $\frac{1}{16}$- and $\frac{1}{8}$-inch steel balls.

RTM: *See* Resin transfer molding.

Rupture strength: The stress at which a material breaks in a tensile, compression, flexure, shear, or other test.

Rust: Iron oxide.

SAE: Society of Automotive Engineers.

Safety factor: The ratio of the yield strength of a material to its maximum expected stress.

Scleroscope hardness: A measure of surface elasticity as determined by the bouncing of a sapphire- or diamond-tipped plummet.

Screed: A board or other device used in leveling concrete or other material.

Screw dislocation: A crystal imperfection in which the atomic planes are rotated slightly from the one below it.

Shear: The application of opposing but slightly offset forces on a material.

Shear modulus: *See* Modulus of shear.

Shear strength: The maximum load per unit area of a material tested in shear. The area is the total surface area.

Significant figures: An indication of the precision of any measurement. The number of significant figures of any calculation is limited to the number of significant figures in the least accurate measurement.

Silica: Silicon dioxide.

Silt test: A method of determining the amount of fine silicas and other very fine materials in an aggregate.

Sintering: The fusing together of particles under pressure at temperatures below their melting point.

Slip (ceramics):   A slurry of ceramic material that can be cast into molds prior to drying and glazing.

Slip (crystals):   The direction that one plane can slide over another.

Slip plane: Planes in the crystal lattice that can slide over each other. There is the least amount of bonding energy between these planes.

Soda-lime glass: Glass made from silica, sodium oxide, and calcium oxide. It is the oldest and still most used glass.

Solid: That state of matter which has a definite long-range crystal structure. Solids are able to hold their own size and shape.

Solidus: In a phase diagram, the line between the solid phase and the liquid–solid two-phase region.

Solubility: The maximum amount of a stated material that can dissolve into a solute at any given temperature.

Specific gravity: The density of a material divided by the density of water in the same system of units. Specific gravity has no units.

Specific heat: The heat capacity of a material divided by the heat capacity of water in the same measurement system. Specific heat has no units.

Spheroidizing: The heat treatment of steel, which causes the cementite to form into small spheres; reduces the brittleness of a steel.

Spherulites: Microscopic spheres of graphite in ductile cast iron.

Stacking fault: A crystal imperfection in which the planes of atoms shift from face-centered cubic to hexagonal close packed and back within a single grain.

Stainless steels: A class of steels that have a high chromium or nickel content that retards or prevents oxidation.

Steel: An iron–carbon alloy that can be heated into austenite. The carbon content must be less than 2%.

Strain: The elongation divided by the original length of a tensile test sample. In other tests, strain is defined as the deformation of the sample.

Stress: The load per unit cross-sectional area applied to a test sample or structural member.

Stress relief: Rounded inside corners that prevent stresses from concentrating at a given point.

Stress risers: Points that concentrate stresses in a structural member or test sample.

Stress–strain curve: A plot of stress plotted against strain for a specimen taken to failure. Stress–strain curves can be done for tensile tests, compressive tests, and shear, torsion, or flexure tests.

Styrene: A vinyl radical with a phenyl group attached. It could be called phenyl ethene.

Styrofoam: A foam formed from polystyrene by use of a catalyst and a blowing agent.

Subjective test: A test in which the operator or person conducting the test can influence the results of the test.

Substitutional: An imperfection in a crystal in which an atom or particle of a material other than the parent matrix is at a lattice point.

Superalloys: Alloys of nickel, chromium, titanium, or other metals with iron which, when properly heat treated, can withstand tensile stresses of up to 300,000 lb/in^2 or more.

Superconductivity: The property of copper and other materials in which the resistivity approaches zero at cryogenic temperatures.

Synergy: A condition resulting from the total value of any property of combined materials being greater than the sum of that property from the individual materials. Synergy can be applied to tensile strengths, flexure strengths, or any other property of composites.

Syntactic foam: A composite made by mixing hollow microscopic spheres of glass or other materials into fluid resin to form a moldable, curable, lightweight fluid mass.

Taconite: An iron ore.

Teflon: A trade name for polytetrafluoroethene.

Tempering: A secondary heat treatment of steel in which the steel is reheated to a temperature about 700°F, then cooled. The purpose of tempering is to relieve internal strains in the steel to make it tougher.

Tensile strength: The highest stress a material can withstand before failure.

Tensile strength/density ratio: A measure of strength per unit weight. Its units are inches or meters.

Tension: The result of forces pulling directly against each other.

Tetragonal system: A system in which the angles between the axes are 90° to each other, with two of the axes having equal lengths and the third axis, a different unit length.

Therm: A unit of heat equal to 100,000 Btu. Natural gas is supposed to have 1000 Btu per cubic foot. When this is the case, a therm would be produced by burning 100 cubic feet of natural gas.

Thermal expansion: The enlargement of a material due to heat.

Thermal properties: The properties of a material related to heat. They include a material's boiling point, melting point, heat capacity, specific heat, coefficient of thermal expansion, heat of vaporization, heat of fusion, heat of combustion, and thermal conductivity.

Thermal tempered glass: *See* Prestressed glass.

Thermite: A mixture of aluminum powder and iron oxide. Upon ignition, thermite will produce iron and aluminum oxide and a tremendous amount of heat. The thermite reaction is used in welding and in incendiary bombs.

Thermoplastic: High polymers that get soft upon heating.

Thermosetting: High polymers that get hard, and char or burn upon heating.

Tiltslab: A method of concrete construction whereby the walls are formed and poured on the ground, cured, then tilted upright into position and anchored in place.

Time–temperature transformation curve: *See* TTT curve.

Torsion: The result of or application of a twisting force to a material.

Toughness: The ability of a material to withstand the application of energy. It involves both the ductility and the strength of a material. *See also* Modulus of toughness.

T-plastic: *See* Thermoplastic.

Transition metals: The group VIII metals of the periodic table: iron, cobalt, nickel, ruthenium, rhodium, palladium, osmium, iridium, and platinum.

Triclinic system: A crystal structure that has no two axes of equal unit length and none of the angles between the axes equal.

Triple bond: Three pairs of shared electrons between two elements in an organic compound.

Tritium: An isotope of hydrogen having two neutrons in the nucleus of the atom.

Tropometer: An instrument that measures the angle through which a force is applied in a torsion test.

T-set: *See* Thermosetting.

TTT curve: A graph of the time and temperatures at which changes from austenite to pearlite or martensite occur during the cooling of a particular steel. The cooling rates at which the various structures of steels are obtained can be taken from this curve.

Tukon hardness: A measure of surface plasticity determined by the indentation of an elongated diamond pyramid indenter called the Knoop indenter.

Turbidity: The property of a glass, liquid, or other semitransparent material to disperse, diffuse, or absorb light.

Twinning: The formation of mirror-image crystals within the same material. It can be microscopic imperfections or large grains.

Ultimate strength: The greatest stress that a material will withstand prior to failure.

Ultimately reinforced thermoset reaction injection: A process in composite manufacture in which a core is cast from a syntactic foam, wrapped with bidirectional graphite fabric, then placed in a heated mold. An epoxy resin is then injected into the mold.

Uniform strain: The even distribution of the strains over the entire length of a test specimen.

Unit cell: The smallest division of a crystal or a solid.

URTRI: *See* Ultimately reinforced thermoset reaction injection.

Vacancy: A crystal imperfection in which an atom or particle is missing from a lattice point.

Vacuum bag molding: A process used in composites and other industries in which a part is put under atmospheric pressure by placing it in an airtight bag and pulling a vacuum on the bag.

Validity: The ability of a test to measure the property it is supposed to measure.

Vapor pressure: The pressure caused by the atoms or molecules escaping or being evaporated from the surface of a material.

Vickers hardness: A measure of surface plasticity as determined by the indentation of a square diamond pyramid.

Vinyl: An ethene radical.

Vinyl chloride: Chloroethene.

Vinylidine chloride: 1,1-Dichloroethene; sold under the trade names Saran, Kynar, and others.

Viscosity: The resistance of a fluid to flowing.

Volt: The unit of electrical force needed to force 1 ampere of current through a resistance of 1 ohm.

Vulcanization: The process of adding sulfur to the polymerization of rubber.

Warp: The longitudinal direction of the reinforcing fabric in composites.

Water/cement ratio: The number of gallons of water mixed with 1 sack of portland cement.

Weft: The transverse direction of the reinforcing fabric in composites.

Well graded: An aggregate in which all particle sizes are present.

Well sorted: An aggregate that includes only one particle size.

White cast iron: Cast iron that contains only 2.5 to 3.5% carbon and a little silicon and chromium. It has good wear resistance, compressive strength, and hardness but is very weak in tension.

White metals: The low-melting metals, consisting of lead, tin, antimony, bismuth, zinc, and cadmium.

Work: A force through a distance ($W = Fd$). If either the force or the distance is zero, no work is done.

Wrought iron: A very low carbon iron made by melting white cast iron and passing it through an oxidizing flame. The wrought iron has a fibrous structure with strands of slag through it in the direction of the rolling.

X-ray tests: Nondestructive tests in which a material is examined for cracks and flaws by means of an X-ray passing through the material and the defects being recorded on a photographic film or other medium.

Xylem: The part of the tree that becomes wood. Xylem lies under the bark and cambium.

Yield (concrete):   The number of cubic feet of concrete obtained from mixing 1 sack (94 lb) of portland cement.

Yield point: The engineering design strength of a material. It is generally found by constructing a line on the stress–strain curve parallel to the elastic portion of the curve and offset by 0.2%.

Young's modulus: The stress divided by the strain in the elastic region of a stress–strain curve. *See also* Modulus of elasticity.

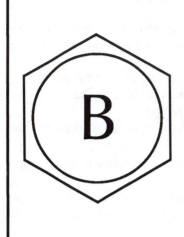

Graphing

Every graph must be self-explanatory. It should stand alone and be easily read and understood without additional notes or accompanying instructions. This can be done if the graph includes *all* of the following elements.

I. Title

The title must be very specific. It should not be vague. Examples of poor titles are

TENSILE TEST

and

STRESS–STRAIN CURVE

A proper title would include all the limits and specific details that the graph covers. An example of a better title would be

STRESS–STRAIN CURVE

for a

1018 Steel Specimen at 20°C

II. Each Axis Must Be Labeled

The labels of each axis must also be specific.

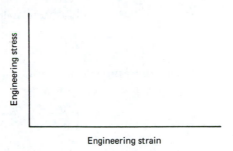

III. Each Axis Must Show the Units that Are Used

Is stress plotted in pounds per square inch or megapascal? Is the strain plotted in inches per inch or meters per meter? For other graphs, velocity could be plotted in miles per hour, inches per second, feet per minutes, kilometers per hour, meters per minute, or paces per picosecond. It makes a difference in interpreting the graph and must be specified.

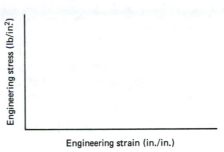

IV. Each Axis Must Show Its Scale

The scale must be:

  a. Uniformly plotted
  b. Easily interpolated
  c. Easily read and interpreted

     1. Do not use a "rubber scale" such as

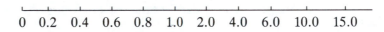

Use

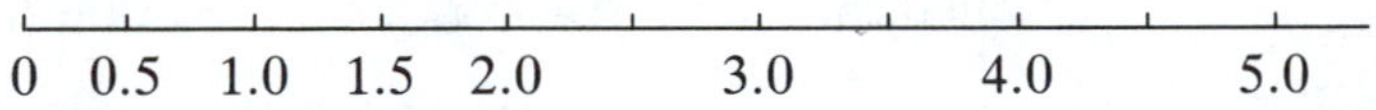

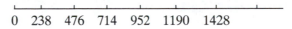

    2. Do not use uninterpolatable intervals such as

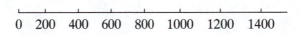

Use

3. Do not use odd or unusual divisions of the graph paper such as

Dividing the graph paper by using three intervals per unit makes it difficult to interpolate and read. Use

4. In determining the scale, divide the maximum less the minimum values of the data to be plotted by the number of divisions on the graph paper for that scale. This gives an approximate number of data units that must be plotted per division on the graph. The intervals can then be adjusted to meet the criteria noted above.

Example:

The maximum stress data obtained was 72,000 lb/in^2. If there are eight divisions on the graph paper for that scale, there should be

$$\frac{72,000 - 0}{8} = 9000 \text{ lb/in}^2 \text{ per division}$$

Since 9000 lb/in^2 is not a very interpolatable division, it would be better to divide the scale into divisions of 10,000 lb/in^2 and let the maximum reading be 80,000 lb/in^2.

V.   Data Points Should Be Plotted and Identified

Plot each point; then put a circle, triangle, square, or other identifying mark around each point. This is especially important if more than one type of data is plotted on the same graph.

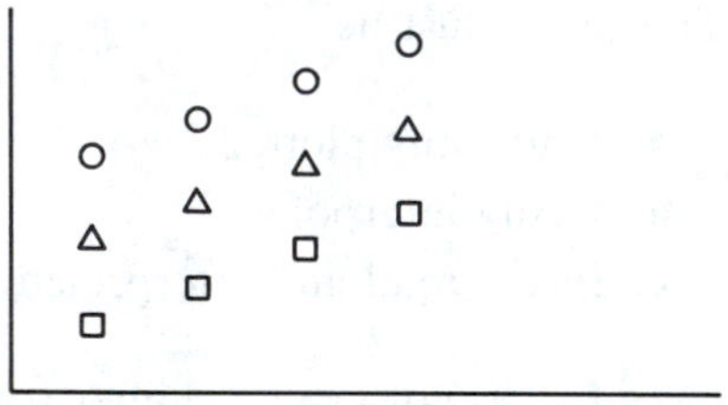

VI.   Plot the Best Fit Line for Each Set of Data Points

Use a straightedge, flexible curve, or French curve to draw the line. Do not "freehand" the curve. Do not plot point-to-point lines unless you are trying to show detailed inflections (such as a stock market fluctuations) or are very sure of the precision of your data.

Poor:

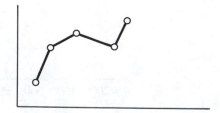

Better:

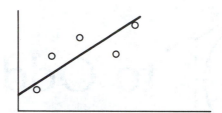

Use dashed, dotted, or other easily identifiable lines for the lines connecting each different set of data points. Then identify each line as to which data are represented by that line.

Example:

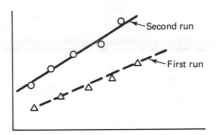

Use drafting instruments and be neat. Type or neatly letter the titles, scales, and remarks. Use black ink only unless instructed or authorized to use other markers. Black ink reproduces better than other drawing media. Colored pencils should not be used since they often do not reproduce or are difficult to see in poor lighting. The use of colored pencil lines can often lead to misreading of the graphs and lead to mistakes in calculations by the person using the graph. An example of a complete, properly drawn graph would be as shown in Figure A-1.

Unless all of the items noted above are included in the graph, it becomes a piece of "artwork," signifying nothing. Do not try to pass it off as a research or engineering document.

CURING CURVE FOR PORTLAND, CEMENT CONCRETE

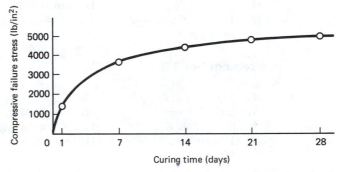

FIGURE A-1   Curing curve for portland cement concrete (compressive failure stress versus curing time) for a water/cement ratio of 5 gallons of water per sack and a 3-in. slump.

Answers to Odd-Numbered Problems

CHAPTER 1

Problem Set 1-1

1. 600 cm	**3.** 39.75 km	**5.** 4,320,000 mg
7. 32,000 mL	**9.** 37 cm is longer	**11.** 2880 m²
13. 3142 cm³	**15.** 2778 cm/s	**17.** 32.7 km/h
19. (a) 32.8 cm³	**19.** (b) 61.44 cm²	

Problem Set 1-2

1. 20.5 lb	**3.** 79 in³	**5.** (a) 2.57 kg
5. (b) 5.66 lb	**7.** 890 N	**9.** 2.1 lb or 0.95 kg
11. (a) 25.9 ft³	**11.** (b) 1618 lb	**13.** 62 mi/hr
15. 1.8×10^{12} furlongs/ fortnight	**17.** 2.64 sec	**19.** 0.33 km/s

Problem Set 1-3

1. 510 N	**3.** 112 lb	**5.** 10.1 N/cm²
7. (a) 637 lb/in²	**7.** (b) 4.4 MPa	**9.** (a) 173,000 lb
9. (b) 770,000 N	**11.** 1.7×10^5 MPa	**13.** 221,000 Pa
15. 1,240,000 Pa	**17.** (a) 2.6 lb/in²	**17.** (b) 18,000 Pa
19. 463 lb/in²		

Problem Set 1-4

1. (a) 0.4095 lb/in³	**1.** (b) 708 lb/ft³	**1.** (c) 11.3 g/cm³
3. 50.4 lb/ft³	**5.** SG = 19.3; if the metal was yellow, it was probably gold	**7.** (a) 56.8 lb/ft³

7. (b) 0.91 kg/L	**9.** 7.85	**11.** (a) 5.5 g/cm³
11. (b) 343.2 lb/ft³	**11.** (c) 5.5	**13.** SG = 7.75, it's iron
15. 108 lb	**17.** 1.82 in.	**19.** (a) 13.54 g/cm³
19. (b) 845.27 lb/ft³		

Problem Set 1-5

1. 20°C	**3.** −40°C	**5.** 621°F
7. 1204°F	**9.** 113°F, yes	**11.** (a) 8132°F;
11. (b) 4773K;	**11.** (c) 8591°R	**13.** 2525°R
15. −297°F	**17.** 85.6°F, yes	**19.** (a) 234.28K;
19. (b) −39.97°F;	**19.** (c) 422°R	

Problem Set 1-6

1. 2280 cal	**3.** 0.021 in.	**5.** 0.1 in.
7. 94.5 Btu/hr	**9.** 1100 cal	**11.** 45.3 s
13. 9.8 in.	**15.** 5000 cal/g	**17.** 2300 cal/g
19. 8.3×10^{-6} cm/cm/°C		

Problem Set 1-7

1. 10 A	**3.** 30 A, no	**5.** 121 Ω
7. $11.88	**9.** 24 Ω	**11.** 3.39 A
13. 36 A, no	**15.** $87.12	**17.** 328 Btu
19. 9.3 Ω		

CHAPTER 2

Problem Set 2-1

1. 15,280 lb/in²	**3.** 0.003 in./in.	**5.** 150,000 lb/in²
7. 0.04 in./in.	**9.** 1326 kg/cm² or 130 MPa	**11.** 0.02 m/m
13. 6.9 MPa	**15.** (a) 200,000 N	**15.** (b) 20,400 kg
17. 0.28 in.	**19.** 6.048 in.	

Problem Set 2-2

1. 1,250,000 lb/in²	**3.** (a) 30,000 lb/in²	**3.** (b) 0.0013 in./in.

3. (c)

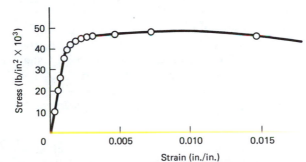

3. (d) 49,000 lb/in²	**3.** (e) 45,500 lb/in²	**3.** (f) 1.5%
3. (g) 33×10^6 lb/in²	**5.** 0.24 mm	**7.** 1.2 in.
9. 0.427	**11.** 0.259	**13.** 255 GPa
15. 0.067 in²	**17.** 77.6 GPa	**19.** 0.26 in.

Problem Set 2-3

1. 1562 lb/in²
5. (b) 2.6 lb/in²
11. 14 cm by 14 cm
17. Load = 4.35 MPa, yes

3. Ave = 3165 lb/in², yes
7. 8 in.
13. 7 columns
19. Need 76,800 lb, no

5. (a) 23.5 lb
9. 38.7 MPa
15. 943 lb/in²

Problem Set 2-4

1. 4685 lb/in²
7. 0.126 in.
13. 143,000 lb/in²
19. 756,000 lb/in²

3. 840 lb/in²
9. 20 rivets
15. 11 × 10⁶ lb/in²

5. 1294 MPa
11. 11 rivets
17. 216,000 lb/in²

Problem Set 2-5

1. 62.5 ft-lb
7. 7,290,000 Pa
13. 0.0013 cm/cm
19. 0.57 in.

3. 98 N-m
9. 78,000,000 Pa
15. Hollow

5. 3975 lb/in²
11. 1.7 cm
17. 1 × 20 has more strain

CHAPTER 3

Problem Set 3-1

1. 321
7. kg/mm²
13. 142
19. 309

3. 109
9. 65
15. 1.93 mm

5. 40.2
11. 322
17. 150

Problem Set 3-2

1. 222
7. HRe
13. 466

3. 454
9. HRb = 69.2
15. 0.55%

5. HRc
11. 287
17. Sceleroscope, Sonodur

19. Advantages: fast, accurate, cheap test to run; easy to train an operator: precise; reliable; valid; recognized in industry; objective
Disadvantages: expensive machine; size limitations; not portable; can be destructive; requires an average of several tests

Problem Set 3-3

1. 1880 ft-lb
7. (a) 1391 in.-lb or 116 ft-lb
11. 157.5 ft-lb
17. 0.28 N-m

3. 147 N-m, no
7. (b) 4200 in.-lb/in³ or 350 ft-lb/in³
13. 18.7 ft-lb
19. 20.6°

5. 1269 in.-lb or 106 ft-lb
9. 168 N-m
15. 5.1 ft-lb

CHAPTER 4

Problem Set 4-1

1.	**(a)** 23	**1.**	**(b)** 51	**1.**	**(c)** vanadium
1.	**(d)** V	**3.**	**(a)** 92	**3.**	**(b)** 143
5.	Mercury	**7.**	13	**9.**	26
11.	1.12×10^{23}	**13.**	4.78×10^{24}	**15.**	9.3×10^{-23}
17.	3.06×10^{21}	**19.**	3.6×10^{24}		

Problem Set 4-2

1.	Fe_2O_3	**3.**	$C_{12}H_{22}O_{11}$	**5.**	$NaHCO_3$
7.	HNO_3	**9.**	N = 2, H = 14, C = 5, O = 7	**11.**	Covalent
13.	$AlBr_3$	**15.**	$CaCO_3$	**17.**	$(NH_4)_2SO_4$
19.	9				

Problem Set 4-3

1.	40	**3.**	132	**5.**	58
7.	78	**9.**	HBr	**11.**	78
13.	1.04×10^{24}	**15.**	46	**17.**	342
19.	32				

Problem Set 4-4

1.	Tetragonal	**3.**	Cubic, tetragonal, orthorhombic	**5.**	Cubic, rhombohedral
7.	Rhombohedral	**9.**	Cubic, orthorhombic	**11.**	(6, 2, 3)
13.	(1, 0, 0)	**15.**	(0, 1, 0)	**17.**	(1, 1, 0), (1, −1, 0), (−1, −1, 0) or (−1, 1, 0)
19.	(2, 4, 1)				

Problem Set 4-5

1.	150°C	**3.**	300°C	**5.**	10%

7.

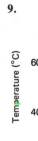

9.

11. 61.9% Sn
15.

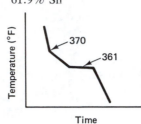

13. 20%
17. α + β

19. 258 sec

CHAPTER 5

Problem Set 5-1

1. Plain carbon (0.4% Mn), 0.4% carbon
3. 0.4–0.7% Ni, 0.4–0.6% Cr, 0.15–0.25% Mo, 0.2% C
5. 0.5–1.15% Cr, 0.15–0.25% Mo, 0.2% C
7. 1.6–1.9% Mn, 0.3% C
9. b-8% C
11. 4340
13. Iron is an element, steel is an iron–carbon alloy < 2% carbon

Problem Set 5-2

1. α iron or pure iron

3. FCC steel

5. The eutectic composition of the Fe–Fe₃C system

7. 4.3% C

9.

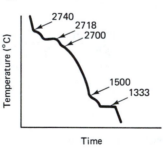

11. Lower temperatures are required, only one process needed

13. 1383°F
15. 2718°F
17. τ and liquid
19. 0.025%

Problem Set 5-3

1. 1-f, 2-c, 3-g, 4-e, 5-b, 6-d, 7-a
3. 414 MPa
5. 207 GPa
7. 16.2 in.
9. No, not acid resistant
11. It could be, but not the best; it's not tough enough.
13. No, cast iron is not easily welded.
15. 201, 202, 301, or 302
17. 405, 430, or 445
19. Maraged is based on a Ni, Co, Mo, Ti low-carbon steel, ausformed is based on a Cr, Mo, V steel

CHAPTER 6

Problem Set 6-1

1.	5−	**3.**	11.5 or 11+	**5.**	5
7.	201	**9.**	6+	**11.**	3+
13.	6−	**15.**	567×	**17.**	Normalizing

19. Induction hardening, flame hardening, carburizing, nitriding, or carbonitriding

CHAPTER 7

Problem Set 7-1

1. 93% Al, 5–6% Cu, 0.4% Si, 0.7% Fe, 0.3% Zn; solution heat treated and cold worked

3. 91% Al, 5.1–6.1% An, 2.1–2.9% Mn, 1.2–2% Cu, 0.5% Si, 0.7% Fe, 0.18–0.4% Cr, 0.27% Ti; solution heat treated and artificially aged

5. 92% Al, 4.5–6% Si, 0.8% Fe, 0.3% Cu, 0.05% Mn, 0.05% Mg, 0.1% Zn, 0.2% Ti; solution heat treated and cold worked

7.	2011-T6	**9.**	22.4%	**11.**	50.8%
13.	Yes	**15.**	99.3% Sn, at 441°F	**17.**	Mixture of $\alpha + \beta'$
19.	1				

Problem Set 7-2

1.	Ni	**3.**	Mg	**5.**	Cu
7.	Cu	**9.**	Ti	**11.**	220

13. MgO, Mg reacts easiest so would be the most difficult to decompose; it will take the heat better than the others

15.	None	**17.**	Lead	**19.**	928 lb

CHAPTER 8

Problem Set 8-1

1.	Soda-lime glass	**3.**	Chemically toughened	**5.**	Cermets
7.	Fused silicates	**9.**	Rare-earth glass	**11.**	13.5°
13.	1.37	**15.**	1.44	**17.**	0.17 in.
19.	0.0014				

CHAPTER 9

Problem Set 9-1

1.	4418 lb	**3.**	5.6 sacks/yd³	**5.**	6.75 sacks/yd³
7.	234 sacks	**9.**	(a) 4.7 gal/sack	**9.**	(b) 3.88 ft³/sack
9.	(c) 6.96 sacks/yd³	**11.**	35 sacks	**13.**	6.4 ft³/sack
15.	212 sacks	**17.**	6 gal/sack	**19.**	5.4 sacks/yd³

Problem Set 9-2

1. 3.1	**3.** Well graded	**5.** 740 lb
7. 195 lb	**9.** $0 + 0 + 0 + 0 + 0$ $+ 0 = 0$	**11.** 3.0, well sorted
13. 3.51	**15.** No. 4 = 286 lb, No. 8 = 286 lb, No. 16 = 428 lb, No. 30 = 572 lb, No. 50 = 142 lb, No. 100 = 142 lb, pan = 142 lb	**17.** 3.54
19. 3.76		

Problem Set 9-3

1. 4527 lb/in²	**3.** 4527 lb/in²	**5.** 113,100 lb, yes
7. (a) 68.2 lb/in²	**7. (b)** 1928 lb	**9.** 3985 lb, yes
11. 4050 lb	**13.** 832,000 Pa	**15.** 1803 kg
17. 556 sacks	**19.** 25 cm	

CHAPTER 10

Problem Set 10-1

1. Hexane	**3.** 3-Hexene	**5.** Toluene
7. 2,2,4-Trichloropentane	**9.** CH_3	

11. $HO—CH_2CH_2CH_2OH$

13. $CH_3CH—CH_2C—CH_3$ with CH_3 groups

15. OH / OH (benzene ring)

17. *para*-Methylamine   **19.** 132

Problem Set 10-2

1. 1-Pentanol	**3.** Ethyl pentanate	**5.** 4-Octanone (butyl propyl ketone)
7. $CH_3CH_2CH_2CH—CH_3$		**9.** 3
11. Alcohol	**13.** Alcohol	**15.** Aldehyde
17. 78	**19.** 94	

Problem Set 10-3

1. Polyethylene, polyvinyl chloride, polystyrene, polycarbonate, Teflon, Kevlar, and Dacron

3. Teflon	**5.** Thiokol	**7.** Polyvinyl chloride
9. Polyethylene	**11.** 28	**13.** Neoprene
15. 194	**17.** 104	**19.** 72.5%

CHAPTER 11

Problem Set 11-1

1. 2.76 GPa	**3.** 3.45 GPa	**5.** 4.28×10^6 in.
7. 0.4 in.	**9.** 0.66 cm.	**11.** 518 GPa
13. 0.0025 in.	**15.** 0.0046 m = 0.46 cm	**17.** 4 cm
19. 0.94 lb		

Problem Set 11-2

1. 23.4 lb	**3.** Yes	**5.** 13 lb
7. Black walnut (3.7×10^5 in. versus 3.19×10^5 in.)	**9.** 7,600 MPa	**11.** 41 kg
13. 6.9 kg	**15.** 31	**17.** 6.35 in. by 6.35 in.
19. 11.6 GPa		

Problem Set 11-3

1. 8 board feet	**3.** 10 board feet	**5.** 64 board feet
7. 1013 board feet	**9.** $436.90	**11.** 318 board feet
13. $273 board feet	**15.** $192	**17.** Hardwood is cheaper.
19. Plywood is cheaper.		

Index